T0329908

**Underwater Inspection and
Repair for Offshore Structures**

Underwater Inspection and Repair for Offshore Structures

John V. Sharp
Cranfield University
Cranfield
UK

Gerhard Ersdal
University of Stavanger
Stavanger
Norway

Registered Offices
John Wiley & Sons, Inc., 111 River Street, Hoboken, NJ 07030, USA
John Wiley & Sons Ltd, The Atrium, Southern Gate, Chichester, West Sussex, PO19 8SQ, UK

Editorial Office
The Atrium, Southern Gate, Chichester, West Sussex, PO19 8SQ, UK

For details of our global editorial offices, customer services, and more information about Wiley products visit us at www.wiley.com.

Wiley also publishes its books in a variety of electronic formats and by print-on-demand. Some content that appears in standard print versions of this book may not be available in other formats.

Library of Congress Cataloging-in-Publication Data

Names: Sharp, John V., 1936- author. | Ersdal, Gerhard, 1966– author.
Title: Underwater inspection and repair for offshore structures / John V.
    Sharp and Gerhard Ersdal.
Description: First edition. | Hoboken, NJ : Wiley, 2021. | Includes
    bibliographical references and index.
Identifiers: LCCN 2020042682 (print) | LCCN 2020042683 (ebook) | ISBN
    9781119633792 (cloth) | ISBN 9781119633822 (adobe pdf) | ISBN
    9781119633815 (epub)
Subjects: LCSH: Offshore structures–Inspection. | Offshore
    structures–Maintenance and repair.
Classification: LCC TC180 .S46 2021 (print) | LCC TC180 (ebook) | DDC
    627/.980288–dc23
LC record available at https://lccn.loc.gov/2020042682
LC ebook record available at https://lccn.loc.gov/2020042683

Cover Design: Wiley
Cover Images: Underwater and above water © SEAN GLADWELL /Getty Images, Oil and gas platform with offshore vessel transporting cargo © Danial_Abdullah / Getty Images, U.S. Navy Diver © Stocktrek Images / Getty Images, A diver inspecting airplane parts in the hold of the Japanese aircraft transport, Fujikawa Maru © A Cotton Photo / Shutterstock, 3d Rendering of a Subsea ROV Inspecting the Seabed © Vismar UK / Shutterstock

Set in 9.5/12.5pt StixTwoText by SPi Global, Pondicherry, India
Printed and bound by CPI Group (UK) Ltd, Croydon, CR0 4YY

C9781119633792_260321

Contents

Preface

All structures deteriorate and experience damage with time, particularly due to fatigue cracking, corrosion and damage from extreme and accidental events. This requires inspections, monitoring and appropriate repair of these structures to be performed to avoid an unsafe condition. Significant research and development work have been undertaken related to typical damage types, inspection and monitoring methods, evaluation of damage, and the need and methods for repair. This book aims at giving the reader an overview of this research and development work in addition to providing current practice in these areas, both to inform the reader about the existence of this work and to avoid unnecessary repetition of research and development.

Since early 1980 the first author of this book, John V. Sharp, has been active in the majority of these research programmes through his role initially in managing the relevant UK Department of Energy research work and later as Head of Offshore Research for the UK HSE. Since early 2000 the second author, Gerhard Ersdal, has had similar roles in managing several research programmes funded by the Norwegian Petroleum Safety Authority. Both authors have specific relevant expertise and have published a number of papers on inspection and repair of offshore structures and both are now actively involved in these areas at the universities of Cranfield and Stavanger, respectively, as Visiting (part-time) Professors.

The authors have had the benefit of working closely with John Wintle, Visiting Professor at the University of Strathclyde and Consultant Engineer at TWI, in preparing this book and his contributions have made significant improvements to the final text.

This book has mainly been written by using online web-conferencing and in its later stages this has been forced upon us because of the Covid-19 lockdown in the UK, Norway and many other parts of the world.

The opinions expressed in this book are those of the authors, and they should not be construed as reflecting the views of the organisations the authors represent. Further, the text in this book should not be viewed as recommended practice but rather as an overview of important issues that are involved in the management of inspection and repair.

The authors would particularly like to thank Mostafa Atteya, Eirik Duesten and Rolf Hanson for carefully reviewing the document and providing many valuable comments. We also want to thank Mick Else and Fu Wu for their efforts in enabling copyrights for essential illustrations. Further, the authors would like to thank Stinger (stinger.com), OceanTech (oceantech.com), BSEE, KBR Energo, Fugro and Atkins (MSL) for allowing us to use their illustrations and Magnus Gabriel Ersdal for drafting some of the figures and valuable input on artificial intelligence and machine learning. The authors would also like to thank the helpful and patient staff at Wiley.

Microsoft Teams, East Hendred, Oxfordshire and Stavanger, July 2020

John V. Sharp
Gerhard Ersdal

Definitions and abbreviations

Accidental limit state (ALS)	Check of the collapse of the structure due to the same reasons as described for the ultimate limit state but exposed to abnormal and accidental loading situations
Acoustic emission	The production of sound waves by a material when it is subjected to stress
ACFM	Alternating Current Frequency Modulated (type of inspection)
ACPD	Alternating Current Potential Drop (type of inspection)
Admixture	Material added during the mixing process of concrete in small quantities related to the mass of cement to modify the properties of fresh or hardened concrete
Ageing	Process in which integrity (i.e., safety) of a structure or component changes with time or use
AIM project	A project undertaken by the US Mineral Management Services (MMS) for assessment, inspection and maintenance, providing guidance on managing the integrity of existing fixed steel platforms in the Gulf of Mexico.
Anomaly	In-service measurement (damage, deterioration, defect, degradation, etc.) that is outside the threshold acceptable from the design or most recent fitness for service assessment
As-built documentation	Documentations that includes as-built documentation collected during in-service
Asset integrity management (AIM)	AIM is the means of ensuring that the people, systems, processes and resources that deliver integrity are in place, in use and will perform when required over the whole life cycle of the asset
Barrier	A measure intended to either identify conditions that may lead to failure or hazardous and accidental situations, prevent an actual sequence of events occurring or developing, influence a sequence of events in a deliberate way, or limit damage and/or loss
Bilge	The area on the outer surface of a ship's hull where the bottom shell plating meets the side shell plating
Caisson	Major part of fixed concrete offshore structure, providing buoyancy during floating phases and the possibility of oil storage within the structure, also used for pipework from topside to underwater typically water lift (intake of firewater, cooling water, etc.) and outlet of wastewater
CFRP	Carbon Fiber Reinforced Polymer

Clamp	A fabricated steel construction encompassing an existing tubular member or a nodal joint. A clamp consists of two or more parts that are bolted together. There are a number of clamp variants depending on whether or not the clamp parts are compressed against the existing member/joint and on whether there is a medium (grout or neoprene) placed between the clamp steelwork and the member/joint. A clamp should not be confused with a guide, which can appear to be superficially similar
Collapse	Total loss of the load bearing capacity of the platform through failure of one or more structural components
CP	Cathodic Protection
Curing	Action taken to maintain favourable moisture and temperature conditions of freshly placed concrete or cementitious materials during a defined period of time following placement
CVI	Close visual inspection (type of inspection)
Defect	An imperfection, fault or flaw in a component
Design service life	Assumed period for which a structure is to be used for its intended purpose with anticipated maintenance but without substantial repair from ageing processes being necessary
DFI resume	A document summarising key information concerned with the design, fabrication and installation
Discontinuity	A lack of continuity or cohesion; an intentional or unintentional interruption in the physical structure or configuration of a material or component
Duty holder	A UK term for the operator in the case of a fixed installation (including fixed production and storage units) and for the owner in the case of a mobile installation
DVI	Detailed visual inspection (type of inspection)
EC, ET	Eddy current testing (type of inspection)
Evaluation	The process of evaluating whether identified changes, defects or anomalies need repair, further inspection or a more detailed assessment
Fairlead	A device to guide a mooring line and to stop it moving laterally before it enters the vessel
False indication	An indication that is interpreted to be caused by a discontinuity at a location where no discontinuity exists
Fatigue limit state (FLS)	Check of the cumulative fatigue damage due to repeated loads or the fatigue crack growth capacity of the structure
Fatigue Utilisation Index (FUI)	FUI is the ratio between the effective operational time and the documented fatigue life
FCAW	Acronym for Fluxed Cored Arc Welding
Fixed structure	Structure that is bottom founded and transfers all actions on it to the sea floor
Flaw	An imperfection or discontinuity that may be detectable by non-destructive testing
FRP	Fiber Reinforced Polymer
FSU	Floating storage units
FSO	Floating storage and offloading units
FPSO	Floating production, storage and offloading units
Flooded member detection (FMD)	Inspection technique that relies on the detection of water penetrating a member by using radiographic or ultrasonic methods
GMAW	Gas Metal Arc Welding

Gross errors	Significant errors, mistakes and omissions in the form of anomaly or defects that may lead to local or global failures
Grout	A mixture of cementitious materials and water, with or without aggregate, to fill cavities and components to form a solid mass when set
GTAW	Gas Tungsten Arc Welding
GVI	General visual inspection (type of inspection)
HAZ	Heat affected zone related to welding
Hazard	Situations with potential for human injury, damage to the environment, damage to property, or a combination of these
High Strength Steels (HSS)	Steels with yield strengths in excess of 500 MPa
Hydrogen induced cracking (HIC) or hydrogen induced stress cracking (HISC)	The process by which hydride-forming metals such as steel become brittle and fracture due to the introduction and subsequent diffusion of hydrogen into the metal
Inspection programme	Scope of work for the offshore execution of the inspection activities to determine the condition and configuration of the structure
Integrity	The state of the structure, ideally being fit for service, and with an acceptable level of safety against failure
Integrity management	Continuous process to manage all changes that will occur during operational life that may affect the integrity of structures and marine systems
Jack-ups	Mobile offshore unit with a buoyant hull and legs that can be moved up and down relative to the hull
JIP	Joint Industry Project, usually in research and development
Kenter shackle	A device for joining two chain links as a repair
KPI	Key performance indicator, measurement of performance against targets
Life extension	The use of structures beyond their originally defined design life
Limit state	A state beyond which the structure no longer fulfils the relevant design criteria
Management of change (MoC)	A recognised process that is required when significant changes are made to an activity or process which can affect performance and risk
Marine fouling (growth)	Seaweed, bacteria and other living organisms in the seawater typically adhering to immersed surfaces such as offshore structures, which may build up to significant thicknesses
Metocean	Syllabic abbreviation of meteorology and (physical) oceanography
Microbiologically induced cracking (MIC)	A form of degradation that can occur as a result of the metabolic activities of bacteria in the environment. The bacteria that cause MIC can accelerate the corrosion process because the conditions that apply already have elements of a corrosion cell
MIG	Metal Inert Gas
Mitigation	Limitation of negative consequence or reduction in likelihood of a hazardous event or condition
MMA	Manual Metal Arc
MPI	Magnetic Particle Inspection (type of inspection)
Mudmat	A structure used to prevent offshore structures from sinking into soft unconsolidated soil on the seabed.
NDE	Non-Destructive Examination
NDT	Non-Destructive Testing

Node	Joining point for brace members in a jacket-type structure
Non-redundant	Structure that fails when the first primary structural member fails
OPB	Out-of-Plane Bending
Partial safety factor	(for materials) factor that takes into account unfavourable deviation of strength from the characteristic value and inaccuracies in determining the actual strength of the material
	(for loads) factor that takes into account the possible deviation of the actual loads from the characteristic value and inaccuracies in the load determination
Passive fire protection (PFP)	Coatings used on critical areas that could be affected by a jet fire. There are several different types, which include cementitious and epoxy intumescent based
Peening	Process of working a metal's surface to improve its material properties, usually by mechanical means
Performance standards	Statement of the performance required of a structure, system, equipment, person or procedure and that is used as the basis for managing the hazard through the life cycle of the platform
Pre-stressing tendons	High strength tendons required to maintain the structural integrity of a concrete structure, particularly in the towers (shafts). These tendons are placed in steel ducts, which are grouted following tensioning
Primary structure	All main structural components that provide the structure's main strength and stiffness
Progressive collapse	The sequence of component failures (from an initial local failure) that will eventually lead to the collapse of an entire structure or large part of it
Push-over analysis	Non-linear analysis for jacket structures used for determining the collapse / ultimate capacity
PWHT	Post Weld Heat Treatment
Redundancy	The ability of a structure to find alternative load paths following failure of one or more components, thus limiting the consequences of such failures
Reserve strength ratio (RSR)	The ratio between the design loading (usually 100-year loading) and the collapse / ultimate capacity
Residual strength	Ultimate global strength of an offshore structure in a damaged condition
Return period	An engineering simplification representing the probability (q) of an event by an assumed average period between occurrences of an event or of a particular value being exceeded. For q less than 0.1 this corresponds approximately to a return period of $1/q$ years
Risk based inspection	Inspection plans developed from an evaluation of the likelihood and consequences of failure associated with a structure to develop the inspection scope and frequency
Robustness	Measure of the ability of the structure to be damage tolerant and to sustain deviations from the assumptions to which the structure originally was designed
ROV	Remotely Operated Vehicle
RT	Radiographic testing (type of inspection)
Safety critical elements (SCE) and Safety and environmental critical elements (SECE)	Systems and components (e.g., hardware, software, procedures) that are designed to prevent, control, mitigate or respond to a major accident event (MAE) that could lead to injury or death. This was further extended in the 2015 version of the UK safety case regulation to include environmental critical elements
SCF	Stress Concentration Factor
Scour	Erosion of the seabed around a fixed structure produced by waves, currents and ice
Secondary structure	Structural components that, when removed, do not significantly alter the overall strength and stiffness of the global structure

Serviceability limit state (SLS)	A check of functionalities related to normal use (such as deflections and vibrations) in structures and structural components
Sleeve	A sleeve is a concentric tubular surrounding a leg or brace member that is several diameters long. The annular gap between the sleeve and member is normally grouted. In the case of an existing member, the sleeve is necessarily split longitudinally and the two halves are joined during installation using short bolts
SMR	Strengthening, Modification and Repair
SMYS	Specified Minimum Yield Stress
S-N curve	A relationship between applied stress range (*S*) and the number of cycles (*N*) to fatigue failure (regarding fatigue failure, *see* fatigue limit state)
Splash zone	Part of a structure close to sea level that is intermittently exposed to air and immersed in the sea
SSC	Ship Structures Committee
SPT	Sacrificial pre-treatment technique (used in adhesive repair)
Station keeping system	System capable of limiting the excursions of a floating structure within prescribed limits
Stress concentration factor (SCF)	Factor relating a nominal stress to the local structural stress at a detail
Structural integrity	A state of being intact and fit for purpose, with an acceptable level of safety against failure
Structural integrity management (SIM)	Means of demonstrating that the people, systems, processes and resources that deliver structural integrity are in place, in use and will perform when required of the whole life cycle of the structure
Structural reliability analysis (SRA)	Method used to analyse the probability of limit state failure of structures
Stud (chain)	Crossbar in the centre of a link of a chain, either welded or mechanically fixed
Studbolt	A threaded rod, generally used in stressed clamps
Subsidence	Settlement of the structure that results, primarily from extraction of reservoir hydrocarbons
Surveillance	All activities performed to gather information required to assure the structural integrity, such as inspection of the condition and configuration, determining the loads, records, and document review (such as standards and regulations)
Testing	Testing or examination of a material or component in accordance with a guideline, or a standard, or a specification or a procedure in order to detect, locate, measure and evaluate flaws
TIG	Tungsten Inert Gas
Topsides	Structures and equipment placed on a supporting structure (fixed or floating) to provide some or all of a platform's functions
Ultrasonic testing	A family of non-destructive testing techniques based on the propagation of ultrasonic waves in the object or material being tested
Ultimate limit state (ULS)	A check of failure of the structure of one or more of its members due to fracture, rupture, instability, excessive inelastic deformation, etc.
Vibration monitoring	Natural frequency monitoring to measure stiffness
Watertight integrity	The capability of preventing the passage of water through the structure at a given pressure head
Wave-in-deck	Waves that impact the deck of a structure, which dramatically increase the wave loading on the structure

1

Introduction to Underwater Inspection and Repair

> *The way you learn anything is that something fails, and you figure out how not to have it fail again*[1].
>
> —Robert S. Arrighi

> *Repair is an exacting, technical matter involving five basic steps: (1) finding the deterioration, (2) determining the cause, (3) evaluating the strength of the existing structure, (4) evaluating the need for repair and (5) selecting and implementing a repair procedure*[2].
>
> —Sidney M. Johnson

> *Before anything else, preparation is the key to success*[3].
>
> —Alexander Graham Bell

1.1 Background

Offshore structures for the production of oil and gas have a long history. The early offshore oil and gas exploration started in the 1940s in the Gulf of Mexico (GoM) and the Caspian Sea[4]. This was followed by the development of the North Sea and Brazil in the 1960s and later activities in the Persian Gulf, Africa, Australia, Asia and other areas. More recently offshore structures for wind energy production have been developed, initially in Denmark in the early 1990s followed by significant growth in several European countries, particularly the UK.

---

1 *Source:* Robert S. Arrighi, "Pursuit of Power: NASA's Propulsion Systems Laboratory No. 1 and 2", 2020.
2 *Source:* Sidney M. Johnson, "Deterioration, Maintenance, and Repair of Structures Modern Structure Series", 1965, McGraw-Hill.
3 *Source:* Alexander Graham Bell.
4 According to the 2020 web-version of the Guinness book of records, the earliest offshore platform was the Neft Daslari in the Caspian Sea 55 km off the coast of Azerbaijan. Construction began in 1949 and it began oil production in 1951. However, other sources report that production in the Gulf of Mexico began in 1946 by the Magnolia Petroleum (now ExxonMobil) platform 18 miles off the Louisiana coast.

Underwater Inspection and Repair for Offshore Structures, First Edition.
John V. Sharp and Gerhard Ersdal.
© 2021 John Wiley & Sons Ltd. Published 2021 by John Wiley & Sons Ltd.

These offshore structures are continuously exposed to:

- a sea-water environment, which can cause corrosion and erosion;
- active and environmental loads, which may cause fatigue cracking and buckling; and
- incidents and accidents causing physical damage such as dents and bows.

Damage of these kinds can cause loss of integrity of the structure and decrease the margin of safety. In addition, many oil and gas structures and the earliest of the wind structures are now ageing and many have been through a life extension process. Nevertheless, there is a continuing requirement to demonstrate that these installations remain safe for the personnel that operate them.

Unfortunately, there have been a number of accidents over the years with considerable loss of life resulting from structural failures related to inadequate inspection or failure to mitigate anomalies. A typical example is the semi-submersible Alexander L. Kielland accident in 1980 resulting in the loss of 123 lives. The cause of the accident was the loss of a brace member from fatigue and fracture leading to overturning and sinking. Fatigue failure initiated at a fabrication defect as a result of a combination of poor welding and lack of in-service inspection, which led to a catastrophic failure.

Other accidents with serious loss of lives have occurred in which structural failure played a part. In the shipping industry the MV Erika and MV Prestige accidents are examples of structural failures in storms where anomalies in these vessels failed to be detected and mitigated. It is clear from these accidents that in-service inspection and repair of structures are vital for the safety of structures. In addition, these are also normally required by regulators and class societies (i.e. ship classification societies, also known as ship classification organisations).

There are numerous incidences of damage and deterioration that had they not been detected by inspection and subsequently repaired or remediated could have led to serious accidents and loss of life. As later shown in this book these types of damage and deterioration include severe corrosion, fatigue cracks, dents and bows from impact loads, and severed members that could have resulted in more widespread structural failure and ultimate collapse of the structure. While these instances are well known to the companies involved and the relevant regulators, they are not necessarily well reported in the public domain. However, they show the importance and value of undertaking inspection and repair in a timely way for the prevention of escalation and maintaining safety. In addition, many offshore structures are now in an ageing phase where inspection and repair are likely to be more important. The authors believe that this is an opportune time to review this previous work on inspection, evaluation, repair and mitigation of such structures.

There are few books on underwater inspection and repair and those that exist are now significantly out of date. However, a significant amount of previous work is available from research and technology developments on the topic providing an extensive expertise accumulated in inspection and repair of structures through many years of offshore operational experience. Unfortunately, many of these reports are presently unavailable in the public domain.

This book is intended to indicate the current practice in these fields for those involved in keeping offshore structures safe, including practicing engineers involved in structural integrity management and also for students in the field.

1.2   Why Do We Inspect and Repair Structures

Although we intend to design, fabricate and install structures for safe operation during their design life, the environment, cyclic loading and accidental events will cause anomalies[5], which if not detected and repaired have the potential to cause failure of the structure. Ageing increases the likelihood of such anomalies being present.

Figure 1 shows the drivers for why structures are inspected. These include factors such as the balance between minimizing the life cycle cost and ensuring safe and functional structures (safe operation) by means of inspection and repair. These two drivers will often be in conflict but will also in some cases coincide as failures that lead to major repairs, loss of functionality or in the worst case, collapse of structures will have a major impact on cost also. An optimal integrity management that ensures safe and functionality at a minimum cost is hence often an important goal in planning inspection and repair of a structure.

Inspections and surveys also provide a means to determine the current condition of a structure and if necessary, timely undertake appropriate and cost-effective mitigation and repair measures to preserve the integrity of the structure. This will be discussed further in the book, especially in Chapter 6 on long-term inspection planning. In addition, regulatory and code requirements need to be met and these may define a minimum inspection level.

The diagram also illustrates the different changes and uncertainties in the current condition of the structure that can be detected by different types of inspection and surveys. The primary goal is to identify any changes, damage and anomalies to the structure but inspections that indicate that no anomalies are present are also important. Such information is vital in reducing uncertainty

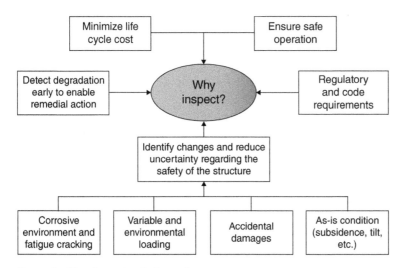

Figure 1   The elements of why we inspect structures.

---

5 Anomalies are in this book used for any deviation in condition (degradation, deformations, defects, damage and deterioration), configuration (change in layout, geometry and weight), design regime (new or updated requirements or practices) and design actions (changes to, e.g., metocean data leading to, e.g., insufficient strength and fatigue life) that may affect the integrity of the structure.

about the condition of a structure and, hence, providing the owner and the responsible engineer with confidence in that the operations remain safe and assumptions are valid when no significant or unexpected anomalies are detected.

The decision to undertake any form of repair or mitigation of an anomaly, detected by inspection, needs to be based on a thorough evaluation and often a more detailed inspection of the anomaly and its effect on the structural safety, based on established standards and knowledge. If repair or mitigation is required, the necessary decisions need to be made on how this can be achieved effectively. Failure to repair or mitigate critical anomalies could cause structural failure with the possibility of significant consequences. Thus, this emphasises the important role of inspection, evaluation, mitigation and repair in maintaining a safe structure.

Structural failures have occurred offshore with significant loss of life. The first of these was the Sea Gem incident in 1965 in UK waters with the loss of 13 lives. The resulting inquiry concluded metal fatigue in part of the suspension system linking the hull to the legs was to blame for the collapse. Fatigue cracking and lack of in-service inspection were significant in the Alexander L. Kielland capsize in 1980 killing 123 people, as already mentioned.

Offshore structures in the Gulf of Mexico have also failed during hurricanes. For example, Hurricane Andrew in 1992 caused significant damage to twenty-two of the offshore regions, with older structures sustaining significant damage. Inspection was needed to determine the extent of the damage and in many cases this information led to the need for repair in order to resume operation. Several examples of hurricane damage in the GoM are reviewed later in this book.

Other offshore accidents have also occurred. Not all of these failures were a result of an anomaly that could be identified by inspection. Some were the results of under-design, underprediction of loading, accidental damage and gross errors. Such failures typically initiate significant subsequent research work providing a better understanding of the cause of failure and appropriate inspection requirements. An example of such is the intensive work that was initiated on fatigue and crack inspection after the Alexander L. Kielland accident.

The reasons for inspection and repair can change over the life of a structure. Ersdal et al. (2019) review the statistics of failure for older offshore structures and show that these structures have a significant failure rate, particularly for floating structures. Figure 2 shows the types of damage to critical hull members; this includes cracking of the hull, corrosion, vibration and other types. This figure also shows the trend for increasing damage with age. This is to be expected knowing that these structures will degrade and accumulate damage, which requires regular inspection and often subsequent repair and mitigation.

As shown in Figure 2, damage rate increases with age, which is also indicated in a traditional bathtub curve as shown in Figure 3. In addition, structures are also known to experience some so-called burn in failures at an early age. These two increases in failure rate are often reflected in a typical bathtub curve as shown in Figure 3. The phases in the life of a structure related to the bathtub curve can be described as:

- an initial phase where anomalies arise from the design, fabrication and installation;
- the maturity phase representing the useful operating life; and
- the ageing and terminal phases representing the first and second part of the end of life.

It is important to recognise that frequency, purpose and method of inspection depend on the phase in the bathtub curve (HSE 2006). In the initial phase, anomalies arising from gross errors in the design, fabrication and installation should be detected by early inspections (baseline inspection). The purpose of these early inspections is to determine the existence of gross errors and to

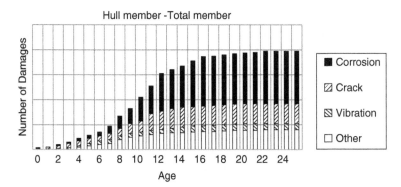

Figure 2 Damage to hull structural members by different causes and ship age for all ship types. *Sources:* Based on SSC (1992) "Marine Structural Integrity Programs (MSIP)", Ship Structure Committee report no 365, 1992; SSC (2000). SSC-416 Risk-Based Life Cycle Management of Ship Structures, Ship Structure Committee report no 416, 2000.

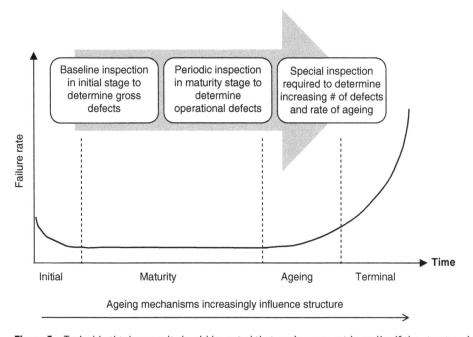

Figure 3 Typical bathtub curve. It should be noted that ageing may set in earlier if the structure is not managed properly. *Source:* Based on HSE (2006). Plant ageing – Management of equipment containing hazardous fluids or pressure, HSE RR 509.

provide confidence about the state of the structure. Any anomalies detected are expected to be rectified and the structure should at the end of this period "enter" the maturity phase.

In the maturity phase a lower failure rate is expected and the purpose of the inspection is to confirm that changes to the structure's condition, configuration and loading are in line with expectation.

In addition, identification of unexpected and unforeseen anomalies and damage is needed, such as unexpected early fatigue cracks and corrosion damage and unforeseen accidental damage and load changes. Inspection in the maturity phase will often be periodic and can be calendar-based, condition-based or risk-based. These strategies for inspection are discussed further in later chapters.

In the ageing and terminal phases the failure rate is expected to increase, although this is seldom recognised by structural integrity engineers. The purpose of inspection in these phases is to provide the basis for evaluation and the need for and prioritisation of repairs to ensure that the structure is still fit for purpose. Standards in some cases also require special inspections to be performed in these phases. The frequency of inspection is expected to increase compared to that in the maturity phase and even more so in the terminal phase.

1.3 Types of Offshore Structures

Inspection and repair depend on the type of structure and the material used. The types of platforms primarily used in the offshore industry are fixed with some variation and floating platforms. Examples of offshore structures are shown in Figure 4 and Figure 5.

Fixed platforms typically are of the type:

- fixed steel structures (jacket structures), which are mostly pile supported or suction anchor supported;
- concrete gravity-based structures;
- jack-ups; or
- monopile.

Floating platforms are in general dependent on their watertight integrity and station keeping in addition to their structural integrity to function. The most used types of floating structures in the offshore industry include:

- semi-submersible platforms (mostly in steel, but one in concrete exists);
- tension leg platforms (mostly in steel, but one in concrete exists);
- ship-shaped platforms and barge-shaped platforms (mostly in steel, but a few in concrete have been made); and
- spar platforms.

Structures must adhere to regional regulations (stationary structures) or to flag state regulations in combination with class society rules (mobile units). So-called mobile offshore units (MOUs) are typically at a location for a limited time and inspection and repair are often performed at a yard at regular intervals. These yard periods allow for inspection and repair to be performed in controlled circumstances, often in dry conditions or benign conditions. In comparison, inspection and repair of stationary structures requires inspection and repair to be performed in an offshore environment and often underwater by an ROV. Hence, inspection and repair are rather different for these two types of structures, both with regards to the regulatory regime and the environment in which the inspection and repair is performed.

1.3.1 Fixed Steel Structures

Fixed steel structures (jackets) consist of a steel spaceframe piled to the seabed (Figure 4 b), supporting a deck with space for drilling rigs, production facilities and crew quarters. Fixed steel

(a) (b)

(c) (d)

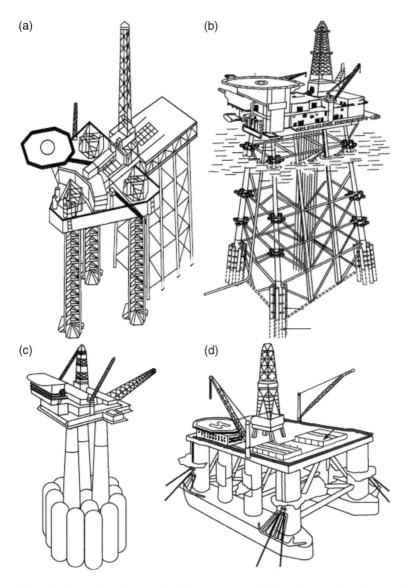

Figure 4 Examples of types of offshore structures: (a) is a jack-up placed alongside a jacket structure, (b) is a jacket structure, (c) is a concrete gravity-based structure and (d) is a semi-submersible unit. *Source:* Sundar, V. (2015), Ocean Wave Mechanics: Application in Marine Structures, © 2015, John Wiley & Sons.

structures are also used as a substructure for wind turbines. Steel jackets are usually made of tubular steel members. A typical six-legged steel platform is shown in Figure 4 b. Historically, the piles were driven directly through the legs and into the seabed. In more recent platforms the piles are typically connected to the legs by pile sleeves where the annulus between the sleeve and the pile is grouted. In a few cases, a specialized suction pile (bucket foundation) has been used, for example the Norwegian platforms Draupner 16/11-E and Sleipner T. In addition to the basic structure there are frames for the conductors, j-tubes, risers and caissons needed for production and operation. J-tubes are typically used to enable small diameter flowlines, electrical cables or pipeline bundles to be connected to the topside facilities. Steel structures are subject to ageing processes such as fatigue and corrosion and hence life extension is a key issue with regard to the structure itself.

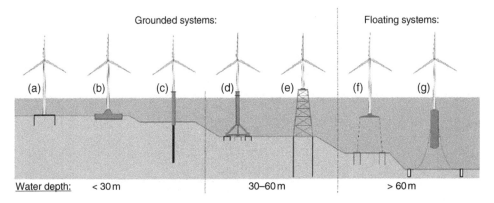

Figure 5 Various forms of wind turbine substructures: (a) suction pile caisson, (b) gravity-based concrete foundation, (c) monopile, (d) tripod, (e) fixed steel structure (jacket), (f) tension leg platform, (g) spar buoy. *Source:* Bhattacharya, S. (2019), Design of Foundations for Offshore Wind Turbines © 2019, John Wiley & Sons.

Jack-ups (Figure 4 a) are self-elevating units with a buoyant hull and several legs which when on location can be lowered to the seabed and raise the deck above the level of the sea thus creating a more stable facility for drilling and/or production. During operation in an elevated situation, jack-ups behave like a fixed platform. Jack-up rigs have been primarily used for exploratory drilling, but there are a few instances where they have also been used for production.

Monopiles (Figure 5 c) support the deck on a single pile / tower. These types of structures are particularly used for offshore wind energy production.

1.3.2 Floating Structures

Semi-submersible units (Figure 4 d) have hulls, together with columns and pontoons, with sufficient buoyancy to enable the structure to float. Semi-submersible platforms change draft by means of ballasting and de-ballasting (changing the water level in seawater tanks). They are normally anchored to the seabed by mooring systems, usually a combination of chain and wire rope. However, semi-submersibles can also be kept on station by dynamic positioning (DP), commonly used for semi-submersibles as drilling units and flotels in deep water. The hull supports a deck on which various facilities for drilling and production can be installed. Semi-submersible platforms are also being used for wind energy production, such as on the Kincardine wind farm in Scotland.

A tension leg platform (TLP) is of similar form to the semi-submersible platform but is vertically moored by tension legs, which decrease the vertical motion significantly. A TLP is designed with excess buoyancy to ensure that the tension in the legs remains in all operational conditions.

A spar platform consists of a single moored large-diameter vertical cylinder that is supporting the topside or a wind turbine. The cylinder is ballasted in the bottom (often by solid and heavy material) to provide stability. Spar-type of structures are increasingly being used for floating wind energy production.

A ship-shaped structure is a floating vessel similar in shape to a conventional ship, primarily used by the production and processing of hydrocarbons. These are often called a floating storage unit (FSU), a floating storage and offloading unit (FSO) or a floating production, storage and offloading unit (FPSO). The steel hull typically consists of plates in a main deck, side shells, turn-of-bilge and a bottom shell in addtion to tank tops and longitudinal bulkheads (wing tank bulkhead and centreline bulkhead) and transverse bulkheads. These main platings together form the ship beam. In most cases these plates are stiffened by girders and stiffeners. The plates are exposed to the loading from hydrostatic and dynamic pressure, which is transferred to the stiffeners and girder beams and into the structure. Ship-shaped structures are usually moored to the seabed by chains or wire ropes, although they can also be held on station using a dynamic positioning system. These mooring systems will require underwater inspection and possibly repair.

Most of the details of floating structures can be inspected from the inside in near dry conditions and are, as such, not a part of this book. However, underwater inspection and repair of the mooring, anchors, outer skin, valves and attatchments are relevant and hence included in the book.

1.3.3   Concrete Platforms

Concrete is used in several different types of offshore structures, which include:

- gravity-based structures, or GBS (for example, the Norwegian platforms Statfjord A-C and UK platform Brent C);
- tension leg platforms, or TLP (the Norwegian platform Heidrun);
- spar platforms, particularly for wind energy production,
- semi-submersible platforms (the Norwegian platform Troll B);
- Jarlan walls enclosing concrete tanks (the UK platform Ninian and the Norwegian platform Ekofisk Tank);
- steel-concrete hybrid (the UK platform Ravenspurn); and
- articulated tower (used as a loading column on Maureen).

The first concrete structure to be installed in the Norwegian sector was the Ekofisk tank in 1973 and the first to be installed on the UKCS was Beryl Alpha in 1975. At the time of writing, there are 22 concrete gravity-based platforms, together with one concrete tension leg platform (Heidrun), one concrete semi-submersible platform (Troll B) and one concrete base for a steel upper section (Harding).

Due to the dominance of GBS-type structures, this book will mainly focus on these, but some aspects will apply for all types of concrete structures. The main parts of a concrete gravity-based structure (GBS) are:

- legs, towers and shafts;
- storage cells (caissons);
- steel-to-concrete transition;
- shaft-to-base junction;
- prestressing anchorages;
- splash zone;
- foundations; and
- cathodic protection system.

Most concrete structures rest on the seabed on a large caisson often with skirts that penetrate the seabed protecting the structure against horizontal and overturning forces (Figure 4 c). The size of the caisson and skirts depends on local soil conditions. The topside steel structure is usually supported on up to four concrete columns/shafts extending from the caisson through the sea surface. The caisson can be divided into cells, which may be used as compartments for oil storage or as ballast. One or two of the columns are usually drill shafts supporting conductor frames and well-conductors. Some of the shafts, which may be water filled or dry, are normally outfitted mechanically.

1.4 Overview of this Book

In this book the important issues discussed so far are addressed by the following topics as indicated in Figure 6:

- regulatory requirements for inspection and repair (Chapter 2);
- where and when a structural inspection is needed (Chapter 6);
- what tools and methods can be used to inspect structures and how inspection tools are deployed (Chapter 4);
- types of damage that are found on offshore structures (Chapter 3);
- what tools and methods can be used to monitor structures (Chapter 5);
- how is the information maintained about the design, fabrication and installation in addition to the history of operation and inspection results (Chapter 6.1.5);
- when is an anomaly in a structure acceptable and when is mitigation or repair needed (Chapter 7); and
- which repair schemes are appropriate and what is the strength and fatigue life of a repaired component (Chapter 8).

In all these areas, the book aims to provide an overview of previous studies relevant to each topic along with what is considered good practice.

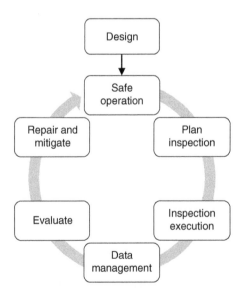

Figure 6 The management process for safe operation by inspection and mitigation.

An earlier book by Ersdal, Sharp and Stacey (Ersdal et al. 2019) addressed the topic of ageing and life extension of offshore structures, which provides a useful addition to this book. The main types of changes identified in this previous book are:

- physical changes (ageing, degradation, loading, etc.);
- technological changes (including compatibility and obsolescence);
- changes to knowledge and safety requirements; and
- structural information changes (e.g., accumulation of inspection data, loss of design data).

The focus of this book is on physical changes. However, the additional changes will also influence the safety of a structural system.

As already mentioned, there has been a significant amount of previous work, including research on each of these key elements related to inspection, evaluation and repair for offshore structures, which are in some cases unavailable to the public. These key reports have been reviewed by the authors and presented in this book with the aim of providing the reader with an awareness of this background, which is important in the context of structural integrity management.

It is important to recognize that choices made in the design will significantly influence inspection and repair of an offshore structure (the need for inspection and how it can be performed). Such choices include:

- the existence of any preinstalled monitoring systems;
- access for inspection and repair;
- cathodic protection system design and coatings;
- material selection;
- design margins, design fatigue life, robustness, redundancy; and
- operational restrictions given by the design.

In addition, the book will discuss the competency requirements for inspectors, structural engineers involved in integrity management and the organisational requirement for integrity management.

1.5   Bibliographic Notes

Section 1.3 is partly based on Ersdal, Sharp and Stacey (2019).

References

Bhattacharya, S. (2019), *Design of Foundations for Offshore Wind Turbines*, Wiley, 2019.

Ersdal, G., Sharp, J.V. and Stacey, A. (2019), *Ageing and Life Extension of Offshore Structures: The Challenge of Managing Structural Integrity*, Wiley, 2019.

HSE (2006). "Plant Ageing—Management of Equipment Containing Hazardous Fluids or Pressure", HSE RR 509.

SSC (1992) "Marine Structural Integrity Programs (MSIP)", Ship Structure Committee report no 365, 1992.

SSC (2000). "SSC-416 Risk-Based Life Cycle Management of Ship Structures", Ship Structure Committee report no 416, 2000.

Sundar, V. (2015), "Ocean Wave Mechanics: Application in Marine Structures", John Wiley & Sons, Ltd.

2

Statutory Requirements for Inspection and Repair of Offshore Structures

It takes less time to do things right than to explain why you did it wrong[1].

—Henry Wadsworth Longfellow

Integrity—the state of being whole and undivided, the condition of being unified or sound in construction[2].

—Oxford dictionary

2.1 Introduction

A number of different regulatory regimes exist for offshore installations for different parts of the world. The regulatory regimes in USA, UK and Norway in many ways took the lead in the development of these, although well-developed regulations can now be found in, for example Brazil, Canada and Australia. In the North Sea, the major change has been the development of risk-based regulations, initially in Norway and since the mid-1990s in the UK. These regulations specify high-level safety requirements (goals) to be achieved through the design, fabrication and operational stages. These regulations are further supported by recognised standards. Some regulatory regimes are more prescriptive with respect to the standards to be used.

Inspection and repair of offshore structures are often regulated as a part of the structural integrity management (SIM) requirements. The purpose of SIM is to identify all types of changes relevant to the safety of a structure in operation, to evaluate the impact of these changes and mitigate (e.g. repair) the impact of these if found necessary. Such changes include anomalies detected during inspections or condition monitoring, changes in loads and configuration and changes to requirements from improved standards. Mitigation will typically be various forms of repair, load reduction or operational restrictions.

A standard or recommended practice for integrity management should guide the responsible party to manage all aspects of safety and functionality for the structure. This will typically be achieved by a major hazard approach based on an established risk picture, describing what can go wrong and how one can protect the structure if such failures should occur. Hence, the structural integrity management should be based on an understanding of the possible hazards to the

---

1 *Source:* Henry Wadsworth Longfellow.
2 *Source:* Oxford dictionary.

Underwater Inspection and Repair for Offshore Structures, First Edition.
John V. Sharp and Gerhard Ersdal.
© 2021 John Wiley & Sons Ltd. Published 2021 by John Wiley & Sons Ltd.

structure and marine systems (for floating structures) and the severity of the consequence of these types of damage. A standard should further include good methods for planning the necessary inspections and surveys to establish an understanding of any change or anomaly influencing the safety of the structure.

Ersdal, Sharp and Stacey (2019) stressed that changes to structures and marine systems may be physical, technological, knowledge based, related to safety requirements in regulations or standards and changes to information about the structure. Although all these changes can be important, physical changes will often dominate the SIM work. The detection of these physical changes results from rather costly offshore structural inspections, structural monitoring, weight monitoring and metocean observations. The other types of change should not be overlooked and these will often include document review, updated engineering methods and standards and maintaining a proper database containing all necessary information about the structure (see Section 6.1.5).

The general principles for the management of physical assets were introduced in PAS 55 (BSi 2008). Although this standard is not sufficiently detailed for structural integrity management, its principles include keeping the asset (in this context structure and marine system) unimpaired and in sound condition throughout the life cycle, whilst protecting health, safety and the environment. This type of integrity management approach has been incorporated in some of the latest structural integrity management standards and recommended practices such as API RP-2FSIM (API 2019a), API RP2-MIM (ISO 2019b), ISO 19901-9 (ISO 2019a) and NORSOK N-005 (Standard Norge 2017b), as discussed later in this chapter.

A major hazard approach is relevant to all types of structures, but for marine systems related to floating structures, many of these hazards and controls relate to factors such as ballast controls and weight distribution, which can change on a daily basis.

UK HSE commissioned Atkins to prepare a report providing an overview of the required structural integrity management for fixed offshore structures (HSE 2009). The aim of the programme was to develop a comprehensive framework, as illustrated in Figure 7, for the structural integrity management (SIM) of fixed jacket structures based on HSE's technical policy in this area. The framework was seen as a mechanism for communicating good industry practice and hence to encourage continual improvement in performance. The inspection strategy and programme, information management, evaluation and sub-sea intervention feature prominently.

In addition, the report provided guidance on how to implement the framework. The document had been developed largely on the basis of existing standards and industry published documents including ISO 19902 (ISO 2007), API RP-2SIM (API 2014b) and PAS 55-1 (BSI 2008).

2.2 Examples of Country Statutory Requirements

2.2.1 Introduction

The US took a lead in the development and regulation of the offshore industry, as the Gulf of Mexico was well developed as early as the 1960s when the first edition of API RP-2A was issued. When other countries were developing offshore structures there was a lack of appropriate local design standards. As a result, API RP-2A became the default used as the basis for design of such structures around the world. By the end of the 1970s, some local standards and codes were introduced. The most notable of these were the issue of the UK Department of Energy Guidance Notes and the Norwegian Petroleum Directorate Rules.

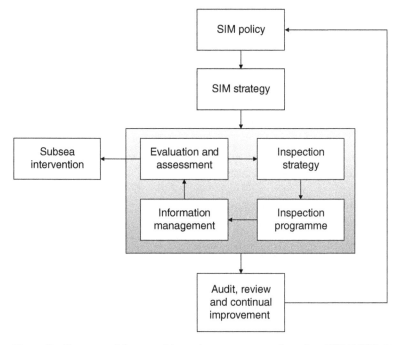

Figure 7 Elements of Structural integrity management, based on HSE (2009). *Source:* Based on HSE (2009), HSE RR684 - Structural integrity management framework for fixed jacket structures, Health and Safety Executive (HSE), London, UK.

In addition, several ship classification societies became involved in the offshore industry and released appropriate rules, some based on API RP-2A. These included American Bureau of Shipping, Det Norske Veritas and Lloyds Register. Many mobile offshore structures have a class certificate provided by one of these leading Classification Societies, where inspection has been an important part of the class system and regulation, which has been developed over many years of experience by owners, regulators and class societies.

2.2.2 Regulation in the US Offshore Industry

The oil and gas activities in the US are regulated by the governmental organisation Bureau of Safety and Environmental Enforcement, BSEE (previously Mineral Management Service, MMS and for a short period, the Bureau of Ocean Energy Management, Regulation and Enforcement, BOEMRE). In general, these have relied heavily on recommended practices by the American Petroleum Institute (API).

API issued in 1969 its first recommended practice of RP-2A for the design of offshore structures (API 1969), with the first set of design procedures for the offshore industry. This recommended practice has been continuously updated since with respect to wave criteria (e.g. moving from a 25-year criterion to a 100-year return period) and design formulae for tubular joints.

The 22nd edition of API RP-2A (API 2014a), issued in 2014, saw the introduction of a supplement named API RP-2SIM (API 2014b) aimed at structural integrity management, which addressed the ongoing integrity management of the structures, an increasingly important matter. In 2019 API in addition published recommended practises on the integrity management of floating systems API RP-2FSIM (API 2019a) and mooring systems API RP-2MIM (API 2019b).

2.2.3   Regulation in the UK Offshore Industry

From 1974 until June 1998, the UK certification regime required the issue of a certificate of fitness every five years by legislation Sl 289, enacted in 1974 (Department of Energy 1974). This was the basis for the regulation of the safety of offshore structures in the UK sector of the North Sea. It was supported by published Guidance Notes, which included specific sections on design, construction, surveys and repairs (Department of Energy 1990). Most of the structures in the North Sea were designed, fabricated, installed and operated in this regime until 1996. The 4th edition of the Guidance Notes had extensive sections on both surveys and repairs. These are reviewed later in this book.

Following the Cullen report on the Piper Alpha disaster (Cullen 1990), a major change in the regulatory regime affecting structural integrity and safety issues for offshore installations was implemented in 1995 with the introduction of the Safety Case Regulations (HSE 2005) in 1995 and the Design and Construction Regulations (DCR) in 1996 (HSE 1996). As a result, all installations operating in the UK sector of the North Sea are now required to have an accepted safety case which needs to demonstrate that all hazards with the potential to cause a major accident have been identified, that the risk has been evaluated and that measures have been or will be taken to reduce the risk to a level that is as low as reasonably practicable (ALARP).

Hazards can arise for structures as a result of inadequate design, poor fabrication or potential failure of materials (e.g. brittle fracture and fatigue) and hence good design and fabrication, as well as effective materials selection, are part of the risk reduction process. In addition, there is a requirement for the provision of a safety management system which demonstrates that the key factors for health and safety management—as identified in, for example, HSG65 (HSE 2013)—are in place.

The DCR introduced the concept of lifetime integrity. An important requirement is to ensure that as far as reasonably practicable, in the event of reasonably foreseeable damage, the installation will retain sufficient integrity to enable action to be taken to safeguard the health and safety of persons on or near it. The retention of residual integrity in the case of damage, such that evacuation can proceed as required, is an important aspect of this. Certain elements in the structure are safety critical and hence are deemed to be "safety critical elements" (SCEs) in the modified Safety Case Regulations HSE (2005).

For North Sea structures on the UK continental shelf (UKCS), preparation of an inspection plan is now a requirement of Regulation 8 of the DCR. This requires that the duty holder ensures that suitable arrangements are in place for maintaining the integrity of the installation, through periodic surveys and assessments and carrying out any remedial work in the event of damage or deterioration. Previously, the guidance (Department of Energy 1990) provided recommendations for the frequency of surveys during the five-year period of the certificate of fitness, including its renewal, on the UKCS. This renewal period tended to determine the inspection schedule.

The Department of Energy Guidance (Department of Energy 1990) and also API RP-2A (API 2014a) have been important sources of design information for a wide range of structural components. From 1998, however, the Department of Energy Guidance is no longer being maintained by HSE and an ISO standard for offshore structures, ISO 19902 (ISO 2007), is now the recognised standard for the UK. The ISO standard includes a substantial new section on structural integrity management and in-service inspection which provides an international framework for this subject. The ISO procedure covers both in-service inspection and structural integrity management, including collation of platform and inspection data and their evaluation to develop an inspection plan.

2.2.4 Regulation in the Norwegian Offshore Industry

The petroleum activities on the Norwegian continental shelf are regulated by the Petroleum Safety Authority, PSA (previously Norwegian Petroleum Directorate, NPD), in the Framework, Management, Facility and Activities Regulations (PSA 2019). Very generic requirements are given to structures in these regulations, mostly indicating the safety level that should be met (structures should be able to withstand all loads at an annual probability of 10^{-4}). The Activities Regulation specifies that the use of structures shall be in accordance with requirements stipulated and any additional limitations identified. Further, the Activities Regulation states that the structures should be maintained so that they are capable of carrying out their required functions in all phases of their life. Maintenance as a term in the PSA regulations includes activities such as monitoring, inspection, testing and repair.

In the guidance to the PSA facility regulation (PSA 2019) reference is made to NORSOK standards N-001 (Standard Norge 2012), N-003 (Standard Norge 2017a) and N-004 (Standard Norge 2013) for how to fulfil design requirements for structures. Further, the PSA regulation refers to NORSOK N-005 (Standard Norge 2017b) for requirements to in-service integrity management and NORSOK N-006 (Standard Norge 2015) for assessment of existing structures.

NORSOK N-001 (Standard Norge 2012) specifies general principles and guidelines for safety of structures. The standard is valid for all types of structures and covers the use of several types of materials such as steel, concrete and aluminium. NORSOK N-001 (Standard Norge 2012) further refers to the following accompanying standards in the NORSOK N-series of standards for specific topics:

- NORSOK N-003 (Standard Norge 2017a) provides the principles and guidelines for calculating loads and load effects;
- NORSOK N-004 (Standard Norge 2013) presents methods for the design and strength evaluations of steel structures;
- NORSOK N-005 (Standard Norge 2019) presents requirements for integrity management and in-service integrity management of structures during operation;
- NORSOK N-006 (Standard Norge 2015) gives recommendations for assessing structures in operation, including recommendations for life extension.

2.3 Standards and Recommended Practices for Steel Structures

2.3.1 Introduction

The SIM process is often defined as; the collection of necessary information about the structure, its condition, its loadings and its environment to enable sufficient understanding of the performance of the structure to ensure that loading limits are not exceeded and that safe operation is assured.

The most recent SIM standards are based on a major hazard approach to the integrity management of structures and other assets, as exemplified in API RP 2FSIM (API 2019a), API RP 2MIM (API 2019b), ISO 19902 (ISO 2007), ISO 19901-9 (ISO 2019a) and NORSOK N-005 (Standard Norge 2017b), which are further discussed in this section. This approach is based on methodology that seeks to understand the hazards to which the structure is exposed, the possible consequences of these and how these should be mitigated. This includes ensuring that suitable safety measures (often called barriers) to protect against the hazards turning into unfavourable consequences are established and maintained throughout the life of the facility. A key element of the major hazard and barrier management regimes is the prioritisation of activities, including inspection and repair for the important safety systems and barriers. Several standards and reports have been published on the principles of SIM, such as HSE RR684 (HSE 2009). These identify a number of key processes which are considered good practice in SIM, together with an appropriate management scheme and documentation.

Inspection is a key element of the surveillance activities and will include in-service inspection of the structure to identify deterioration, degradation and damage. The process of identifying such anomalies will often be the most extensive and costly part of the surveillance activity (both strategy and programme).

Repair is the necessary mitigation activities in order to retain the safety of the structure if inspection has revealed anomalies of significant concern. The repair method of choice is typically based on output from a structural evaluation or assessment.

Several standards have recently been developed to guide the operators of floating facilities in their integrity management including inspection, such as ISO 19904-1 section 18 (ISO 2006) and NORSOK N-005 appendices F, G, H and I (Standard Norge 2017b). Further, a document prepared by Oil & Gas UK (O&GUK 2014) on the management of ageing and life extension for floating production installations provides detailed information relevant for inspection of ageing floating installations. Inspection features as a key control measure in managing the ageing mechanisms such as fatigue and corrosion, although little detail is given on the actual recommended inspection tools and technique. The main topics in the Oil & Gas UK (O&GUK 2014) are the integrity management and inspection of:

- hull (structural and watertight integrity);
- marine system, including ballast system, control system, cargo system, inert gas system and marine utilities (pumps, generators, etc.);
- station-keeping systems (mooring and DP).

NORSOK N-005 (Standard Norge 2017b) builds upon this Oil & Gas UK (O&GUK 2014) report, breaks down the non-structural systems into a number of components and identifies whether inspection based on class rules or based on generally accepted maintenance standards such as NORSOK Z-008 (Standard Norge 2011) is appropriate.

Standards and recommended practices for SIM are given in API RP-2SIM (API 2014b), API RP-2FSIM (API 2019a), API RP-2MIM (API 2019b), ISO 19902 (ISO 2007), ISO 19901-9 (ISO 2019a), NORSOK N-005 (Standard Norge 2017b) and HSG65 (HSE 2013). These do to a large extent include the SIM process described above.

2.3.2 API RP-2A and API RP-2SIM (Structural Integrity Management)

API RP-2A has since 1969 been providing guidance for offshore structure, including surveys (inspections) during operation. The required surveys range from a minimal annual inspection (level I) to more extensive inspections at longer intervals (levels II–IV), depending on manning levels and exposure (API 1993). This approach has since been taken up by other standards. A brief overview of the required surveys required for manned and unmanned platforms is shown in Table 1.

API RP-2SIM (API 2014b) was developed from the API RP-2A (API 2000), ISO 19902 (ISO 2007) and industry best practices. The aim was to provide a stand-alone recommended practice (RP) for structural integrity that clarified the link between platform data, risk categorisation, fitness-for-purpose (FFP) assessment and inspection. The risk categorisation depends on the likelihood of

Table 1 Inspection Intervals According to API RP-2A (API 1993).

	Interval years for manned platforms	Interval years for unmanned platforms
Level I inspection • Effectiveness of CP system • Above water visual survey In the event of suspected underwater damage, a level 2 survey was stated to be required.	Annual	Annual
Level II inspection • General underwater visual survey • Damage survey • Debris survey • Marine growth survey • Scour survey • Anode survey • Cathodic potential The detection of significant structural damage is stated to be the basis for a level 3 survey.	3–5 years	5–10 years
Level III inspection • Visual inspection of preselected areas or areas with known or suspected damage • Cleaning of marine growth The detection of significant structural damage is stated to require a level 4 survey	6–10 years	11–15 years
Level IV inspection • Underwater NDT of preselected areas or areas with known or suspected damage • Detailed inspection and measurement of damaged area	Dependant on outcome of level III inspection	Dependant on outcome of level III inspection

failure and the consequences from failure, in terms of manning level and environmental exposure. The risk categories are expressed as L-1 to L-3, with L-1 being the higher risk platforms (typically manned or high environmental consequence).

The RP builds upon the required survey intervals as shown in Table 1 and includes a simplified risk-based methodology for inspection planning. Further, the RP details engineering practices for the evaluation, assessment and inspection of existing fixed offshore structures to demonstrate their fitness-for-purpose.

The RP describes the recommended SIM process, with details of each aspect of the process, as shown in Figure 8. As can be seen, the damage evaluation, assessment process, risk reduction (e.g. repair), the inspection plan and scope are of particular relevance to this book.

The RP requires a baseline underwater inspection to be undertaken to determine the as-installed platform condition. The baseline underwater inspection should include a visual survey of the platform for structural damage and the presence and condition of anodes and installed appurtenances. Further, the RP requires routine underwater inspections to be conducted to provide the information necessary to evaluate the condition of the platform and appurtenances. A plan for these routine inspections, depending on the risk evaluation, is provided in API RP-2SIM (API 2014b) giving the intervals and scope for the inspections, the tools and techniques to be used and the methods of deployment.

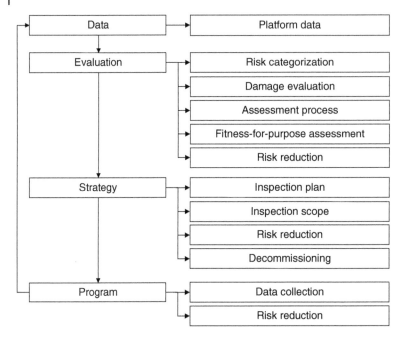

Figure 8 Main SIM process of API RP 2SIM (API 2014b) and follow-on steps. *Source:* Modified from API (2014b), API RP-2SIM Recommended Practice for Structural Integrity Management of Fixed Offshore Structures, American Petroleum Institute, 2014.

A simplified risk-based approach to inspection was introduced in API RP 2SIM (API 2014b). Intervals for level II inspection were stated depending on risk levels (manning level and likelihood of failure). The risk-based inspection programme should also specify if level III or level IV inspections are required. Damage or deterioration found during a level II inspection may trigger a level III or level IV inspection. This risk-based approach is used to redefine the inspection intervals as shown in Table 1, allowing for longer intervals for low likelihood of failure L-1 platforms but also requiring shorter intervals for high likelihood of failure L-2 platforms. The setting of inspection intervals for level II inspections greater than 10 years for L-1 platforms requires the operator to demonstrate a number of factors, including that the platform is unmanned and that the annual level I CP readings are performed and are acceptable.

API RP-2SIM (API 2014b) states that flooded member detection (FMD) can provide an acceptable alternative to close visual inspection of preselected areas in a level III survey. Engineering judgment should be used to determine the optimum use of FMD and close visual inspection of joints. Close visual inspection of pre-selected locations for corrosion monitoring should be included as part of the level III survey.

Inspection tools and techniques are covered in general terms in API RP-2SIM (API 2014b), including details of the standard methods as covered in Chapter 4 of this book. Evaluation of damaged components and repair is addressed in the recommended practice in line with Chapters 7 and 8 of this book. The standard also addresses the structural assessment process, outlining a series of approaches. Four different levels of assessment are developed, which include:

- simple methods, e.g. comparison with a similar platform;
- design methods, e.g. linear (elastic) methods to check the platform member-by member, similar to the approach used for the design of new platforms;

- ultimate strength method, e.g. using a non-linear or equivalent linear method to determine platform performance on a global basis;
- alternative methods, e.g. using historical performance of the platform or explicit probabilities of survival of the platform for the assessment.

Personnel qualifications are listed for personnel responsible for conducting the evaluation and developing the inspection strategy, as further discussed in Section 4.4.5.

2.3.3   API RP-2FSIM (Floating Systems Integrity Management)

API RP-2FSIM (API 2019a) is a Recommended Practice (RP) intended to be used by owners and engineers in the development, implementation and delivery of a process to maintain system integrity of floating production systems (FPSs), which includes tension leg platforms (TLPs). The specifications, procedures and guidance provided within the RP are based on internationally recognized industry standards and on global industry best practices. This RP does not cover moorings, which are separately covered by API RP-2MIM (API 2019b).

API RP-2FSIM (API 2019a) states that an inspection plan should define the scope and frequency of inspection, the tools and techniques to be used and the deployment methods. This plan should be developed for the FPS structural and system components. The main objectives of the inspection plan are to identify areas of corrosion, coating deterioration, damage due to overloading, impact or abrasion, as well as areas affected by marine growth and debris. The plan should also measure or confirm the CP potentials, sensors and alarms, marine system functions and mooring equipment functions. The inspection plan should consist of a set of scheduled work scopes to be performed over the service life of the FPS. In addition, unscheduled surveys should follow an unexpected event such as exposure to a near-level design event (e.g. hurricane).

API RP-2FSIM (API 2019a) states that underwater surveys should be performed on the submerged areas of the hull and this should include external marine system components and the mooring system hull attachments. These surveys should also confirm that the corrosion protection system of the external hull is functioning adequately. It is recommended that external hull surveys can be performed on a continuous cycle where a specified percentage of the hull is inspected over a period and at the end of the period all of the hull is covered. A default inspection program for the hull structure is included with calendar-based intervals and the extent of structure to be inspected by different inspection methods (GVI, CVI, ultrasonic testing and weld inspection).

Monitoring of the structure and marine systems are recommended in API RP-2FSIM (API 2019a) in addition to inspection. Typical monitoring methods for floating structures are further discussed in Chapter 5.

API RP-2FSIM (API 2019a) describes an assessment process and acceptance criteria, similar to other relevant API RPs, including the assessment of fatigue damage. The fatigue assessment should assess the fatigue loading behaviour and analyse the remaining fatigue life of critical locations such as welded joints and details in highly stressed areas. If the predicted fatigue life exceeds allowable levels, API RP-2FSIM (API 2019a) states that improvement options may be considered, including local structural modifications, increased inspection frequency, change of loading conditions and using improved inspection techniques.

Methods for the repair of damage and modifications are described in API RP-2FSIM (API 2019a), including completely renewing a locally damaged area by means of a structural repair. This may involve the addition of a new structure (e.g. stiffening to reinstate the capacity in a damaged area). Modification of the structural details may be used to minimise the potential for future damage, particularly fatigue (e.g. by grinding).

2.3.4 ISO 19902

The development of the international standard ISO 19902 (ISO 2007), concerned with fixed off-shore structures began in 1993, including a section on structural integrity management in the early drafts. This standard was first issued formally in 2007, but the early drafts were used as the basis for structural integrity management in the period from 1993 to 2007. In 2019 the SIM part of this standard was transferred to a new ISO standard, ISO 19901-9 (ISO 2019a), which focused on structural integrity management and assessment of existing structures.

The ISO 19902 (ISO 2007) took into account all the issues and methods that had been used to inspect, monitor, evaluate and assess fixed offshore steel structures over the previous years. ISO 19902 (ISO 2007) described four different types of inspection of the structure as shown below:

- baseline: inspection soon after installation and commissioning to detect any defects arising from fabrication or damage during installation;
- periodic: inspection to detect deterioration or damage over time and discover any unknown defects;
- special: inspection required after repairs or to monitor known defects, damage or scour;
- unscheduled: inspection undertaken after a major environmental event such as a severe storm, hurricane, earthquake or after an accidental event such as a vessel impact or dropped objects.

ISO 19902 (ISO 2007) stated that "the purpose of in-service structural inspection is to determine, with a reasonable level of confidence, the existence and extent of deterioration, defects, or damage. Data collected during an inspection are needed to verify the integrity of the structure".

ISO 19902 (ISO 2007) described the process of managing the integrity of structures with four major elements as shown in Figure 9, where inspection is the key element of both the long-term surveillance plan and particularly the surveillance execution. The remaining two processes include the storage of inspection data and the evaluation of possible anomalies found during inspection.

Assessment was introduced in ISO 19902 (ISO 2007) as an option following engineering evaluation if anomalies identified during inspection require a more detailed analysis.

A possible outcome of the evaluation or the assessment is the implementation of prevention measures which include structural strengthening and reduction of loads on the structure or

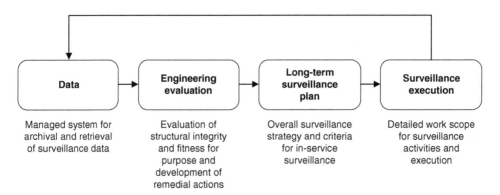

Figure 9 Structural integrity management cycle (surveillance includes inspection of the condition, determination of loading, review of documents, etc., needed to determine changes that may affect the safety of the structure). *Source:* Based on ISO (2007), ISO 19902:2007 Petroleum and natural gas industries—Fixed steel offshore structures, International Standardization Organization.

mitigation actions; these might include measures such as operational restrictions and possible de-manning.

2.3.5   ISO 19901-9

This relatively new standard (at the time of writing) focusses on structural integrity management (ISO 2019a) based on other standards such as ISO 19002 (ISO 2007) and NORSOK N-005 (Standard Norge 2017b) and N-006 (Standard Norge 2015). It includes a section on developing an inspection strategy (plan) to be prepared from the engineering evaluation which should determine the likely existence and extent of any deterioration and damage. The inspection strategy should periodically be reviewed and updated throughout the service life based on new data and information being received from, for example inspection reports and structural evaluation and assessments. The types of inspection proposed follow that in other standards (e.g. API RP-2A) with both scheduled and unscheduled inspections. The scope of work should include standard inspection methods such as GVI, CVI, FMD and NDE. Selection of appropriate inspection methods should be determined by qualified personnel based on knowledge and availability of the different methods. Inspection intervals for periodic inspections are selected according to the structural integrity strategy using a risk-based approach. However, if an operator does not want to use a risk-based approach, an alternative is a consequence-based method developed from world-wide experience which provides pre-determined inspection intervals.

The ISO standard also addresses inspection methods and identifies specific areas for inspection, including underwater cathodic protection, coatings and air gap. The survey should include dimensional measurements to measure such quantities as damage size and geometry, member out-of-straightness, crack length and depth and depth of corrosion. Focussed inspection is proposed for several failure modes and degradation mechanisms such as impact damage from ships, debris and dropped objects, fatigue failure and damage to protective devices such as riser guards. The standard notes that structures are usually designed with a corrosion allowance in the splash zone, which should be monitored.

At the time of writing, this standard is in the process of being updated as a result of concerns expressed about its approach, particularly on the assessment part of the standard. In addition, structural integrity management is a rapidly evolving field given the pressures of the extending life of many platforms and the subsequent cost of inspection and repair.

2.3.6   NORSOK N-005

NORSOK N-005 (Standard Norge 2017b) was published in an updated version in 2017 and incorporated a major hazard-based approach to the integrity management of structures. This approach is also implemented in other newer standards for structural integrity management such as ISO 19901-9 (ISO 2019a). The approach is based on risk analysis methodology, understanding the hazards the structure is exposed to and minimising the likelihood of these hazards and possible unfavourable consequences occurring. In major hazard management this will typically include implementation of suitable safety measures (often called barriers or safety critical elements) and are established and maintained throughout the life of the facility. This current approach to SIM is based on the Norwegian and UK Major Hazard regulatory regimes.

As-is surveillance as described in this standard includes activities to determine anomalies and changes of all types relevant for the safety of the structure. The most important of these surveillance activities is the inspection of the structure's condition and the determination of

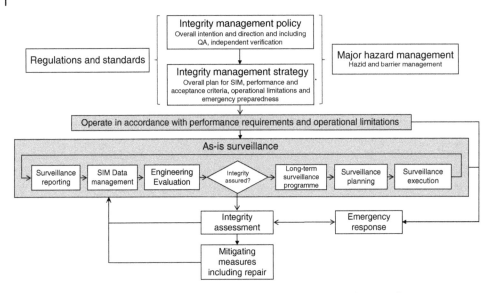

Figure 10 An example of a typical SIM process based on a major hazard approach.

loadings. The in-service inspection process described in this standard is very similar to the process described in Figure 10.

2.4 Standards and Recommended Practices for Mooring Systems

2.4.1 Introduction

Several recommended practices and standards address inspection and integrity management of mooring systems. API RP-2I (API 2008) and API RP-2SK (API 2005) have developed over many years and have been the industrial standards over this period. Recently, API issued a separate recommended practice (RP) for integrity management of mooring systems, API RP-2MIM (API 2019b), which extends the guidance from these earlier RPs. NORSOK N-005 (Standard Norge 2017b) and ISO 19901-7 (ISO 2013) also address inspection of mooring systems.

Inspection intervals for chain mooring systems for floating structures are recommended in, for example API RP-2I (API 2008) and NORSOK N-005 (Standard Norge 2017b). The maximum interval between major inspections is linked to the age of the chain in years. For relatively new chains (i.e. 0–3 years) the recommended interval is 3 years; for slightly older chains (4–10 years) it is 2 years and for chains older than 10 years the interval is reduced to only 8 months. This short interval is very demanding and costly and hence chains are normally replaced before they reach the 10-year criterion. Inspection of mooring systems can be undertaken visually by an ROV, but this has significant limitations. A more detailed inspection, e.g. by MPI, requires removal of the mooring and dockside inspection with significant cost and operational implications. More details of mooring line inspection are given in Section 4.9.

ISO 19901-7 (ISO 2013) is accepted as the recommended code for offshore moorings in many parts of the world. HSE (HSE 2019) states that the following aspects of Annex B.2 of ISO 19901-7 (ISO 2013) are considered important by HSE and should be met where this is reasonably practicable:

- increased wear and corrosion allowances in splash zone for permanent moorings;

- assessment of a simultaneous double failure (two-line failure) for permanent moorings; and
- increased safety factors which take account of the operational state of the installation when connected to risers or when in proximity to other structures.

The Norwegian Maritime Directorate (NMD) has issued local regulations, including the Anchoring Regulations (NMD 2009), concerning positioning and anchoring systems on mobile offshore units. The 2013 version of Annex B.2 of the ISO 19901-7 (ISO 2013) has been updated to reflect in part the NMD regulation (2009). A Canadian section is included in Annex B, which states the same technical requirements as for Norway.

The following codes and standards are now widely accepted in the offshore industry:

- API RP RP-2SK (API 2005);
- DNVGL-OS-E301, which superseded POSMOOR '96 (DNVGL 2018);
- Lloyds Register Rules for Classification Floating Units at Fixed location – Part 3 (Lloyds Register 2019).

2.4.2 API RP-2MIM (Mooring Integrity Management)

API RP-2MIM (API 2019b) is a recommended practice (RP) providing guidance for the integrity management of mooring systems connected to permanent floating production systems (FPS). The scope covers the system from the anchor to the connection to the floating unit. The guidance provided covers inspection, monitoring, evaluation of damage, fitness-for-service assessment and mitigation planning. This RP includes guidance from API RP-2I (API 2008) and API RP-2SK (API 2005) and expands on both of these. The RP allows risk-based principles to be used to develop an appropriate strategy, with higher risk moorings requiring more frequent inspections than lower risk moorings.

The integrity management plan defines two approaches, which are prescriptive and risk based. For the prescriptive method the operator is referred to the requirements specified in API RP-2I (API 2008) or API RP-2SK (API 2005). A risk-based approach should be developed in terms of the potential risk of damage to the component being inspected. The inspection plan outlines the frequency and scope of the inspections, the tools to be used and how these will be deployed. Inspection techniques are listed in an annexe to the RP and include a comprehensive set covering all aspects of the mooring system, including the anchor, CP system, chain, coatings. The techniques listed include GVI, CVI, high resolution 3D photography, physical measurements and the use of an inclinometer.

API RP-2MIM (API 2019b) recommends monitoring of the mooring system to continouosly verify its condition and performance. This may include direct measurements of the mooring response (e.g. line tension, strains) or measures than can indirectly predict mooring response such as vessel offset. Data from a position-monitoring system can be used to ascertain that the mooring system is meeting design requirements. Environmental monitoring may also be used to provide metocean data to support the position monitoring. Monitoring for individual line performance is also useful to detect a line failure or loss of tension in the mooring system.

API RP-2MIM (API 2019b) recommends determining assessment initiators, based on a change of condition outside the limits of the original design basis or from a recent in-service assessment. Possible assessment initiators could result from, for example, damage, changes in station-keeping performance and stability. To demonstrate an acceptable risk level for the mooring system, the likelihood of failure can be estimated by the use of a suitable risk assessment method for several different cases. These range from as-originally-designed to the as-is-condition using current design criteria. For the risk acceptance criteria, pre-requisites are listed including defining mooring system failures and acceptable risk levels and demonstrating acceptable levels for the as-is or planned configurations using the results from the previous analyses. If a mooring system does not pass the fitness for service criterion, risk reduction measures that reduce risk to an acceptable level are required. These measures can include changed operational procedures and modifications to the system. In addition, modifications to inspection and monitoring programs can be implemented to identify any further deterioration and reducing the likelihood of any undesired consequences.

Unlike other integrity management standards, API RP-2MIM (API 2019b) addresses emergency response planning based on unexpected failures of a mooring system. Single line failure can escalate to multiple line failure quickly, requiring a range of options to be considered to develop a response plan. API RP-2MIM (API 2019b) recommends that at least a short-term emergency response plan should be in place before operations commence. In addition, there is a requirement for an immediate operational response plan. This immediate response plan needs to provide operators with a set of operating procedures that address a range of reasonably foreseeable situations that could occur both under normal and severe weather conditions. Actions arising from the plan could include ceasing production or limiting operations with defined limits.

API RP-2MIM (API 2019b) recommends that a conservative approach is adopted until the nature of the failure is understood and any escalation ruled out. Inspections can be undertaken to provide information of the degradation mechanism to be identified. This then allows an assessment to be made of the potential for multiple degradation mechanisms to exist, which would affect the emergency response planning.

The short-term plan should include procedures to address situations of substantial or imminent loss in mooring capacity, requiring emergency repairs to be made to the degraded mooring system sufficient to create the necessary time for a more permanent repair.

The medium-term emergency response could entail a complete repair of the mooring system, eliminating the recorded degradation mechanism. It is recognised that there may be an intervening period between the inspection finding and the short-term repair response during which the mooring system could operate in a degraded state, where accepted procedures for stopping operations and production shutdowns depending on the observed degree of degradation are needed.

Overall, API RP-2MIM (API 2019b) is a comprehensive document relating to the integrity management of mooring systems and the only (to the knowledge of the authors) integrity management standard that includes a proper guidance for incident response planning and execution.

2.4.3 IACS Guideline for Survey of Offshore Moorings

A set of guidance for surveys of the moorings of floating units has been produced by IACS (2010). It states the minimal requirements for an annual survey, the scope of which are mooring components adjacent to the winch or windlass. It should be conducted with the vessel at operational draft with the position mooring system in use. This is to be supplemented by a more extensive survey if typical damage is present such as:

- chain: reduction in diameter exceeding 4%, missing studs, loose studs in grade 4 chain;
- wire rope: obvious flattening or reduction in area, worn cable lifters causing damage to the rope, severe wear or corrosion, broken wires.

Special periodical surveys are also required at approximately 5-year intervals. These special surveys are quite demanding and should include the following:

- close visual examination of all chain links, with cleaning if required;
- enhanced NDE sampling of:
  - 5% of the chain
  - 20% of chain which has been in the proximity of fairleads in the last 5 years
  - all connecting links
- dimensional checks, including lengths over five links;
- checking looseness and pin-securing arrangements of joining shackles;
- measurement of thickness (diameter) of approximately 1% of all chain links distributed through the working length of the chain; and
- dismantling and MPI of all Kenter-type shackles which have been in service for more than 4 years.

2.5 Standards and Guidance Notes for Concrete Structures

2.5.1 Introduction

The international recognised standard for designing, fabricating and maintaining offshore concrete structures is the ISO 19903 (ISO 2019b), originally introduced in 2006 and updated in 2019. The ISO standard was based on the Norwegian standard NS 3473 with input from other relevant standards. Previously, the Department of Energy introduced a set of Guidance Notes for offshore concrete structures in 1974 which was significantly updated in the fourth edition, published in 1990 (Department of Energy 1990) and supported by a background document on concrete (HSE 1997). The emphasis of these documents, however, was on design and construction of offshore concrete structures with limited sections dealing with inspection and repair. In addition, the Norwegian standard NORSOK N-005 (Standard Norge 2017b) has some international status. NORSOK N-005 includes a special annexe covering inspection of concrete structures. The relevant sections of ISO 19903, the Department of Energy Guidance Notes and NORSOK N-005 are discussed below.

2.5.2 ISO 19903 – Concrete Structures

ISO 19903 (ISO 2019b) includes an inspection programme for concrete structures similar to that in, for example, ISO 19902 for steel structures, requiring an initial (baseline), periodic and special inspection to be performed. The initial inspection is required as soon as possible after installation to verify the original design and that all the major parts of the installed structure have no obvious defects. Following this initial inspection, inspection and condition monitoring of the structure shall be carried out regularly in accordance with the established programme. Special inspections should be conducted after direct exposure to an accidental or design environmental event (wave, earthquake, etc.). These inspections should encompass the critical areas of the structure. However, following an accidental event such as a boat collision and a dropped object, the inspection may, in certain circumstances, be limited to the local area of damage. In addition, the standard requires that measures should be taken to maintain the structural integrity appropriate to the circumstances in an event such as change of use, life extension, major modifications and when inspection has revealed damage or deterioration.

Similar to the standards already discussed for steel structures, assessment of the condition of the structure should be carried out following the inspection activities. This assessment should include a summary evaluation being prepared at the end of each programme for inspection. It is also stated that the data gathered from each periodic inspection should be compared to data gathered from previous inspections, with the purpose of establishing any data trends that can indicate time-dependent deterioration processes.

The typical inspection methods used for a concrete structure include:

- GVI to detect obvious or extensive damage such as impact damage, wide cracks, settlements and tilting. Prior cleaning of the inspection item is not needed for this type of inspection.
- CVI to detect less extensive damage. CVI requires direct access to the inspected area and prior cleaning of the item to be inspected is normally needed.
- Non-destructive testing involves close inspection by electrical, electrochemical or other methods to detect hidden damage such as delamination. Prior cleaning of the inspection item is normally required.
- Destructive testing such as core drilling which is used to detect hidden damage or to assess the mechanical strength or parameters influencing concrete durability.

In the standard, structural monitoring methods are proposed to be used in areas with limited accessibility or for monitoring of, for example, action effects and corrosion development. The sensors needed for this monitoring, such as strain gauges, pressure sensors, accelerometers, corrosion probes should preferably have been fitted during construction.

The structure may also have been instrumented in order to record data relevant to pore pressure, earth pressure, settlements, subsidence, dynamic motions, strain, inclination and possibly temperature in oil storage. These monitoring systems are required by the standard to be tested and inspected regularly.

Required locations for inspection on a concrete structure include a survey of the atmospheric zone, the splash and the tidal zones and the important areas of immersed concrete. It is generally recognised that the splash zone is most vulnerable to corrosion. To determine the extent and frequency of inspection for different structural parts it is recommended that the exposure or vulnerability to damage is considered. ISO 19903 (ISO 2019b) states that the inspection of the internal parts of the structure (e.g. inside the legs) should focus on:

- detecting any leakage;
- biological activity;
- temperature, composition of sea water and pH values in connection with oil storage;
- detecting any corrosion of reinforcement; and
- concrete cracking.

Special areas for inspection:

- Splash zone; it can experience damage from impact of supply vessels and is also sensitive to corrosion of the reinforcement. It can also deteriorate from ice formation with ensuing spalling in surface cavities where concrete has been poorly compacted. Ice abrasion and freeze-thaw cycling can also lead to early deterioration even with high quality concrete. This deterioration can lead to subsequent loss of cover over the reinforcement steel leading to potential steel corrosion. ISO 19903 (ISO 2019b) recommends that repairs to these damaged surfaces should be made as soon as possible to prevent further deterioration and structural overload.

- Construction joints in the concrete structure; these represent potential structural discontinuities where water leakage and corrosion of reinforcement are possible negative effects. As a minimum, the monitoring programme should identify construction joints located in high stress areas and monitor the performance with respect to evidence of leakage, corrosion staining or local spalling at joint faces (which indicates relative movement at the joint). In addition, evidence of poorly placed and compacted concrete, such as aggregate pockets and delaminations and joint cracking or separation, are areas for investigation.

- Embedment plates; these may create a path for galvanic corrosion to the underlying steel reinforcement. The main concerns are corrosion and spalling around the plates. Galvanic corrosion is a serious possibility where dissimilar metals are in a marine environment and could lead to deterioration of the reinforcing steel, which is in contact with the embedment plates.

- Repair areas and areas of inferior construction; ISO 19903 (ISO 2019b) states that concern is associated with areas that provide a permeable path through which a flow of seawater can take place as the continuous flow of saline and oxygenated water can cause corrosion of the reinforcement. In addition, attention is recommended to be given to the surface and the perimeter of patched areas for evidence of shrinkage cracking and loss of bond to the parent concrete surface.

- Penetrations; these are areas of discontinuity and are therefore prone to water ingress and spalling at the steel to concrete interface. Any penetrations added to the structure during the operational phase are particularly susceptible to leakage resulting from difficulties in achieving high-quality concrete in the immediate vicinity of the added penetration. ISO 19903 (ISO 2019b) states that all penetrations in the splash and submerged zones require frequent inspections (but not specified).

- Steel transition ring to concrete interface; this interface is the main load transfer point between the concrete towers and steel topsides. ISO 19903 (ISO 2019b) recommends that the steel transition ring should preferably be examined annually. The examination should include the load transfer mechanism (flexible joints, rubber bearings, bolts and cover) and the associated ring beam. In addition, the concrete interface should be inspected for evidence of overstress and corrosion of embedded reinforcing steel. Corrosion-potential surveys can be used to detect ongoing corrosion that is not visible by visual inspection alone.

- Debris; it can cause structural damage through impact, abrasion or by accelerating the depletion of cathodic protection systems. It can also limit the ability to inspect various parts of the structure (e.g. tops of storage cells). Drill cuttings can build up on the cell tops and/or against the side of the structure and should be assessed for lateral pressures exerted by the cuttings, and whether they cause an obstruction to inspection. As a result of this assessment the removal of drill cuttings needs to be assessed accordingly.

- Scour; this is the loss of foundation-supporting soil material and can be induced by current acceleration round the base of the structure or by "pumping" effects caused by wave-induced dynamic rocking motion. It can lead to partial loss of base support and unfavourable redistribution of actions. GVI is the recommended method.

- Drawdown (differential hydrostatic pressure); a level of drawdown in many concrete structures is necessary for structural integrity. Structural damage or equipment failure can lead to ingress of water and loss of hydrostatic differential pressure. Special inspection may be required if loss of drawdown is detected.

- Temperature of oil sent to storage; the storage cells are designed for a maximum temperature differential. In cases where differential temperatures have exceeded these design limits, special inspections may be required.

- Sulphate-reducing bacteria (SRB); these can occur in anaerobic conditions where organic material is present (such as hydrocarbons). The bacteria produce H_2S (hydrogen sulphide) as their natural waste which could cause a lowering of pH value of the cement paste in the concrete. Favourable conditions for SRB growth can be present in unaerated water in, for example, the water-filled portion of shafts and cells. An acidic environment can cause concrete softening and corrosion of reinforcement. However, it is recognised that an inspection of a concrete surface likely to be affected by SRB activity is difficult to undertake. Some guidance can be obtained by adequate monitoring of SRB activity and pH levels (HSE 1990).
- Post-tensioning; the tendons are usually contained within ducts which are grouted. Post-tensioning anchorage zones are commonly areas of complex stress patterns. Because of this, considerable additional reinforcement steel is used to control cracking. Anchorages for the post-tensioning tendons are generally terminated in prestressing pockets in the structure, which are also vulnerable. Inspection of tendons is, however, very difficult using conventional inspection techniques. In many cases, the reinforcing steel is very congested and this condition can lead to poor compaction of concrete immediately adjacent to the anchorage. The recess is fully grouted after tensioning. These conditions expose the critical tendon anchors to the marine environment, causing corrosion of the anchor and additional spalling and delamination of concrete and grout in the anchorage zone. Regular visual inspection of the anchorages is recommended.

Experience has shown that the anchorage zones are vulnerable to distress in the form of localized cracking and spalling of grouted anchorage pockets. Where evidence exists for this type of damage, ISO 19903 (ISO 2019b) recommends a more detailed visual inspection supplemented by impact sounding for delamination. The visual inspection should focus on corrosion staining, cracking and large accumulations of efflorescence deposits.

Partial loss of prestress is generally recognised as leading to local concrete cracking resulting from redistribution of stress. This should be investigated upon discovery. In addition, design documents should be reviewed by the inspection team to establish the arrangement and distribution of cracking that could be expected to result from partial loss of prestress.

Durability of concrete and the corrosion protection system are addressed in terms of inspection and focus on:

- Concrete: factors that may change with time and may need to be surveyed regularly, such as chloride profiles, chemical attacks, abrasion depth, freeze and thaw deterioration and sulphate attack in petroleum storage areas.
- Corrosion protection: Periodic examination with measurements should be carried out to verify that the cathodic protection system is functioning within its design parameters and to establish the extent of any material depletion of the anodes. Cathodic protection is also provided for the protection of the reinforcing steel, which is important for the structural integrity of the concrete. In this respect the level of adequate potential should be monitored. In general examination shall be concentrated in areas with high or cyclic stress utilization, which need to be monitored and checked against the design basis. Heavy unexpected usage of anodes should be investigated.

Examination of any coatings and linings is normally performed by visual inspection to determine the need for repairs. A close visual examination will also disclose any areas where degradation of coatings has allowed corrosion to develop to a degree requiring repair or replacement of structural components.

It is noted in ISO 19903 (ISO 2019b) that several techniques have been developed for the detection of corrosion in the reinforcement in land-based structures. These are mainly based on electro-potential mapping, for which there is an ASTM standard (ASTM 2015). These techniques are useful for detecting potential corrosion in and above the splash zone but have limited application under water because of the low resistance of sea water. ISO 19903 (ISO 2019b) also notes that it has been established that under many circumstances underwater corrosion of the reinforcement does not lead to spalling or rust staining. The corrosion products are of a different form and can be washed away from cracks and hence leave no evidence on the surface of the concrete (see Section 3.5). However, when the reinforcement is adequately cathodically protected, any corrosion should be prevented. In cases where cathodic protection of the reinforcement is limited, the absence of spalling and rust staining at cracks in the concrete cover should not to be taken as evidence for no corrosion.

2.5.3   Department of Energy Guidance Notes

The "Survey" section of the fourth edition Guidance Notes (Department of Energy 1990) focussed mainly on steel structures with regard to the requirements for re-certification. At the time of preparing the fourth edition Guidance Notes, concrete structures were relatively new and a comment was made that experience to date had shown little requirement for maintenance and repair. However, specific mention was made of the lower elements of a GBS-type structure, which were only likely to suffer significant damage if erosion or uneven settlement had taken place. It was noted that many units of this type have built-in instruments to record the state of the foundations. With regard to corrosion of the steel reinforcement, it was noted that this is less likely in permanently submerged areas than in the splash zone. Hence, if the splash zone was found to be in good condition, then only a limited number of checks needed to be made at lower levels, except where there were sudden changes of section or high stress concentrations.

The Background document to the Guidance Notes (HSE 1997) supported the limited problems to date but commented that this was probably a result of limited life, particularly for evidence of any corrosion of the steel reinforcement. Some problems with draw-down had been found which were mentioned, due both to failure of the draw-down system and accidental impact damage. Several platforms had suffered ship impact or impacts from dropped objects, with several cases where the damage had been minimal. Where damage had occurred, it had resulted in holes in cell roofs or a leg and in each case a repair had been undertaken. It was noted that the degree of marine growth had varied substantially between structures, some structures with almost no growth, others with extensive growth requiring cleaning.

The "Repairs" concrete section of the Guidance Notes stated that the accepted materials for repair offshore were concrete, cement grouts and mortars and epoxy resins. The importance of the ability of the repair material to bond to concrete, reinforcement or pre-stressing ducts needed to be considered. The Guidance Notes also noted that when selecting a repair material for protection of the reinforcement against corrosion, the properties of the material to be considered should include permeability to water, presence of chloride ions and resistivity. An important factor is the durability of the repair material in the marine environment.

Repair of cracking in a concrete structure was addressed, with low viscosity epoxy resins being the usual material and the repair procedure should be established by trials.

Repairs to steel reinforcement were considered where the reinforcement has been damaged or cut away. The new reinforcement should either be lapped on to the existing reinforcement with a full bond length or connected by a suitable coupler.

Repairs to prestress were also addressed making the point that it is very difficult to establish the effectiveness of any damaged prestress, as this is to a large extent dependent on the bond with the grout in the cable duct. The effectiveness of any repairs to tendons which have been broken would require re-stressing of the new or repaired tendons, which is not easy in a repair situation.

Repair of cracking in a concrete structure should utilise low viscosity epoxy resins, and the repair procedure should be established by trials.

2.5.4   NORSOK N-005 – Concrete Structures

NORSOK N-005 (Standard Norge 2017b) has a special annexe covering inspection of concrete structures. This standard makes the point that concrete structures are normally damage tolerant because of their large dimensions and planning of inspections should take this into account. However, it also points out that ageing concrete structures in air may experience increasing amount of chloride ingress that may lead to corrosion of reinforcement in areas of poor-quality concrete.

Specific requirements for data collection and management relating to concrete structures include:

- operational criteria for caisson cells operated with differential hydrostatic pressure (drawdown) or normal hydrostatic pressure in cells;
- operational criteria for cells used for oil storage and/or water ballast;
- any reported repairs or discontinuities in the concrete from the fabrication phase. Specifically, repair of reported leakages in the concrete;
- description of mechanical systems for operating the ballast water, oil filling and draw down systems in the storage cells; and
- settlement and tilt monitoring.

Regarding evaluation of changes NORSOK N-005 (Standard Norge 2017b) states that the following items are noted for concrete structures:

- the condition of the concrete surface. Particularly the concrete surface above seawater level, due to weather exposure and chloride penetration, including core testing or dust samples of the concrete;
- condition of coatings and linings, if present;
- the cathodic protection of the concrete reinforcement; and
- monitoring of crack development in highly utilized areas (e.g. near topside footings/bearings).

NORSOK N-005 (Standard Norge 2017b) notes that the probability of failure of concrete structures is most often related to corrosion and less to fatigue damage. In addition, failure below the waterline is minimised by the presence of anodes providing corrosion protection to the steel reinforcement. As a result, large areas of a concrete structure such as caisson walls and shaft walls below water have relatively low probability and consequence of failure. Hence it is considered that a lower level of inspection for these structural components may influence the level and content of the inspection programme.

NORSOK N-005 (Standard Norge 2017b) states that special attention should be given to areas of major importance for the structural integrity and for the atmospheric and splash zones, particularly where the risk of deterioration is highest. Areas of major importance are listed as complex connections such as wall-dome connections, cell wall joints, columns to caisson connections and load bearing areas in the top of shafts. Construction joints, penetrations and embedments should also be considered.

NORSOK N-005 (Standard Norge 2017b) notes that for inspection planning the potential failure mechanisms listed below should be taken into consideration. These can lead to typical types of failure, which include reinforcement corrosion, cracking, spalling and delamination and damaged coatings and abrasion:

- construction-related imperfections;
- cracks due to formwork adhesion;
- lack of reinforcement cover;
- lack of concrete compaction;
- unplanned construction joints;
- chloride ingress;
- corrosion of reinforcement behind cast-in plates;
- subsidence and seabed scouring;
- internal defects in conjunction with oil storage such as leakage and biological activity;
- leakage;
- impact damage; and
- freeze/thaw damage and ice abrasion.

In executing inspection of concrete structures NORSOK N-005 (Standard Norge 2017b) states that programmes should be developed after the DFI resume has been delivered and updated periodically. A general statement is made that inspection planning should focus on verification of the condition of the structural components so that structural integrity can be maintained. It also states that taking into account the likely types of defects and damage which may develop, appropriate inspection methods should be developed. An example is given indicating that because delamination is often not visible on the concrete surface, hammer tapping may be needed to detect such areas.

NORSOK N-005 (Standard Norge 2017b) notes that corrective maintenance should be implemented to prevent minor damage escalating into major damage. It is important to maintain steel reinforcements and pre-tensioning to ensure structural functionality during the lifetime of the structure.

2.6 Discussion and Summary

As shown in this chapter, the first standards and recommended practices relating to structural integrity management were developed in the late 1960s and early 1970s. These standards were in practice inspection standards and often included inspection plans based on routine intervals inspired by ship rules. The parts of the structures that were regarded as major or important were inspected typically every fifth year. The offshore industry began to consider more cost-effective and optimised inspection strategies in the late 1980s and methods for condition-based and risk-based inspection planning were developed and started to be used in practice. A broader view on integrity management was included in the development of the international standard ISO 19902 (ISO 2007), which commenced in 1993 and included a section on structural integrity management in early drafts. This standard was first issued in 2007, but the early drafts were used as the basis for structural integrity management in the interim. The process of inspection, data management, evaluation of findings and assessing the safety of the structure were included.

HSE report RR 684 (HSE 2009) states that an inspection strategy should be:

- consistent with the structural integrity strategy and SIM processes including structural evaluation, assessment, maintenance and information management requirements;
- able to determine with reasonable level of confidence the existence and extent of deterioration, defects and damage;
- able to address the motives for inspection listed by ISO 19902 (ISO 2007);
- developed and maintained by a competent person, using appropriate experience, data and, where required, analysis
- able to specify the tools and techniques to be used; and
- documented.

The report (HSE 2009) further states that an inspection programme should be developed from the integrity management policy and strategy. It requires schedules, budgets, personnel profiles, inspection procedures for implementation. Guidance on developing an inspection programme is given in ISO 19902 (ISO 2007) and API RP-2SIM (API 2014b).

A structural integrity management framework is proposed as shown in Figure 10, based on what the authors consider to be the best practice from the various standards reviewed in this chapter. Each element of the process is to varying degrees included in each of these standards and recommended practices. Inspection is a major feature of "as-is surveillance" and repair is a key factor in "mitigating measures".

Most standards in detail include the middle "as-is surveillance" process with the six sub-processes and also include guidance on "integrity assessment" and "mitigation measures".

However, very few of the standards give guidance on hazard identification and establishing the risk picture of the structure and marine system. Two exemplary exceptions are API RP 2MIM (API2019b), which includes an annex on "causes of failure" and API RP 2FSIM (API 2019a), which includes an annex on "Damage and failure modes", both with valuable information for hazard identification and risk management.

Many standards and recommended practices also implicitly include information and guidance on operational limitations, such as the design events, the size of vessels assumed to be visiting the platform and the importance of knowing the weight of the topside and the inventory. However, the standards and recommended practices do not necessarily give guidance on its importance for the operation of the platform or how to operate in situations outside the operational limitations. For floating structures, several relevant issues are often included in the marine operational manual.

Emergency preparedness and emergency response planning should be performed for reasonably foreseeable situations that may occur in normal and in severe conditions. This seems to be an area where the structural and marine engineers rely on the traditional emergency response discipline to understand structures and marine systems. As a result, only one standard is providing guidance on this topic. This exemplary exception is API RP-2MIM (API 2019b), which includes an annex on "incident response planning".

As a result, it can be concluded that for some of the important factors, the standards and recommended practices have limited information. This book attempts to bridge these gaps.

References

ABS (2017), "Inspection Grading Criteria for the ABS Hull Inspection and Maintenance Programme (HIMP)", American Bureau of Shipping, 2017.

API (1969), RP-2A *Recommended Practice for Planning, Design and Constructing Fixed Offshore Platforms*. API Recommended practice 2A, 1st edition, American Petroleum Institute, 1969.

API (2005), *API RP-2SK Design and Analysis of Stationkeeping Systems for Floating Structures*, Third Edition, American Petroleum Institute, 2005.

API (2008), *API RP-2I In-service Inspection of Mooring Hardware for Floating Structures*, Third Edition, American Petroleum Institute, 2008.

API (2011), *API RP 2FPS Recommended Practice for Planning, Designing and Constructing* Floating Production Systems, American Petroleum Institute, 2011.

API (2014a), *API RP-2A Recommended Practice for Planning, Design and Constructing Fixed Offshore Platforms*. API Recommended practice 2A, 22nd Edition, American Petroleum Institute, 2014.

API (2014b), *API RP-2SIM Recommended Practice for Structural Integrity Management of Fixed Offshore Structures*, American Petroleum Institute, 2014.

API (2015), *API RP-2T Recommended Practice for Planning, Designing and Constructing Tension Leg Platforms*, Third Edition, American Petroleum Institute, 2015.

API (2016), *API RP 579-1 / ASME FFS-1, Fitness-For-Service*, Third Edition, American Petroleum Institute 2016.

API (2019a), *API RP-2FSIM Floating System Integrity management*, American Petroleum Institute, 2019.

API (2019b), *API RP-2MIM Mooring Integrity Management*, American Petroleum Institute, 2019.

ASTM (2015), *ASTM C876-15 Standard Test Method for Corrosion Potentials of Uncoated Reinforcing Steel in Concrete*, ASTM International.

BSI (2008), *PAS55 Asset Management*, British Standardisation Institute.

Cullen, The Hon Lord (1990). The Public Enquiry into the Piper Alpha Disaster, *HMSO*, London, UK

Department of Energy (1974), *The Offshore Installations (Construction and Survey) Regulations, HMSO*, London, UK.

Department of Energy (1990), *Offshore Installations: Guidance on Design, Construction and Certification*, Fourth Edition, HMSO, London.

DNVGL (2018), *DNVGL-OS-E301 Position Mooring, DNVGL*, Høvik, Norway.

HSE (1990), *Effect of Bacterial Activity on North Sea Concrete, Undertaken by Khoury G.A, Sullivan P.J.E., OTH 90* 320, HSE Books, London, UK.

HSE (1996). *The Offshore Installations and Wells (Design and Construction, etc) Regulations, Health and Safety Executive (HSE) SI* 1996/913.

HSE (1997) "Guidance for Concrete structures – Background Report", HSE OTO Report 96 050.

HSE (2005), *Offshore Installations (Safety Case) Regulations, Health and Safety Executive (HSE)*, HSE Books, London, UK.

HSE (2009), *"HSE RR684—Structural Integrity Management Framework for Fixed Jacket Structures", Health and Safety Executive (HSE)*, London, UK.

HSE (2013), *"Managing for Health and Safety, HSG 65", Health and Safety Executive (HSE)*, London, UK.

HSE (2019), "HSE Offshore Information Sheet 4/2013 Offshore Installation moorings Rev.3", Health and Safety Executive (HSE), London, UK.

IACS (2010). "Guidelines for the Survey of Offshore Chain Cable in use, No. 38, 2010", International Association of Classification Societies (IACS).

ISO (2006), "ISO 19904:2006 Petroleum and Natural Gas Industries—Floating Offshore Structures", International Standardisation Organisation, 2006.

ISO (2007), "ISO 19902:2007 Petroleum and Natural Gas Industries—Fixed Steel Offshore Structures", International Standardisation Organisation.

ISO (2013), "ISO 19901-7:2013 Petroleum and Natural Gas Industries—Specific Requirements for Offshore Structures—Part 7: Stationkeeping Systems for Floating Offshore Structures and Mobile Offshore Units", International Standardisation Organisation, 2013.

ISO (2019a), "ISO 19901-9 Petroleum and Natural Gas Industries—Specific Requirements for Offshore Structures—Part 9: Structural Integrity Management", International Standardisation Organisation, 2019.

ISO (2019b), "ISO 19903 Petroleum and Natural Gas Industries—Fixed Concrete Offshore Structures", *International Standardisation Organisation*, 2019.

Lloyds Register (2019), "Lloyds Register Rules for Classification Floating Units at Fixed location—Part 3", Lloyd Register. London. UK.

NMD (2009), "Anchoring Regulations of 10 July 2009, No.998", Norwegian Maritime Directorate.

O&GUK (2014) "Guidance on the Management of Ageing and Life Extension for UKCS Floating Production Installations", Oil and Gas UK, London, UK.

PSA (2019), *Framework, Management, Facilities and Activities Regulations*, Stavanger, Norway: Petroleum Safety Authority.

Standard Norge (2011), "*NORSOK Z-008 Risk Based Maintenance and Consequence Classification - Rev. 3, June 2011*", Standard Norge, Lysaker, Norway.

Standard Norge (2012), "*NORSOK N-001: Integrity of Offshore Structures, Rev. 8, September 2012,*" Standard Norge, Lysaker, Norway.

Standard Norge (2013), "*NORSOK N-004: Design of Steel Structures, Rev. 3, February 2013*", Standard Norge, Lysaker, Norway.

Standard Norge (2015), "*NORSOK N-006 Assessment of Structural Integrity for Existing Offshore Load-Bearing Structures, Edition1; March 2009*", Standard Norge, Lysaker, Norway.

Standard Norge (2017a), "*NORSOK N-003: Actions and Action Effects, 3e,*" Standard Norge, Lysaker, Norway.

Standard Norge (2017b), "*NORSOK N-005 In-Service Integrity Management of Structures and Maritime Systems, Edition 2, 2017*", Standard Norge, Lysaker, Norway.

3

Damage Types in Offshore Structures

If you don't know what it is you're looking for you're never going to find it.
You have to be clear on what it is you're seeking[1].

—Germany Kent

3.1   Introduction

3.1.1   General

Damage of various types occur in all offshore structures as a result of the environment in which they operate and accidents occurring. The types of damage that the structure is likely to experience will to some extent depend on the type of structure and the material involved. Cracks, corrosion and accidental damage such as dents, bows and gouges dominate in fixed and floating steel structures. Whereas for concrete structures, corrosion of the steel reinforcement leading to spalling and cracking in addition to accidental damage dominate. In mobile structures also the mooring system is exposed to wear, fatigue cracking and corrosion.

A main role of inspection is to detect, locate and characterise damage with respect to its type and size in order to enable actions to be taken if required. A good understanding of the possible causes, likely locations and form of these damage types is essential to select the appropriate inspection technique. Examples of causes, types and effects are illustrated in Figure 11.

This chapter aims to provide an overview of the types of damage, causes and effects. Due to the prevalence offshore of steel structures compared to concrete structures, the number of damage types that have occurred is dominated by damage to steel structures. However, this chapter attempts to give a balanced review of damage to both steel (fixed and floating) and concrete structures, giving due regard to previous work on these topics.

The most common damage types for steel structures are corrosion, cracking (most often from fatigue) and various forms of impact damage. For concrete structures, the reinforcement corrosion often leading to spalling and impact damage are dominant. These are introduced below and along with several other damage types are briefly explained below.

---

1 *Source:* Germany Kent, "You Are What You Tweet: Harness the Power of Twitter to Create a Happier, Healthier Life", Star Stone Press, LLC, 2015.

Underwater Inspection and Repair for Offshore Structures, First Edition.
John V. Sharp and Gerhard Ersdal.
© 2021 John Wiley & Sons Ltd. Published 2021 by John Wiley & Sons Ltd.

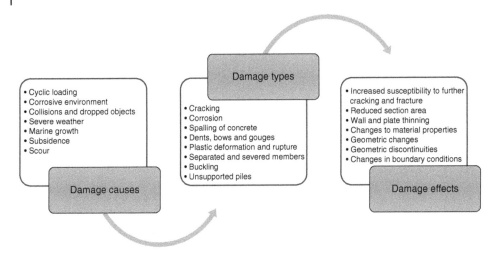

Figure 11 Examples of damage cause (why the damage occurs), types (what they look like) and effects (their consequence on strength).

3.1.2 Corrosion

Corrosion is a result of a chemical or electrochemical reaction between a metal and its environment that leads to a generalized or local loss of material and sometimes degradation of its properties. The following basic conditions must be fulfilled for corrosion to occur:

- a metal surface is exposed to a potentially damaging environment (e.g. bare steel in physical contact with seawater);
- presence of a suitable electrolyte able to conduct an electrical current (e.g. seawater containing ions); and
- an oxidant able to cause corrosion (e.g. oxygen, carbon dioxide).

External corrosion of steelwork can occur in seawater with the presence of absorbed oxygen leading to loss of material and reduced load-carrying capacity. The rate of corrosion is dependent on the level of oxygen and the temperature of the seawater. To make it more complex, the oxygen solubility decreases with increased temperature.

In the splash zone there is a plentiful supply of oxygen which together with the presence of seawater can lead to high corrosion levels. In the North Sea the seawater at all depths is normally saturated with oxygen at approximately 6 ml per litre. In contrast, in more temperate regions such as the Gulf of Mexico the temperature will be higher which could result in more severe corrosion. In areas with less mixing of water masses through the depth, the oxygen level in deep water may be much less and corrosion would be inhibited.

External corrosion is usually mitigated by the use of a cathodic protection (CP) system usually involving sacrificial anodes and in some cases by the use of external corrosion coatings. The design of the CP system is dependent on the design life of the structure and the type and quality of the external coating system. There are typically recommended levels of cathodic protection in the range between –800 to –950 mV vs Ag/AgCl. On some installations there are more negative levels of protection and this over-protection can lead to the production of hydrogen with embrittlement and other adverse effects on the steelwork. High-strength steel (with yield strength greater than 500 MPa) are more susceptible to this and more stringent requirements are recommended for the level of cathodic protection (HSE 2003). The CP system in not effective in the splash zone and alternative means of protection are required, typically coatings and a corrosion allowance. A number of coating systems have been used offshore with epoxy-based systems more typical. Bahadori (2014) is a useful guide to corrosion protection systems for the oil and gas industry.

Coatings are applied for corrosion control as they act as barriers that isolate the steel from the corrosive environment. Coatings can be applied in the construction yard or offshore although the latter is more expensive and often less efficient. Typical coatings used offshore include coal tar epoxy (as noted later for the West Sole WE platform) and epoxy resins. Coatings should be a part of the inspection plan and programme and may require repair or replacement depending on the level of damage. Coatings and CP systems are also a feature of corrosion protection for ballast tanks in floating structures. Breakdown of these coatings can lead to localised corrosion and loss of watertight integrity. Proper re-instalment of coatings in such areas is an important mitigating measure.

As a result, a key objective of the inspection process is checking:

- the state of the corrosion protection system, whether CP readings are within the accepted range;
- the condition of anodes;
- the condition of coatings; and
- any evidence of corrosion.

If the inspection reveals anomalies related to any of these factors, then evaluation or possibly mitigation measures are required. The types of structure and the material involved will determine what corrosion can occur and at which location in the structure.

During the first decades of oil and gas production, fixed steel structures (jacket structures) were primarily used. Below water these were protected with both coatings and cathodic protection systems. Although many of these structures have been in operation far beyond their original design life, there has been a minimum of corrosion experienced underwater. At present, jacket structures are designed in accordance with recognised standards where a cathodic protection system underwater and coatings both above and below water are required. In addition, a corrosion allowance (extra thickness of steel) should be included in the splash zone.

In concrete offshore structures the continuing integrity of the steel reinforcement is an essential requirement. The steel which is embedded in concrete should normally be protected from corrosion for long periods, provided there is a good depth of high-quality cover over the steel (typically 70 mm in the splash zone and 45 mm in the underwater zone). Concrete is a permeable material and hence the chlorides in seawater will penetrate to the steel reinforcement in the longer term. Activation of the reinforcement with loss of passivity can occur when sufficient chlorides reach the steel surface and if sufficient oxygen is available this will usually lead to corrosion. This is usually the case for the splash and air zones, where over a period of time, corrosion will progress leading to expansive products which usually cause spalling of the concrete cover. Following spalling of the cover corrosion usually occurs more rapidly unless repaired. This type of corrosion is very typical of many marine structures.

Semi-submersible platforms and tension leg platforms are normally built without corrosion allowances, except in the splash zone. The coating system needs to be inspected and maintained during the operation phase and in some cases it can be demanding to obtain adequate corrosion protection.

The first FPSOs were in use in the 1980s and the use of these production vessels increased throughout the 1990s. These were often designed without a corrosion allowance and were therefore vulnerable to thickness loss resulting from corrosion (DNVGL 2019). The ballast tanks were typically protected with both a coating and cathodic protection, where the latter is only effective for areas underwater. It has been found that the coating and the workmanship in some cases were not of sufficient quality. Consequently, there has been some corrosion in ballast tanks on FPSOs. Storage tanks for oil have been less susceptible to corrosion as they are exposed to a less corrosive medium. It has been common to apply coatings at the bottom and top of the tank to protect against corrosion from water accumulating at the bottom and condensation at the top. In general, the experience for storage tanks is good (DNVGL 2019). The degree of corrosion damage varies as the corrosion process is sensitive to temperature and the concentration of, for example, sulphates. When referring to corrosion of steel connectors and components, six levels of corrosion are described by ABS (2017), which are shown in Table 2, and are useful for describing and evaluating corrosion. The design of FPSOs has changed since the 1990s and FPSOs are now designed with a corrosion allowance in hulls, tanks and decks, which gives additional robustness.

A decommissioned offshore installation allows the state of anodes to be examined in more detail. One example was the recovery of the West Sole WE platform in 1978 after 11 years in the sea. A detailed examination of components was undertaken including the anodes, as shown in Figure 12. Weight loss of the anodes was found to be only 40%, less than the expected value. Protection potentials were measured before recovery and were in the range −0.87 to −0.93 V (compared with Ag/AgCl), which is more positive than set at the design stage. The splash zone coating was a coal tar epoxy and it was found that this protection had broken down from wave and spray action. It was noted that this could have been minimised by more painting during operation. Isolated pitting and metal wastage were seen on members in the splash zone, up to a depth of 3 mm in places, which could have become more serious if the platform had been in the sea for a longer period.

3.1.3 Cracking Due to Fatigue

Fatigue is characterised by cumulative material damage caused by numerous loading cycles during the structure's life, resulting in crack initiation and propagation. Fatigue cracks often occur from discontinuities and pre-existing defects in areas with cyclic stress. Welded joints with high stress concentrations are the areas where fatigue cracks typically initiate, after which they will grow incrementally. In design standards and recommended practises, fatigue failure is normally

Table 2 Levels of Corrosion (ABS 2017).

Level	Description
1	More than 90% of primer paint is intact, surface rust local to primer damage. No loss of material
2	Surface rust over more then 50% of H link with no loss of material on plates or hardware
3	Level 2 corrosion on plates, but with also significant degradation or loss of securing hardware such as spacer pins, handling shackles
4	Deep local rist blooms on H-link plates and /or securing hardware
5	Widespread loss of material on main plates. Gauging of plates show losses to be within ABS allowances
6	Widespread loss of material on main plates. Gauging of plates shows losses exceeding ABS allowances

Figure 12 Aluminium anode recovered from West Sole WE platform showing less wastage than expected, with an average weight loss of 40%. *Source:* JV Sharp.

considered to have occurred when a crack has grown through the wall thickness. However, as will later be described in this book, there is often still a certain amount of life remaining until severance.

Fatigue cracking is a time-dependent and accumulative degradation mechanism and hence cracks normally should be expected to occur late in the life of a structure. However, there is also evidence that cracking can occur at a much earlier stage. Evidence of this has been seen when significant defects from the original fabrication that were missed at the fabrication inspection remain when the structure enters service. This is even a greater likelihood of significant defects occurring and missed by fabrication inspection when modifications and repairs are performed after the structure has entered service. An example of this is the fatigue crack that formed in the Alexander L. Kielland flotel and eventually caused the capsize of this semi-submersible in 1980 from a defect.

Fatigue failure is a significant hazard to offshore structures subjected to cyclic loading in harsh environmental conditions, typically identified by sea-states such as those found in the North Sea and similar regions. As a result, a key element of the inspection process in such regions should include checking for any signs of fatigue cracking. If the inspection reveals any cracks or analysis indicates insufficient remaining fatigue life, then evaluation or possibly mitigation measures, such as repairs, are required.

3.1.4 Dents, Bows and Gouges Due to Impact

Impact by ship collision, dropped objects and swinging loads during lifts by cranes constitutes an important hazard scenario for damaging offshore structures. Offshore structures as

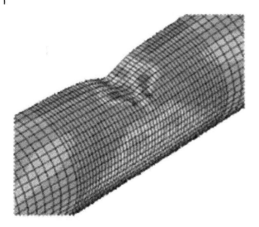

Figure 13 Dented member. *Source:* Courtesy of KBR Energo.

a result accumulate dent and bow damage during service. In some cases, cracks are also a result of these impact incidents in addition to the dent and bow damage.

Figure 13 shows a typical dented member. Such dents and often associated bows can have a significant effect on the ultimate capacity. Similarly, the presence of cracks associated with dents and bows could have a significant impact on the fatigue life of the structure.

3.1.5 Cracking Due to Hydrogen Embrittlement

Hydrogen embrittlement (HE), or hydrogen-induced stress cracking (HISC), is the result of ingress of hydrogen atoms into the metal. When these hydrogen atoms re-combine in voids in the metal matrix to form hydrogen molecules, they create pressure from inside the cavity and this pressure can reduce ductility and tensile strength up to a point where cracks open. In marine environments the principal sources of hydrogen in steel are from corrosion and cathodic protection (CP). Welding can also cause high hydrogen contents if sufficient care is not taken during welding, for example, by applying preheat and drying consumables. HE is a less common mode of failure in offshore structures than corrosion fatigue, but it was found to be the cause of cracking that occurred in the leg chords and spud cans of jack-up drilling rigs operating on the UKCS in the late 1980s (HSE 1991). High-strength steels structures are particularly vulnerable to cathodic over-protection leading to HE. For such structures, a strict monitoring of the CP levels is required during inspection.

3.1.6 Erosion, Wear and Tear

Erosion can be defined as physical removal of surface material due to numerous individual impacts of solid particles, liquid droplet or implosion of gas bubbles (cavitation). One example of erosion is the result of the interaction between a structure and moving ice floes in arctic waters. Erosion is a time-dependent degradation mechanism but can sometimes lead to very rapid failures. In its worst form, considerable material loss can occur. Loss of material will reduce strength and increase potential for fatigue cracking. As a result, inspection should include determining any signs of erosion.

Wear has been observed to cause substantial degradation in mooring chains. Significant loss of material can occur, sometimes in short periods of time measured in months, as a result of the rubbing of chain link surfaces. There is a subsequent loss of load-bearing capacity as the cross-sectional area of mooring chain links becomes too low to carry the applied load. Chains and wire ropes are vulnerable to erosion and wear due to contact at the fairlead, contact between individual chain links and also due to impact damage occurring on the seabed, particularly if the seabed is rocky (see Figure 21). The corrosive nature of seawater may lead to synergistic wear rates that are higher than the sum of wear and corrosion as independent processes. Wear can remove protective deposits enhancing the rate of corrosion and corrosion may then again increase surface roughness.

Wire ropes are more subject to seabed damage and for this reason the seabed section of a mooring line is usually chain, which is less vulnerable. Erosion and wear can reduce the diameter of chain links and the outer wires of a wire rope, in both cases reducing overall strength. Loose studs can also arise from localised erosion.

3.1.7 Brittle Fracture

Brittle fracture is in general not an issue with structural steels. However, with the increasing need to develop mooring lines to handle larger loads as vessel sizes increase, higher strength steels have been used. For example, R5 chain has a 30% increase in yield stress compared to the earlier R4 chain. R5 is typically a quenched and tempered steel with a yield stress of typically ~760 MPa. This is classified as a high-strength steel in most definitions. When first introduced, this high-strength chain had some brittle fracture failures in service, but more careful manufacturing methods have led to a decrease in this problem.

3.1.8 Grout Crushing and Slippage

Grouted connections are used in fixed offshore structures and in offshore wind turbines to connect parts of the structure. Offshore wind turbines typically consist of a mono-pile connected by a grouted connection to the upper tower section. Similarly, in a fixed offshore steel structure the piles are connected to the jacket structure by a grouted pile-sleeve connection. Fixed offshore steel structures in very deep waters have also used grouted connections to combine an upper and lower section of the jacket. In addition, grouted connections are also used for repairs.

Loss of friction capacity between the grout and the steel sleeve can lead to slippage and loss of strength in the connection. As a result, weld beads are often used to increase the capacity of such connections. However, crushing of the grout between adjacent weld beads can also occur and will lead to insufficient capacity of the connection.

This type of damage has been observed in offshore wind turbines but less so for fixed offshore steel structures. This slippage, should it occur, is very difficult to detect and repair.

3.2 Previous Studies on Damage to Offshore Structures

Many studies on damage to offshore structures and their frequency of occurrence have taken place. Many of these have been reviewed in this section to provide the reader with appropriate background on damage to offshore structures and their likelihood of occurrence. These studies include more details than reported here and should be visited for further information. Table 3 gives an overview of the reviewed reports and their relevance to the types of offshore structures.

Table 3 Overview of Previous Work on Damage to Offshore Structures.

Report	Sect.	Fixed steel structures	Floating structures	Concrete structures
MTD Underwater inspection of steel offshore structures	3.3.1	X		
MTD Review of repairs to offshore structures and pipelines	3.3.2	X		X
PMB AIM project for MMS	3.3.3	X		
HSE Causes of damage to fixed offshore structures	3.3.4	X		
Single-sided closure welds	3.3.5	X	(X)	
MSL Rationalization and optimisation of underwater inspection	3.3.6	X		
Studies on hurricane and storm damage	3.3.7	X		
D.En. Defects in semi-submersibles	3.4.1		X	
SSC reports on damage types to ship-shaped structures	3.4.2		X	
TSCF Defect types for tanker structure components	3.4.3		X	
Semi-submersible flooding	3.4.4		X	
API RP-2MIM	3.5.2		X	
HSE Studies on mooring systems	3.5.3		X	
Corrosion of mooring systems	3.5.4		X	
Fatigue of mooring systems	3.5.5		X	
Concrete in the ocean	3.6.1			X
Durability of offshore concrete structures	3.6.2			X
PSA Ageing of Offshore Concrete Structures	3.6.3			X
Marine growth reports	3.7	X	X	X

3.3 Previous Studies on Damage to Fixed Steel Structures

3.3.1 MTD Underwater Inspection of Steel Offshore Structures

A report was produced in 1989 by MTD (1989) from a joint industry-funded project, involving a wide range of industrial participants and the Department of Energy. The review lists detected anomalies and damage to 21 platforms in the North Sea installed between 1971 and 1978 and based on records of inspections up to 1984. It is likely that the data as presented in Table 4 are focussed on issues to do with design since they were taken primarily from a UEG report (UEG 1978) on underwater inspection prepared to give advice to designers.

As can be seen from Table 4, cracks were reported in 12 of the platforms and dents in 15. The most common causes were reported to be:

- direct design deficiencies such as incorrect estimation of wave loading and hot spot stresses;
- indirect design deficiencies such as insufficient access for proper fabrication;

Table 4 Summary of Underwater Inspection Findings for 21 Steel Platforms in the North Sea.

Type of defect	Number of platforms with the defect
Crack confirmed	12
Propagating crack	3
Dented member	15
Deflected member	11
Low CP potentials	10
Missing anode	9
Defective anode	4
Loose anode	7
Pitting corrosion	20
General corrosion	2
Burn mark	4
Heavy marine growth	13
Wire chafe	18
Scour	1
Debris	21

- construction deficiencies such as faulty weldments; and
- accidental events such as collisions and dropped objects.

The review also noted that pitting corrosion was reported in 20 cases (most of the platforms) and it noted some surprise given the age of the platforms. However, there was no indication whether these were underwater or on the topsides (which is more likely). Debris from drill cuttings was reported for all 21 platforms and these accumulated around the base of the platform applying additional loads to the members at or close to the seabed. The review commented that insufficient attention had been paid during design for the disposal of these cuttings. One case was reported where member collapse had resulted from the loads from this type of debris.

The report highlighted corrosion and fatigue and for corrosion four types were noted:

- general corrosion (spread evenly over a steel surface);
- pitting corrosion (localised and deep corrosion that may be obscured by corrosion products);
- galvanic corrosion (occurs when two different metals are placed in contact in an electrolyte such as seawater with the chemically more reactive one corroding); and
- corrosion fatigue (simultaneous occurrence of cyclic loads and corrosion, pitting or localised corrosion were reported to have had the most effect on fatigue life).

The report noted that corrosion should not be a significant problem with offshore installations as generous corrosion allowances were designed into members, particularly in the splash zone and the underwater section of the structure was protected by a CP system. However, problems had occurred when the system proved inadequate or had broken down. Table 4 shows 20 incidents of missing, defective or loose anodes and also 10 cases of low potentials. The report also noted that metallic debris, such as scaffold poles in contact with the structure on the seabed, could place an increasing demand on the CP system as well as creating a hazard for divers.

The number of these incidents seems to be surprisingly high and may reflect the early stage of off-shore industry in UK waters. However, also in more recent studies (MTD 1990) issues with corrosion protection have been reported, although in the overprotection range. The survey showed that 65% of the platforms surveyed had potentials more negative than the recommended level of –900 mV (Ag/AgCl), and 30% of those surveyed had operating potentials more negative than –1050 mV (Ag/AgCl). This highlights the difficulty in getting the corrosion protection system to be efficient in operation and avoiding both under- and overprotection.

The MTD (1990) review also noted that steels with high hardness microstructures were more susceptible to corrosion. This could arise from poor welding conditions such as low preheat, low heat input welding or uncontrolled post-weld cooling.

For fatigue the review noted that fatigue damage was linked to areas of high stress concentration. It also noted that early North Sea structures suffered from poor fatigue design and lack of fabrication control, which led to premature cracking. The location of fabrication defects was more likely to occur at weld repairs, closure welds and welds with poor access. The report also noted several fatigue cracks being found at single-sided butt welds which occurred mainly at brace-to-stub connections or chord-to-can connections (see Section 3.3.5). Fatigue problems at these locations are difficult to detect as the fatigue crack can grow from inside of the member outwards if not designed correctly. The design of these closure welds was improved by the 1990s, but the problem remains in many earlier designs.

3.3.2   MTD Review of Repairs to Offshore Structures and Pipelines

One of the first major reviews of damage to fixed offshore structures requiring repair was undertaken by MTD (MTD 1994). This report analysed both steel and a smaller number of concrete platforms over the period up to 1991 for structures on the North-West European continental shelf.

The types of damage to both steel and concrete structures requiring repair are shown in Table 5. It should be noted that this list of damage does not necessarily include all the damage to platforms in this time period. In particular, there would have been damage not needing repair and therefore

Table 5   Damage to Steel and Concrete Structures Requiring Repair.

Steel platforms		Concrete platforms	
Damage	**No. incidents**	**Damage**	**No. of incidents**
Fatigue	39	Corrosion of steel components	3
Vessel impact	36	Fatigue of steel components	1
Fabrication fault	12	Construction fault	3
Installation fault	12	Dropped objects	2
Design upgrade	11	Vessel impact	1
Corrosion	10	Design fault	1
Design fault	9	Other	2
Operating fault	4		
Other or unknown	11		

not included. Details of the repairs which were undertaken for these damage are further discussed in Chapter 8.2.2.

For concrete structures, corrosion to steel components and construction faults were the most frequent causes of damage. Accidental damage due to dropped objects and vessel collisions also occurred with a significant frequency over the period. Particular examples of damage to offshore concrete structures include:

- vessel impact requiring repair on the Brent C installation with a 2500-tonne supply ship;
- impact from a 0.5 tonne crane block on the cell roof of Beryl A platform, which damaged only the concrete protective layer; and
- an oil storage cell on a concrete platform was penetrated by a section of steel pipe in 1981 which fell as a result of a lifting tackle failure; this led to loss of draw-down pressure and loss of production and required repair.

3.3.3 PMB AIM Project for MMS

The US Mineral Management Services (MMS) undertook a project to develop a methodology to manage the integrity of existing fixed steel platforms in the Gulf of Mexico that were reaching the latter part of their design lives. The project was carried out in four phases by PMB Engineering Group as a Joint Industry Project (PMB 1987, 1988 and 1990). The reports from this project include assessments of the effects of damage on ultimate strength and evaluation of relevant repair measures required to preserve safety. The aim of the Joint Industry Project was to develop a documented procedure to address the performance of existing platforms, the potential for life extension, current inspection procedures and repair procedures.

Phase I of the programme developed a general engineering approach to requalifying platforms. Phase II (PMB 1987) focussed on two particular platforms, a 5-legged K-braced fixed steel platform in 42.7 m water depth installed in 1963 (A) and a fixed jacket structure installed in 1959 (B). At the time of this phase taking place, the unmanned platform A was shown to have a wide variety of damage and defects following inspection as seen in Table 6 and Figure 14, but it had a remaining economic life of 12 years. Platform B was also unmanned and had a remaining economic life of 5 years and a condition survey had shown limited damage.

K-braced fixed steel structures as shown in Figure 14 are rarely used at present and X-braced and XH-braced configurations as shown in Figure 78 are more commonly used at the time of writing this book.

This phase II report applied the phase I requalification (assessment) approach to these existing platforms. Part of this involved engineering analyses of platform capacities and ultimate limit state resistances. Assessments were made on the basis of calculations of platform failure rates and a cost-benefit analysis. For platform A four options were investigated including leave as-is, repair all damage to the platform, repair damage and grout legs and finally raise the deck and carry out all the repairs. On a cost-benefit analysis the most attractive solution was to leave the damage in place but inspect on a more regular basis. Assessing both initial and future costs, the most effective alternative was to repair and raise the deck. For platform B several different options were evaluated including repair, but on a cost analysis the most attractive solution was to remove boat landings and periodically remove marine growth.

The phase II report (PMB 1987) concentrated on structural assessment and cost-benefit analyses with less emphasis on detailed inspection and repair options. A particular problem identified was the lack of a process for justifying and communicating acceptability of platforms for suitability of service, with special concerns for the public regulatory process.

Table 6 Damage to Platform A.

Item no.	Location	Damage
1	Vertical interior diagonal B2 at −27.5' (−8.4 m) to centre leg at −65' (−19.8 m)	Missing
2	Vertical interior diagonal B1 at −27.5' (−8.4 m) to centre leg at −65' (−19.6 m)	Completely separated from leg at B1
3	Vertical face diagonal Row B midpoint B1 to B2 at −27.5' (−8.4 m) to B1 at −65' (−19.8 m)	Completely separated from horizonal member B1 to B2 at −27.5' (−8.4 m)
4	Vertical face diagonal Row B midpoint B1 to B2 at −65' (−19.6 m) to B1 at −102.5' (−31.2 m)	Completely separated from horizonal member B1 to B2 at −65' (−19.6m)
5	Vertical interior diagonal A2 at −27.5' (−8.4 m) to centre leg at −65' (−19.8m)	Cracked at A2 from 12:00 to 5:00. Crack length = 40" (1016 mm)
6	Horizontal interior diagonal −65' (−19.8 m) B2 to centre leg	Cracked at B2 from 9:30 to 2:30. Crack length = 14" (355.6 mm)
7	Horizontal interior diagonal −28' (−8.5 m) B2 to centre leg	Cracked at B2 from 4:00 to 8:30. Crack length = 12.75" (323.9 mm)
8	Horizontal interior diagonal −28' (−8.5 m) A2 to centre leg	Cracked at A2 at 4:30. Crack length = 12.5" (317.5 mm)
9	Horizontal face member row B, B1 to B2 at −65' (−19.8 m)	Cracked at B2 at 3:00, 5:00, 9:00. Crack length 7.25" (184 mm) total

Source: Based on PMB (1987) AIM II assessment inspection maintenance, Published as MMS TAP report 106ak.

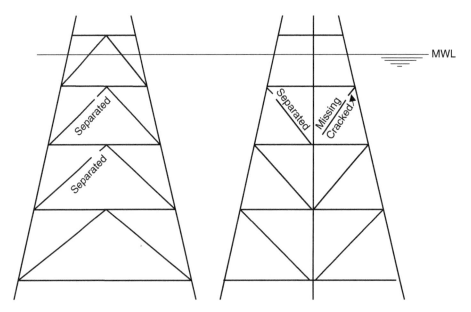

Figure 14 Illustration of location of damage to platform A in Table 6. *Source:* Based on PMB (1987) AIM II assessment inspection maintenance, Published as MMS TAP report 106ak.

The problems inherent in the platforms studied in phase II (PMB 1987) such as inadequate original design standards, damaged members and a need for life extension are typical of many early generation platforms still in operation.

Phase III (PMB 1988) of this project studied the damaged condition of an eight-legged platform installed in 1970. The typical damage types that were assessed included:

- bent and dented members near waterline caused by workboat impacts;
- completely severed members near the base of the jacket caused by dropped objects from topside operations;
- damage to interior horizontal braces and transverse diagonals near platforms' midpoints;
- general corrosion and localised corrosion damage in the splash zone; and
- under-driven piles.

The consequences of these damage types were also evaluated as shown in Section 7.2.6.

Phase IV of this study included an assessment of damage types resulting from inspections. An additional task was to identify any trends. These trends were then used in a later task to review inspection approaches. The findings were based on 40 reported incidents from the Gulf of Mexico representing significant damage. The number of cases of a particular damage type is shown in Figure 15. As can be seen, cracks were the most frequent damage type followed by holes and then dents and bows. Surprisingly, corrosion only appeared as a type of damage in three cases. However, as indicated in Figure 16 which shows the *causes* of damage, corrosion and fatigue and accidental damage dominate.

The failure database review confirmed that many platform failures could have been predicted by a simple engineering review of the wave criteria and deck elevation. This was particularly relevant to those structures designed prior to the use of the 100-year design wave.

Figure 15   Frequency of damage types. *Source:* PMB (1987) AIM II assessment inspection maintenance, Published as MMS TAP report 106ak.

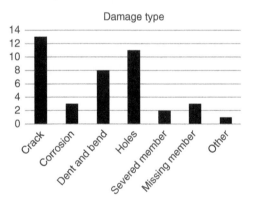

Figure 16   Frequency of causes of damage. *Source:* PMB (1987) AIM II assessment inspection maintenance, Published as MMS TAP report 106ak.

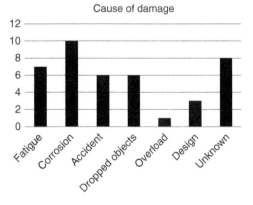

3.3.4 HSE Study on Causes of Damage to Fixed Steel Structures

There are a number of sources providing an overview of damage types and causes to offshore structures. The key ones are briefly reviewed in Section 3.3. As an example, an overview of major structural damage to fixed structures in the UK sector of the North Sea was reported by Sharp et al. (1995). This information had been taken from a database developed by one of the UK certification societies. This contained information on 170 incidents in the period 1972 to 1991 from a review of 174 platforms. The following data were included; severances, known and suspected through-thickness cracks, dents deeper than 50 mm and bows with a maximum deflection greater than 100 mm.

Analysis of the data showed that:

- out of the total of 179 damages, 105 repairs were undertaken;
- 19 severances were reported of which 10 were caused by boat impact, 6 by fatigue and 3 by dropped objects. Six of the severances were on non-redundant members, but only 3 members were repaired;
- 79 through-thickness cracks were reported of which 21 were on non-redundant members and 15 members were repaired;
- 58 damages were by fatigue, 12 of which were due to fabrication defects; and
- 71 damages were caused by boat impact.

The frequencies and causes of the various types of damage recorded in the database are summarised as shown in Table 7 and Figure 17.

Further analysis of the data in Table 7 showed that there is large variability in the interval between the discovery of major damage and repair. Large intervals indicate that the structure may be at risk during this interval. Further, the review indicated that not all 'major damage' found was considered to have significant consequences and only approximately 50% of the major damage was repaired. This implies that the definition used for major damage was too broad. The damage shown in Table 7 may place the structure at risk but did not necessarily lead to structural collapse.

Table 7 Cause and Frequency of Damage to Fixed Offshore Structures.

CAUSE	Damage Type							
	Severance	Through crack	Dent	Bow	Tear	Hole	Crease	Total
Boat impact	10	13	22	23	1	2	-	71
Cyclic loading (fatigue)	5	41	-	-	-	-	-	46
Cyclic loading (fatigue) from fabrication defect	1	11	-	-	-	-	-	12
Installation	-	10	2	2	-	2	1	17
Dropped object	3	2	6	-	1	-	-	12
Mud cuttings mound	-	1	-	-	-	-	-	1
Unknown	-	1	13	1	1	4	-	20
Total	19	79	43	26	3	8	1	179

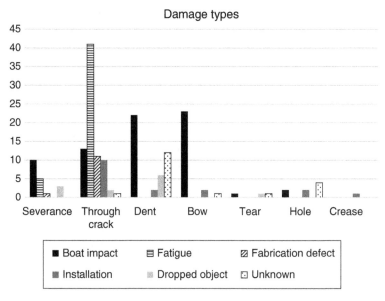

Figure 17 Cause and frequency of damage to fixed offshore structures.

3.3.5 Single-Sided Closure Welds

A number of joints in offshore structures have been fabricated using single-sided welds (also called circumferential butt welds). Typically, a brace in a fixed steel offshore structure should be connected to the leg by a double-sided weld to obtain optimal fatigue life. However, due to the lack of access during construction, a short stub or can close to the leg is often used allowing a double-sided weld to the leg. This requires that the brace is then welded to the stub or can by a single-sided closure weld. A possible mitigation is to cut out an access or closure window to allow for internal double-sided welding of the brace to stub or can connection. However, this access window has to be closed using single-sided welding.

The potential for root defects during single-sided welding is high and there is a greater possibility of these root defects remaining undetected during fabrication because of the difficulties of inspection. This inspection is limited to methods such as UT and radiographic testing which are less successful than more conventional methods used for detecting external surface defects.

There have been several cases of fatigue failure of offshore members containing single-sided closure welds. Failure at single-sided closure welds was, for example, detected on the Chevron Ninian platform in the mid-1980s (Stacey et al. 1997). Wet radiography and ultrasonic methods were used to perform NDE at the access windows. These indicated that fabrication defects were present and that fatigue cracks had initiated at some of these defects. As a result, a horizontal member fractured and two through-thickness cracks were also detected at similar access windows. These members were repaired in 1984 and 1985. In many of these cases cracking has initiated inside the member propagating outwards and when discovered externally, the crack had developed to a through-thickness crack.

A review of these welds was undertaken by Stacey et al. (1997), which included an assessment of fabrication procedures for two platforms. The review further investigated the types of defects left after inspections, mainly using MPI and UT. In one of the platforms inspected also by radiography, 55% of the defects were found to be located at the weld root, whilst in the second platform inspected only by UT and MPI, only 7% were found. This difference was attributed to the use of radiographic inspection for the first structure, providing more efficient inspection than the use of MPI and UT.

The use of backing strips for the closure weld was also addressed by Stacey et al. (1997) with a lower percentage actually needing repair at the fabrication stage. Access windows were also inspected underwater after installation on the Chevron Ninian Southern platform and compared to the results obtained during its fabrication. It was found that the underwater inspections tended to slightly underestimate the defect height (depth).

The paper also reported on fatigue testing at The Welding Institute, Cambridge, UK, (TWI) on three single-sided butt-welded pipes. This included checking the inspections made during their fabrication using CVI, UT and radiography. It was found that both UT and radiography proved the most successful in detecting root defects.

The capability of inspection methods to locate root defects is limited, at the time of writing. Stacey et al. (1997) recommended that research was needed to develop PoD and POS data for the detection of root defects using ultrasonic methods. This requirement still remains although some work was done by TWI in detecting defects at internal ring stiffeners using ultrasonics.

3.3.6   MSL Rationalization and Optimisation of Underwater Inspection Planning Report

A Joint Industry Project titled 'Rationalisation and optimisation of underwater inspection planning consistent with API RP2A section 14' (MSL 2000) was undertaken for the Mineral Management Services and several US offshore companies. The aim of the project was to provide industry with the data to enable optimised inspection-planning without compromising safety, based on both ISO 19902 (ISO 2007) and API RP-2A (API 2014) practices. As a result, an industry-wide database was compiled from the inspection data amassed by the industry over the period 1950 to 1999. The objectives of this investigation were also listed as answering the following questions:

- What defects were found on platforms in the Gulf of Mexico?
- Where were these defects occurring and what components were affected?
- What were the causes of the defects?
- Which platforms were susceptible to defects and why?

This database contained platform, inspection and anomaly (damage) data from over 2,000 Gulf of Mexico platforms in water depths ranging from less than 6 m to over 600 m, collated from the 12 major operators involved. Overall details of almost 3,000 underwater inspections were catalogued and ~5,000 anomalies were recorded. Sixty-four percent of the platforms had been designed either to versions of API RP-2A (API 2014) before 1972 or prior to the introduction of this standard.

The database of anomalies found in GoM platforms could be divided into four categories, as shown in Figure 18 and the break-down of these categories is shown in Table 8. It can be seen that mechanical damage consisting of dents, bows, gauges and separated members was responsible for over half of the damage recorded. The dominating type of mechanical damage was dents.

It was found that 29% of the platforms surveyed had some form of mechanical damage and of these, 3% had major damage (dent depth greater than 30% of the nominal member diameter or 10 times the nominal wall thickness). For those defects where the cause was known, 53% were from vessel impact and 47% from dropped objects.

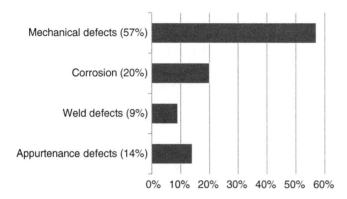

Figure 18 Reported defects by category. *Source:* Based on MSL (2000). Rationalisation and optimisation of underwater inspection planning consistent with API RP2A section 14. Published as MMS TAP report 345aa.

Table 8 Categories of Platform Defects.

Mechanical damage	Corrosion	Weld and joint defects	Appurtenance defects
Dents	Uniform (general)	Crack indications	Anodes
Bows	Pitting	Fabrication defects	Risers and conductors
Gouges	Holes	Overload defects	Boat landings
Separated members	Crevice		Intake caissons
	Fretting		Other

Source: MSL (2000). Rationalisation and optimisation of underwater inspection planning consistent with API RP2A section 14. Published as MMS TAP report 345aa.

The JIP database indicated that weld and joint defects were responsible for 9% of the reported defects. These included crack indications, fabrication defects and defects resulting from overload. These defects were more prominent in platforms installed before 1980.

Corrosion damage observed consisted of pitting, holes, crevice corrosion, fretting and general uniform corrosion. Pitting corrosion and holes were the most common corrosion defects recorded. Most of it was observed in or above the splash zone. Inspection data showed that CVI was a reliable indicator of the general corroded state of the platform. It was also noted that the more heavily corroded platforms had an increased susceptibility to general fatigue damage.

The report shows the number of platforms (208) with ineffective cathodic protection systems (at the time of their inspection) based on anode surveys and dropped cell measurements. Also shown is the number of platforms within the MMS damage database reported to be unprotected at the time of inspection.

As the drop cell readings were available for each of the platforms in the database it was possible to identify those platforms where at least 40% of the recorded readings were outside the range −850 mV to −1150 mV. These were then analysed in connection with fatigue crack indications. It was found that none of the platforms designed to modern versions of API RP-2A had any fatigue cracks. For those platforms designed to earlier versions of the API recommended practice eight platforms had fatigue crack indications. The figures appear to indicate that excessive corrosion increased a platform's susceptibility to fatigue cracking for the early-RP-2A vintage platforms. However, these platforms were also older and hence fatigue and corrosion would have been more likely. Fatigue damage in early-RP-2A platforms was dominated by damage to the conductor guide frame at the first elevation below the water surface.

It was found that marine growth measurements in excess of the API recommended design levels were widespread in the Gulf of Mexico, mainly in water depths above 12 m. It was noted that the marine growth thickness established a stable level after a few years but with annual and seasonal variations in thickness. Measured thicknesses indicated that a variable marine growth profile was preferable during design to the constant thickness of 3.8 cm (1.5") beyond a water depth of 46 m, as recommended in API RP-2A (API 2014).

It was further found that seabed scour was not a concern for the large majority of Gulf of Mexico platforms due to the generally cohesive nature of the soils. However, there was evidence that temporary seabed movements occurred during severe storms or hurricanes.

The JIP reported that flooded members were generally the result of through-thickness fabrication flaws, corrosion, mechanical damage and fatigue. The database contained records from almost 300 FMD surveys containing over 5,500 FMD readings on 244 platforms. As would be expected, the number of platforms with positive flooded member readings increased with the age of the platform. For the earliest platforms (designed to earlier versions of API RP-2A), 33% of those surveyed showed member flooding. The percentage was even higher for those platforms designed prior to API RP-2A with 62% showing positive readings using FMD. For the platforms designed to the more modern versions of API RP-2A only 5 of the 44 platforms (11%) surveyed showed positive readings for flooded members.

The JIP findings supported the existing industry position on Flooded Member Detection as stated in API RP-2A (API 2014) that the appropriate use of FMD generally provides an acceptable alternative to close visual examination and may sometimes be preferable.

Debris typically results from objects dropped or discarded overboard during operations and throughout the service life. Debris surveys are routinely carried out during API level II platform inspections. The database contained over 14,000 items of debris that had been recorded. Of these less than 4% were ever recovered or removed from the structure. The database indicated that debris associated with platforms was not generally detrimental to the structural integrity. One exception to this was where discarded wires, cables and grout-lines were found, which could lead to fretting corrosion. In addition, metallic objects in contact with the structure could also interfere with the corrosion protection system. The report noted that these types of debris should be removed.

3.3.7 Studies on Hurricane and Storm Damage

Most structural failures have occurred in the Gulf of Mexico (GoM) during hurricanes. A review of hurricane damage to fixed offshore structures and wellhead platforms was performed for the HSE (1995). A database of hurricane damage was developed, which contained information on a total of 295 incidents: 131 incidents for jackets (of which 50 have toppled) and 164 satellite, wellhead or caisson platforms (of which 36 have toppled).

Hurricane and storm damage will typically result in plastic deformation, buckling, cracks and ruptures, severed members in the worst cases leading to leaning or global collapse of the platforms. However, due to the de-manning procedures in place in the Gulf of Mexico where the platforms are evacuated before the hurricanes arrived, very little loss of life occurred. An example of a leaning platform is shown in Figure 19 and an example of a damaged leg is shown in Figure 20. A summary of platforms damaged by hurricanes in the Gulf of Mexico is presented in PMB

Figure 19 Gulf of Mexico platform leaning damaged in a hurricane. *Source:* photo provided to Gerhard Ersdal.

Engineering (1993), HSE (1995), Puskar et al. (2004), Energo (2006), Energo (2007), Energo (2010) and Ersdal et al. (2014). The main conclusions of these studies were that:

- fixed structures designed before 1977 were less resistant to hurricane loading than those designed after 1977 (the early structures were designed using the 9th or an earlier edition of API RP 2A with only a 25-year design wave recipe);
- the majority (85%) of the failures have occurred from Hurricane Andrew in 1992 and after.

Several platform and satellite and wellhead structures of more modern design (post 9th edition of API RP 2A) were damaged by Hurricane Andrew and the later hurricanes, indicating the severity of these later hurricanes. Most damage to platforms in the later hurricanes resulted in leaning (between 1° and 45°) or toppling (collapse), as a result of member or joint failures and in many cases multiple component failures.

Sharp et al. (1995) reported that the number of fixed platforms damaged by Hurricane Andrew in August 1992 included both platforms and satellites that were toppled or leaning as illustrated in Table 9.

A later study by Kareem et al. (1999) included similar numbers to those in Table 9 and indicated that 43 platforms suffered irreparable structural damage and 100 platforms suffered significant but repairable damage.

Of the ten platforms that were completely toppled, only one had been designed post-1977. However, failure of this more modern structure was initiated by collision with a drifting mobile platform. The remainder were found to be leaning significantly or to have sustained significant topside's damage, which would have placed offshore workers at significant risk if evacuation had not taken place prior to the hurricane.

Table 9 Damage Caused by Hurricane Andrew (from Sharp et al. 1995).

Damage	Quantity
Platforms toppled	10
Platforms leaning	26
Satellites toppled	25
Satellites leaning (total)	120
Satellites leaning ($\leq 5°$)	77
Pollution events	11
Fires	2
Drifting rigs	5
Pipeline segments damaged	454

Offshore: Risk & Technology Consulting Inc. (2002) carried out a study for MMS on damage to MODUs in Hurricane Lily and earlier hurricanes (including Andrew). It is reported that in Hurricane Andrew, 5 MODUs broke adrift and 2 fixed platforms were toppled as a result of drifting MODUs. The storm snapped seven of the eight anchor chains on one of the MODUs and drove the unit some 40 miles to the north. After breaking loose from its location, it collided with 2 platforms. The anchors from another MODU dragged along the bottom for approximately 4 miles and ruptured a large pipeline resulting in a significant oil spill. In addition, there were 16 pipeline failures from the anchors of MODUs which drifted from their initial position during the storm.

MMS commissioned a review of damage types after each of the major hurricanes, in particular the damage from Andrew in 1992 (PMB Engineering 1993) and onwards to Hurricane Ike in 2008 (Puskar et al. 2004, Energo 2006, Energo 2007, Energo 2010). Particular examples of damage from hurricanes include destruction to platforms, and buckled braces, cracked joints, cracked legs primarily due to strength overload.

The general types of platform damage below water both to the main structure and the secondary structure were as follows (Energo 2007):

- braces: buckles, dents, holes, cracks, tears, out-of-plane bowing, severed members;
- legs: buckles, dents, holes, cracks, tears, pancake leg severance;
- joints: cracks at welds, cracks into chords, cracks into braces, punch-through of braces, pull out of braces (including a piece of the leg material, leaving a hole in the leg);
- conductor trays: cracks at joints (typically at 6 and 12 o'clock), conductor torn loose from guide; and
- risers: broken water caissons, broken riser standoffs.

In several hurricanes pancake leg severances have been observed and according to Energo (2010) this type of damage has been reported to have occurred in 21 cases. This type of damage has been called "pancake leg" due to the flattening of the leg in the damaged area, see Figure 20. The damage is believed to develop as a result of the significant stiffness change between the thin-walled section of the jacket leg and the thicker joint-can section at the horizontal elevations. According to Energo (2010) in most cases it occurs for D/t ratios over 60 and an average thickness transition between join-can and nominal leg of 1.6 mm. The majority of the platforms that experienced this type of damage were older 60s and 70s vintage platforms.

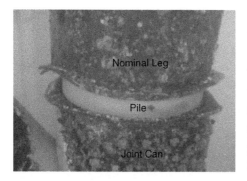 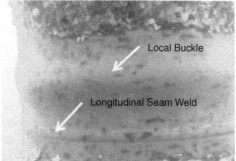

Figure 20 Pancake leg damage—fully severed leg and initial damage configuration. *Source:* Courtesy of US Bureau of Safety and Environmental Enforcement (BSEE).

3.4 Previous Studies on Damage to Floating Steel Structures

3.4.1 D.En. Studies on Semi-Submersibles

A review of defects detected in semi-submersibles is reported by the Department of Energy (1989a). Eleven rigs conforming to four basic designs were considered. The ages of the structures ranged from 3 to 17 years and defects were found in all of the structures, necessitating repairs and in some cases, major structural modifications. The defect data enabled the identification of the weld details with the highest propensity for fatigue cracking. It was found that the majority of defects were associated with particular weld details, namely:

- bracing to column joints;
- abrupt profile gusset terminations on bracings; and
- fillet welded ring reinforced penetrations on bracings.

It was noted that there is a lack of 'consistent systematically recorded survey data' and consequently this causes difficulties with the interpretation of the data. A further consideration is the lack of information on fabrication defects which may have been missed during inspection and hence were part of the initial damage in the structure.

3.4.2 SSC Review of Damage Types to Ship-Shaped Structures

In the Ship Structure Committee review SSC-337 (SSC 1990), case studies of failures of ship structures were undertaken, with the aim of determining the modes of serious damage in ship structures. Fatigue cracking was observed or reported in 11 of the 16 cases examined. In addition, fatigue cracking followed by brittle failure was found in nine of these cases. Overall brittle fracture was seen in 11 of the 16 cases, all of these originating at design or fabrication anomalies. Most of the brittle factures examined were in steel grades A and B (as given in typical classification society rules). Arrest of brittle facture was due to riveted construction in three cases and in one case this was due to available redundancy. Overall, the main conclusion from the studies reported was that fatigue and corrosion were the most common types of damage seen in ship structures.

The SSC-381 report (SSC 1995) attempted to review damage to ship structures for the purpose of evaluating their effect on residual strength. Data were collected based on literature reviews, interviews with ship owners and operators and searches in available databases. However, SSC (1995) reports that the majority of the ship owner and operators were reluctant to release or disclose damage records of their fleet. The focus of the review was on the impact the damage had on the overall residual strength of the structure as a function of its extent, mode of failure and its relative shipboard location. As an example, it was recognised that small cracks (also named nuisance cracks) often will hardly have any effect on the overall strength, while cracks of considerable length on the main deck or side shell can seriously affect the structure's residual strength. Further, a structure experiencing a brittle failure hardly possesses any reserve strength and the failure can lead to total collapse of the structure. Although it is recognised that the primary causes of failure to ship-shaped structures were corrosion and fatigue cracking, this report primarily evaluated damage and its effect on structures (i.e. not incorporating corrosion). The modes of failure mentioned in SSC 391 is listed in Table 10.

Table 10 Modes of Failure for Ship-Shaped Structures.

Type of failure	Extent	Possible locations	Remarks
Yielding	Local	At discontinuities and joints. In plating under pressure. Near concentrated loads.	May not be damaging unless it occurs repeatedly.
	Global	In structures under axial tension such as beams or grillages under lateral load.	Resulting gross distortions cannot be accepted at loads below collapse load. Energy absorption will also depend on ductility.
Buckling	Local	In thin plating between stiffeners such as deep webs in shear or compression and pillars.	Elastic local buckling may not be damaging unless it overloads the remaining structure.
	Global	In stiffened panels in compression or shear.	Will generally involve yielding, hence, unacceptable permanent distortions. Final collapse strength may also depend on ductility.
Fracture - ductile	Local		Unlikely in view of the high strains required.
	Global		Unacceptable. Design will be governed by general yielding load, but safety depends on ultimate tensile strength.
Fracture - brittle	Local	At discontinuities or where ductility is reduced by triaxial stresses or metallurgical damage.	Undesirable though not serious if propagation prevented by fail-safe devices and remaining material is sufficiently tough.
	Global		Unacceptable but hardly calculable. Good material properties, detail design and workmanship must be ensured.
Fracture - fatigue	Local	At stress concentrations such as joints.	Undesirable but not serious if material prevents development of a brittle crack. Generally unacceptable in longitudinal material.
	Global		No known cases. Preventive action should be possible[2] before crack propagates generally.

Source: SSC (1995), SSC-381 Residual strength of damaged marine structures, Ship Structure Committee, Washington, D.C., US. © 1995, SHIP STRUCTURE COMMITTEE.

---

2 It should be noted that this SSC report predates the MV Erika and MV Prestige accidents (1999 and 2002 respectively).

The report (SSC 1995) evaluated a total of 41 instances of damage where a complete description of the damage in terms of location, cause, mode of failure, extent of damage and more was seen as relevant for the project. The evaluation of this damage indicated that for tankers (which is assumed to be most relevant for offshore structures) the side shell longitudinal and secondary connecting structures (16%), brackets (13%) and flat bar stiffeners (10%) were found to be the components most often damaged. Among primary longitudinal members the side shell longitudinals (33%) and the main deck plating and associated longitudinals (33%) were found to be most often damaged. Among primary transverse members the web frames (55%) were found to be the most affected due to the presence of lap joints. Among secondary connecting structures, brackets accounted for 50% of the damage. For tankers most of the longitudinal damage occurred in the middle cargo block. However, transverse damage was found to be evenly distributed among the port, starboard and the centerplane areas. The predominant form of local failure was found to be fracture (mostly fatigue but also brittle).

The Ship Structure Committee report SSC-416 (2000) states that the primary damage types for ship structures were fatigue and corrosion and these, if not properly repaired or rectified, could potentially lead to catastrophic failure. However, also general wear and tear and deformation defects were mentioned. The review indicates that for the commercial fleet classified by the Japanese classification society Nippon Kaiji Kayokai (NKK), corrosion damage accounted for more than half the total damage.

3.4.3 Defect Type for Tanker Structure Components

The SSC report "SSC-421 Risk-informed inspection of marine vessels" (SSC 2002) provides an overview of:

- current inspection methods and degradation mechanisms of ship structures;
- proposed methodology and guidelines for risk-informed inspection;
- demonstration of these guidelines and the methodology using a case study and examples; and
- a software development plan.

The typical degradation mechanisms of ship structures are also assumed to be relevant for floating offshore structures and particularly for ship-shaped offshore structures. A summary of tanker structural elements and their components that are prone to damage categories is presented in Table 11 based on the TSCF (Tanker Structure Cooperative Forum) 1997, "Guidance Manual for Tanker Structures" (TSCF 1997).

3.4.4 Semi-Submersible Flooding Incident Data

HSE undertook a study (HSE 1999) to record and review incidents that had led to semi-submersibles losing stability or buoyancy. These events were caused by either internal or external factors.

The internal events included:

- accidental flooding of normally dry spaces or tanks;
- loss of control of the ballast system;
- failure of pumps or valves;
- mal-operation; and
- cracks leading to leakage.

The external events included:

- damage by anchors;
- collision;

Table 11 Typical Defect Types for Tanker Structural Components (SSC 421).

	Category of damage susceptibility		
	Corrosion	**Fatigue cracking**	**Buckling**
Longitudinal elements	• Upper deck plating and longitudinals • Welds between structural elements (especially deck longitudinals to deck plating) • Scallops & openings for drainage • Webs of longitudinals on longitudinal bulkheads (high rates & localized 'grooving') • Flanges of bottom longitudinals (pitting) • Bottom plating (pitting, erosion near suctions) • Longitudinal bulkhead plating (relatively thin)	• Discontinuities • Openings & notches • Connections with transverse elements	• Upper deck plating and longitudinals • Bottom plating and longitudinals • Longitudinal bulkhead plating (middle & upper) • Deck & bottom girders
Transverse web frames	• Upper part, connection to deck • Just below top coating • Flanges of bottom transverses	• Connections with longitudinal elements • Scallops in connection with longitudinals • Bracket toes • Holes & openings • Crossing face flats	• Web plate (shear buckling) • Brackets, flanges and cross-ties
Transverse bulkheads	• Upper part, connection to deck • Stringer webs • Close to openings in stringers • Highly stressed regions (e.g. around bracket toes)	• Connections with longitudinal elements • Connections between girder systems • Bracket toes	• Horizontal stringers, web plate (shear) • Girder/stringer brackets • Vertical girders, web plate (shear buckling) • Corrugated bulkhead plating
Swash bulkheads	• Upper part, connection to deck • Stringer webs • Close to openings in bulkhead plating • Highly stressed regions (e.g. around bracket toes)	• Connections with longitudinal elements • Connections between girder systems • Bracket toes • Openings in bulkhead plating	• Horizontal stringers, web plate (shear) • Vertical girders, web plate (shear buckling) • Girder/stringer brackets • Openings in bulkhead plating

Source: SSC (2002), SSC421 Risk-informed inspection of marine vessels, Ship Structure Committee, Washington, D.C., US. © 2002, SHIP STRUCTURE COMMITTEE.

Table 12 Incident Frequencies for Semi-Submersibles for the Period 1970–1997 (HSE 1999).

Area	Country	No. incidents	Rig years	Incident frequency per rig year
North Sea	UK	20	918	0.022
	Norway	7	493	0.014
	Other	3	25	0.12
Total, North Sea		30	1463	0.021
Outside North Sea		13	2610	0.005
Total		43	4073	0.011

Source: Modified from HSE (1999). Semi-Submersible Flooding Incident Data, Report OTO 99 016, Health and Safety Executive (HSE), HMSO, London, UK.
The total also includes numbers for The Netherlands.

- grounding; and
- severe weather.

The study (HSE 1999) confirmed the significant presence of fatigue cracks in semi-submersible structures. To the knowledge of the authors, more than 100 cracks had been detected yearly requiring repair in these worst cases for older floating offshore structures. In such extreme cases permanent repair teams would be required on site. This has to be expected for the oldest platforms if these are to be kept in operation.

The trawl of data sources resulted in 43 relevant incidents world-wide in the period 1970–97. Fourteen of these were caused by external impact or from other external events and the remaining 29 were caused by internal failures. Thirty of the incidents were recorded from operations in the North Sea. The report includes details of the individual incidents, as indicated above.

The report included details of damage and incident frequencies for different parts of the world as shown in Table 12. The review was mainly concerned with the variability in available databases and noted that the data for incident frequencies depended on which database was used, with UK values varying by a factor of two. One recommendation was for a common database to be set up. The report makes no comment on the differences in the data for different locations, with, for example, the incident frequencies being four times higher for the North Sea compared to those for operations outside the North Sea.

3.5 Previous Studies on Damage Types to Mooring Lines and Anchors

3.5.1 Introduction and Damage Statistics for Moorings

Typical floating production systems have a spread mooring system where the mooring lines are often a combination of chain, wire rope and synthetic rope. Chains are most frequently used in shallow waters. In deeper waters wire rope and synthetic rope are used due to their lower self-weight compared to chain. Synthetic ropes (e.g. high modulus polyethylene) are often combined with chains to minimise contact of the synthetic rope with the seabed. These ropes are often provided with a protective outer sheath. A mooring system usually consists of a minimum of six lines.

Mooring line failures have occurred on a fairly regular basis at intervals substantially less than the design life of most units. Deterioration of the lines due to fatigue, corrosion, abrasion and wear over time can increase the likelihood of single or multiple line failures. An illustration of a mooring line and the typical locations of different types of damage is given in Figure 21.

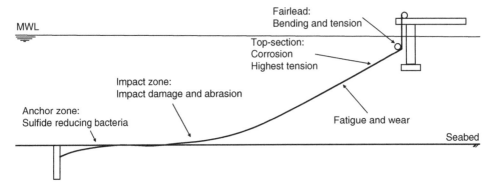

Figure 21 Mooring line elements and typical damage zones.

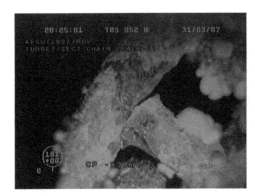

Figure 22 Partially loose stud, from corrosion or mechanical damage (HSE 2017b). *Source:* HSE (2017b). HSE RR1091 Remote Operated Vehicle (ROV) inspection of long-term mooring systems for floating offshore installations - Mooring Integrity Joint Industry Project Phase 2. Prepared by the Joint Industry Project Steering Committee for the Health and Safety Executive (HSE). HMSO, London, UK.

Multiple line failures have resulted in a floating production unit drifting creating a likelihood of the unit causing a collision and unless disconnected from the wells a possibility of serious oil leakage and a blow-out. There are examples where FPSOs have suffered a line failure without it being discovered for a period of at least several weeks (HSE 1999). During this time the installation could suffer further line failures and possibly more severe consequences.

Mechanical damage can occur in chains and wire ropes either from poor handling, use of anchor chasers or from problems at the fairlead. This damage can lead to loose studs (see Figure 22), bent links or gouges. All of these reduce the load-carrying capacity of a mooring line. Mechanical damage to wire ropes can result in twisted or loose strands, reducing strength.

3.5.2 API RP-2MIM Overview of Damage Types to Mooring Lines

The recommended practice API RP-2MIM (API 2019) provides an analysis of causes of failure of mooring lines. On the basis of an industrial survey the recommended practice (RP) notes that the failure rate per line has remained approximately constant at ~0.015 failures per exposed year. This is significantly above the assumed design reliability of 1×10^{-3} per year, which is implicit in codes. The RP also lists failure causes for both chain and wire rope as shown in Table 13. Fatigue and corrosion dominate for chains, making up 64% of the total. For wire ropes installation is apparently the highest cause of damage with fatigue coming second. API RP-2MIM (API 2019) also shows the vulnerability of different types of mooring system components to potential causes of failure. The two causes addressed are brittle fracture and ductile failure.

Table 13 Causes of Failures for Both Chains and Wire Ropes (API 2019).

	Chain	Wire rope
Causes of failure	%	%
Corrosion	20	
Design	6	8
Fatigue	17	19
Fatigue/corrosion	19	
Out-of-plane bending fatigue	8	
Installation	8	38
Manufacturing defect	8	
Mechanical	8	12
Multiple causes	3	4
Unknown	3	15
Overload		4

Source: Modified from API (2019), API RP-2MIM Mooring Integrity Management, American Petroleum Institute, 2019.

API RP 2MIM (API 2019) also shows the location of damage. For chains the fairlead/chain stopper dominates as the location for damage whereas for wire ropes the dominant location is near the socket.

API RP 2MIM (API 2019) details inspection objectives for different causes divided into three categories. These are spiral strand wire rope (unjacketed), wire rope (spelter socket termination) and chain. For each of these three categories the typical anomalies that can be detected during inspection are listed together with causal mechanisms, known locations and potential degradation rates.

3.5.3 HSE Studies on Mooring Systems

HSE report RR219 (HSE 2004) was the result of a Joint Industry Project and lists causes of failure of moorings. For chains it is stated that the primary cause of line failure is with the connecting shackles or with links that have been mechanically damaged. Common causes of chain failure are given as:

- mechanical damage to links;
- missing or loose studs;
- failure of connecting links; and
- brittle fracture of links (less common with better manufacturing control).

For wire ropes the three main causes listed are:

- mechanical damage to the wire;
- corrosion and wear; and
- fatigue.

HSE report RR444 (HSE 2006) lists the main causes of line failure as:

- overload and overstress;
- fatigue in a catenary, at a sheave or connection;
- brittle fracture;
- corrosion;
- wear and abrasion; and
- mechanical failure of the mooring line handling system.

As can be seen, there are differences between the causes listed in API RP-2MIM (API 2019) and HSE reports RR219 (HSE 2004) and RR444 (HSE 2006). This is particularly the case for wire ropes, where installation dominates in API RP-2MIM (API 2019). It is possible that poor installation led to the causes listed in RR219 (HSE 2004).

3.5.4   Studies on Corrosion of Mooring Systems

Chains are vulnerable to corrosion in seawater reducing strength as cathodic protection (CP) is normally not implemented on mooring chains and connecting components. However, cathodic protection systems on the floater itself can reach some way down the chain. The reach will depend on the electrical resistance between chain links and hence the level of protection will decrease with distance from where the chain is electrically connected to the floater.

Corrosion rates are approximately proportional to the level of dissolved oxygen in the water. The oxygen content is influenced by temperature and the degree of mixing. As earlier mentioned, in the North Sea the oxygen level in seawater is typically close to the saturated level due to the constant mixing at all depths. A corrosion allowance (additional sacrificial thickness) is usually part of the design specification. Currently industry practice is to increase the chain diameter by 0.2 mm to 0.4 mm per service year in the splash zone where oxygenated water tends to accelerate corrosion. ISO 19901-7 (ISO 2013) states for chain in the splash zone or in contact with a hard bottom seabed, the diameter should be increased by 0.2 mm to 0.8 mm per year of the planned design service life.

Bare steel surfaces are more anodic (more prone to corrosion) than surfaces covered with deposits and biofilms. Formation of calcareous deposits on the steel surface (see Figure 23) is considered to be one of the strongest features to limit the corrosion rate. Deposits of corrosion products (such as iron hydroxides) will also contribute to limiting corrosion rates. Thus, damage to surface deposits can lead to enhanced corrosion and possibly tendencies to pitting corrosion.

Corrosion pitting has also been observed, with consequent reduction in strength. Pitting in shackles has been observed in the North Sea which was higher than expected and believed to be due to galvanic corrosion (dissimilar metals) or sulphate-reducing bacteria (HSE 2006). Chain links with studs can suffer corrosion at the junction of the stud with the link itself, particularly if the stud is welded in place. In addition, loose studs have been implicated in crack propagation and fatigue.

Wire ropes are also subject to corrosion and protective coatings are normally used to minimise corrosion. However, if corrosion occurs, it can reduce the rope strength and discard criteria for wire ropes are usually based on the degree of outer corrosion of individual wires (HSE 2006), see Figures 24–26.

3.5.5   Studies on Fatigue of Mooring Systems

Chain links can suffer from fatigue with cracking. For studded links the location of maximum tensile cyclic loading is a position on the inner shoulder of the link where cracking has been observed (see Figure 27). For non-studded links the location of maximum cyclic tensile loading is

Figure 23 General corrosion on a recovered floating production unit mooring line after 16 years' service (HSE 2006). *Source:* HSE RR444 Floating production system - JIP FPS mooring integrity. Prepared by Noble Denton Europe Limited for the Health and Safety Executive (HSE). HMSO, London, UK.

Figure 24 Example of moderately severe corrosion in wire rope (HSE 2006). *Source:* HSE RR444 Floating production system - JIP FPS mooring integrity. Prepared by Noble Denton Europe Limited for the Health and Safety Executive (HSE). HMSO, London, UK.

Figure 25 Rope in Figure 23 after thorough cleaning (HSE 2006). *Source:* HSE RR444 Floating production system - JIP FPS mooring integrity. Prepared by Noble Denton Europe Limited for the Health and Safety Executive (HSE). HMSO, London, UK.

Figure 26 Example of moderately severe external wear in wire rope (HSE 2006). *Source:* HSE RR444 Floating production system - JIP FPS mooring integrity. Prepared by Noble Denton Europe Limited for the Health and Safety Executive (HSE). HMSO, London, UK.

at the crown where neighbouring links are joined together (see Figure 28). Fatigue cracking can lead to failure of the link and the chain itself.

Fatigue at the links close to a fairlead is a typical problem for mooring lines which is easily solved by moving the line at suitable intervals such that the same section is not exposed to the bending stresses for longer periods.

3.6 Previous Studies on Concrete Structures

3.6.1 Concrete in the Oceans Project

Concrete in the Oceans (Department of Energy 1989b) was a major UK programme started in 1976 and completed in 1986. It was initiated due to the lack of knowledge on several important aspects of offshore concrete, given the development of several offshore concrete installations at that time. It was jointly funded by the UK Department of Energy & the offshore industry and coordinated by the Underwater Engineering Group (UEG) of CIRIA. The total cost was £1.31M. It included 19 individual projects in 2 phases, covering:

- corrosion;
- structural aspects and design;

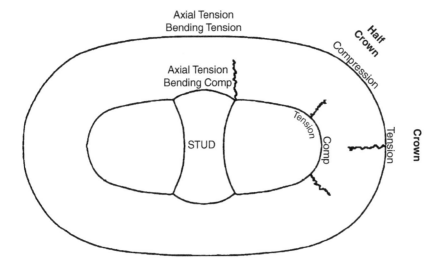

Figure 27 Load regimes for a studded link showing failure locations, terminology and areas of high tensile stress around the link. *Source:* HSE (2017a). HSE RR1093 An assessment of proof load effect on the fatigue life of mooring chain for floating offshore installations - Mooring Integrity Joint Industry Project Phase 2. Prepared by the Joint Industry Project Steering Committee for the Health and Safety Executive (HSE). HMSO, London, UK. Licensed under Open Government License.

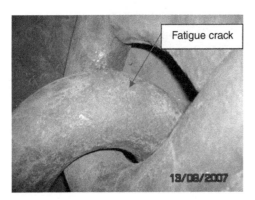

Figure 28 Fatigue damage at the crown of a non-studded chain (HSE 2017a). *Source:* HSE RR1093 An assessment of proof load effect on the fatigue life of mooring chain for floating offshore installations - Mooring Integrity Joint Industry Project Phase 2. Prepared by the Joint Industry Project Steering Committee for the Health and Safety Executive (HSE). HMSO, London, UK.

- fatigue; and
- inspection & maintenance.

Several test facilities were used, including deep water (Loch Linnhe in Scotland), splash zone (Portland on the south coast) and also tank testing.

Several actual concrete structures were surveyed in the programme. These included:

- Tongue Sands Tower (World War II fort in Thames estuary, 35 years old at the time of survey);
- Mulberry Harbour Unit from World War II, situated at Portland (36 years old at time of survey);
- Royal Sovereign Lighthouse (11 km south east of Eastbourne, 10 years old at time of survey); and
- Shoreham Harbour Breakwater on the Sussex coast (22 years old at time of survey).

More information on these surveys is given in the chapter on inspection. A large number of reports were published, including a coordinating report OTH 87 248 (Department of Energy 1989b).

The Concrete in the Oceans project collected information on defects, damage and deterioration from actual offshore concrete structures and provided photographic examples of these. These areas are listed in six groups which are covered in detail in Section 4.2.10.

The major types of deterioration listed were:

- impact damage;
- cracks in the concrete cover (structural cracks);
- corrosion of reinforcement and spalling of the cover;
- deterioration of surface treatment; and
- abrasion and erosion.

3.6.2 Durability of Offshore Concrete Structures

The Norwegian Research Council initiated a review of the durability of concrete structures in the late 1990s. Part 2.4 of this review concerned offshore concrete structures (Norges Byggforskningsinstitutt 1999). The review used the results of visual inspections, measurement of concrete cover, electro-chemical potential and core tests to map the chloride ingress and other anomalies on 6 offshore concrete structures in and above the splash zone. Four of these were GBS, one semi-submersible and one concrete tank. All these were provided with a cathodic protection system to protect mechanical equipment. An overview of the identified damage types featured in this report is shown in Table 14.

Table 14 Overview of Identified Damage on Concrete Structures.

Structure	Year of installation	Year in operation at inspection	
Statjord A	1977	16	Was provided with an epoxy membrane on the outer surface of the shafts which seemed to have a significant effect in reducing the chloride ingress
Gullfaks A	1986	7	Moderate chloride ingress (0.1% at 10 mm depth)
Gullfaks C	1989	5	Moderate chloride ingress (0.1% at 25 mm depth)
Oseberg A	1988	8 and 9	Moderate chloride ingress (0.1% at 25 mm depth)
Troll B	1995	2	Low chloride ingress (0.1% at 6–10 mm depth)
Ekofisk Tank	1973	17 and 20	High chloride ingress (0.1% at 40–45 mm depth)

When measuring chloride ingress there is a difference whether the amount is quoted as a percentage by weight of cement or concrete. The critical level of chloride ingress is in the Concrete in the Oceans project given as 0.4% by weight of cement, while in this report it is reported as 0.1% of weight of concrete.

Visible rust stains were identified in areas with low concrete cover for all structures, particularly for Oseberg A. A very low degree of carbonation was seen on all structures.

3.6.3   PSA Study on Damage to Offshore Concrete Structures

As part of PSA's initiative on ageing and life extension a study was undertaken by Ocean Structures (Ocean Structures 2009) to investigate ageing aspects for concrete structures. The typical damage types and defects reported were:

- deformations and structural imperfections;
- cracks typically occurring in tension stressed areas;
- corrosion of reinforcement due to inadequacy in the corrosion protection system (loss or depletion of anodes);
- damage to coatings;
- damage at post-tension cable anchor points;
- effects due to alternating wetting and drying of the surface (splash zone);
- freeze and thaw damage (splash zone);
- spalling and de-laminations normally due to corrosion of the reinforcement;
- local impact damage due to vessel collisions and dropped objects; and
- strength loss of the concrete material due to sulphate-producing bacteria (storage cells).

In addition, as for offshore steel jacket structures, changes to the seabed and foundations will influence the capacity of a structure. These include:

- scouring of the seabed under, or in the immediate vicinity of the installation;
- build-up of cuttings or sediments particularly if such build-up covers significant parts of the structure;
- movement in bottom sediments; and
- settlement.

The various parts of a concrete structure typically have their own particular damage types.

- The legs, towers and shaft of a concrete offshore structure are characterised by very large general featureless areas. The typical damage in these areas arises from corrosion of the reinforcement, impact damage from ships or possibly chemical deterioration.
- For storage cells the most likely form of damage is spalling of the concrete from dropped objects.
- The pre-stressing anchorages and tendons are required to maintain the structural integrity of a concrete offshore structure. These tendons are placed in ducts which are usually grouted following tensioning. The degree to which grouting was effective, given the long ducts and in some case their horizontal orientation, has led to concerns that seawater can penetrate into the ducts and cause corrosion of the very high strength tendons. A review of the durability of prestressing components (HSE 1997) concluded that the first tranche of concrete offshore structures (up to 1978) was more vulnerable to corrosion of the prestressing tendons, as later platforms benefited from improved grouting materials and procedures. It was also considered that there would need to be significant loss of prestress (~40%) in a leg before it would fail under typical design wave loading. These failures would also need to be in the same section area to be a danger. In land-based structures failures have tended to occur near anchorages or construction joints.

- Foundations of most offshore concrete structures are massive, as they are gravity-type structures. At the installation stage the foundation is usually grouted to the seabed, providing a solid base. As structures age, this grouting may be damaged by scour.
- Cathodic protection (CP) systems in most offshore concrete structures are primarily present to protect the attached steelwork and also protects the reinforcement. In the early structures there were attempts at the construction stage to isolate the reinforcement from this system. In most cases this failed as there was unintended electrical connectivity to flowlines and pipelines as well as to other external attachments. This led to a higher than planned drain on the sacrificial anodes, and in some cases, these had to be replaced. However, the availability of cathodic protection has the advantage that it protects the reinforcement to some extent, minimising the level of corrosion in situations where seawater has permeated to the steel. Current design criteria for CP systems recommend or require a minimum of $1\ mA/m2$ for the reinforcing steel. Inspection and maintenance of the CP system is therefore a basic requirement to minimise the corrosion reaction. This is usually carried out both visually and by monitoring the CP potentials using a probe operated from an ROV.

The several forms of deterioration listed above can affect different parts of the structure. Table 15 shows the correlation between these parts and the primary degradation mechanisms.

Table 15 Damage Types and Typical Locations in Offshore Concrete Structures (Ocean Structures 2009).

	Legs / Towers / shafts – general	Storage cells	Steel concrete transition	Shaft / base junction	Prestressing anchorages	Splash zone	Foundations
Chemical deterioration	✓	✓		✓		✓	
Corrosion of steel reinforcement	✓	✓		✓		✓	
Corrosion of prestressing tendons	✓	✓		✓	✓	✓	
Fatigue cracking			✓	✓			
Damage from ship impact	✓				✓	✓	
Damage from dropped objects		✓			✓		
Bacterial degradation	✓	✓		✓		✓	
Cracking from thermal effects		✓		✓			
Loss of pressure control		✓		✓			
Loss of air gap			✓				
Scour & Settlement							✓

3.7 Previous Studies on Marine Growth (Marine Fouling)

Offshore installations in general, unless protected by effective ant-fouling measures, will become fouled with marine growth. As part of the MTD report on underwater inspection of steel offshore installations (MTD 1989) a review of marine fouling was included. The effects of fouling can be described as shown in Table 16.

Corrosion associated with marine fouling was stated to be not clearly understood, as fouling can both act as a barrier to oxygen to a steel surface and hence reduce corrosion but could also prevent the CP anodes working properly. Marine growth can also provide conditions in which microorganisms capable of damaging steel can survive. The most important of these are the sulphate-reducing bacteria (SRB) which are present in seawater but only become active in anaerobic conditions, such as below marine fouling. In addition, the environment below layers of drilling mud, particularly oil-based mud, can be conducive to rapid growth of SRBs. The presence of an active SRB population on a steel surface can lead to biological corrosion, which could create pitting. The most likely forms of fouling to lead to active SRBs and potential corrosion at the steel surface are kelps, hydroids, mussels and barnacles (MTD 1989).

Robinson and Kilgallon (HSE 1998) reviewed the effects of the presence of SRBs on steel surfaces underwater. The surface of steel when exposed to seawater quickly develops a thin biofilm. This film will grow as marine fouling species develop and can restrict the access of dissolved oxygen in the seawater to the metal surface. These biofilms and fouling can provide anaerobic conditions at the metal surface and under these conditions SRBs can grow. Hydrogen uptake by steel in a seawater environment has been shown to be strongly influenced by the combined effects of cathodic protection and the presence of SRBs. The CP process generates hydrogen on the steel surface and its absorption is then enhanced by biogenic sulphides which are produced by the bacteria. The absorbed hydrogen can lead to hydrogen embrittlement and enhanced rate of fatigue crack growth, particularly in high-strength steels. The problem increases as more protective potentials are being used and hence there should be a prescribed limit for CP of high-strength steels to $-0.8V$ vs Ag/AgCl.

Corrosion on a section of a steel chain mooring due to SRBs close to or below the seabed has been reported by Gabrielson et al (2018). The main corrosion was observed on chain in the top 2m of sediment on the seabed, with an observed corrosion attack of 3-4mm depth of varying size from

Table 16 Effects of Fouling on a Structure.

Effects of fouling:	Description
Obscured surfaces	There is a need to remove fouling before detailed inspection is possible.
Static loading	Fouling may add weight to a structure and the most likely types of fouling to add to weight are those with hard calcareous shells, plates or tubes as the soft fouling is mostly self-buoyant. The most important fouling to contribute to weight are mussels.
Hydro-dynamic loading	Both soft and hard growths contribute to hydro-dynamic loading by increasing member dimensions, added mass of members and increasing the surface roughness.
Physical damage to steel	Cleaning work to remove growth can potentially lead to damage of the steel surface.
Corrosion	Corrosion under fouling may occur potentially from sulphate-reducing bacteria.

a few millimetres to larger areas. The presence of these pits was seen to reduce the fatigue capacity of the chain.

As a general rule, the development of fouling and its ultimate thickness decrease with increasing distance from the shore and with increasing depth. Marine fouling limits the opportunity to inspect either a steel or concrete offshore structure. There are many different types of fouling, both hard and soft. Typical groups of fouling are listed below. Table 17 shows an overview of common marine fouling species and their characteristics. Examples of marine fouling are easily available on the internet.

Table 17 Growth Characteristics of Common Marine Fouling Species (HSE 2002).

Type	Settlement season	Typical growth rates	Typical coverage (%)	Typical terminal thickness	Depth range (relative to MSL)	Comments
Hard fouling						
Mussels	July to October	25 mm in 1 year 50 mm in 3 years 75 mm in 7 years	100%	150 to 200 mm	0 to 30 m	But faster growth rates are found on installations in the southern North Sea
Solitary tube-worms	May to August	30 mm (in length) in 3 months	50–70%	About 10 mm (tubeworms lay flat on the steel surface)	0 to mudline	Coverage is often 100%, especially on new structures 1 to 2 years after installation. Tubeworms also remain as a hard background layer when dead
Soft fouling						
Hydroids	April to October	50 mm in 3 months	100%	Summer: 30 to 70 mm winter: 20 to 30 mm	0 to mudline	A permanent hydroid 'turf' may cover an installation and obscure the surface for many years
Plumose anemone	June to July	50 mm in 1 year	100%	300 mm	−30 m to −120 m (0 to −45 m on platforms in southern North Sea)	Usually settle 4 to 5 years after installation and can then cover surface very rapidly. Live for up to 50 years
Soft coral	January to March	50 mm in 1 year	100%	About 200 mm	−30 m to −120 m (0 to −45 m on platforms in southern North Sea)	Often found in association with anemones
Seaweed fouling						
Kelp	February to April	2 m in 3 years	60–80%	Variable, but up to 3 m	−3 m to −15 m	May be several years before colonisation begins but tenacious holdfast when established. Present on some installations in northern and central North Sea

Source: HSE (2002). OTO 01 010 Environmental considerations. Health and Safety Executive (HSE), HMSO, London, UK. Licensed under Open Government License.

Seaweed requires sunlight for existence and is therefore confined to surface zones, typically limited to about 15 m below sea level. Examples include brown kelp which can reach lengths of more than 3 m but can be removed by cutting through the stalk that attaches the kelp to the surface.

Hard fouling possesses a hard external skeleton protecting the fleshy body within. The most important of these are mussels, which usually settle in the intertidal zone offshore and are therefore limited to shallow depths. Other hard fouling includes shallow water barnacles, which are found offshore from the splash zone to a depth of about 20 m. Barnacles tend to settle close to each other creating a dense area, limiting inspection. Deep-water barnacles are also typical of most offshore platforms typically in the depth range from 30 m to deep water. The third group are tube-worms, which live in hard calcareous tubes cemented to the surface. These are found throughout the depth range of the North Sea. These can require vigorous cleaning to remove.

Soft fouling has no hard external skeleton. The most common type offshore is the hydroid, which forms soft plant like growths. Different types can reach lengths from 50 mm to 150 mm. These are also found throughout the depth range of the North Sea. They are one of the first species to foul an installation. On older platforms the fouling may change in character and hydroids can be replaced by other species. Below 100 m hydroids may continue to form the dominant species as others prefer shallower waters.

Anemones tend to dominate in the mid depth range from 30 to 100 m. The anemone is anchored by a disc attached to the substrate. There are also sea squirts which can be found in deep waters. They prefer sheltered conditions such as protected areas on offshore platforms.

3.8 Summary of Damage and Anomalies to Offshore Structures

3.8.1 General

The studies listed in section 3.2 provides a significant review of the damage that has occurred for different types of offshore structures and their causes. These types can be divided into several different categories, depending on whether the installation is a fixed steel structure, a floating platform or a concrete platform.

Although it is recognized that some of these studies are now old, it is expected that the type and pattern of damage will still be similar. With the continued improvement of design and fabrication standards for offshore structures, it would be expected that the likelihood of these types of damage now should be less. However, no studies at time of writing have confirmed this to date.

In addition, a brief overview of damage types to offshore structures is given in Section 3.1, covering fatigue, corrosion, impact damage, hydrogen embrittlement, erosion, wear and tear, brittle fracture and mechanical damage.

3.8.2 Damage Types Specific to Steel Structures

The previous studies presented in Section 3.3 provide a good overview of damage types and effects for steel structures. The conclusions to be drawn for these studies are given below.

For fixed steel structures the main damage types are shown in Table 18. In addition, in severe hurricanes and storms the damage can be more severe and include toppling (collapse) and leaning.

For floating steel structures the main damage types are also shown in Table 18. Fatigue cracking to hull structures is generally more frequent compared to fixed steel structures, particularly in

Table 18 Categories of Potential Causes of Damage for Three Different Types of Structure.

Cause of damage	Fixed steel and hull structures	Floating platforms - mooring	Concrete platforms
Fatigue	Fatigue cracks particularly in welded structures with potential for through-thickness cracks, member severance (separation) and missing members	Fatigue crack in chain link, link failure, loss of mooring line, drifting units	Fatigue of reinforcement (infrequently observed)
Corrosion	General corrosion will result in reduced wall thickness and in some cases patch corrosion with potential for holes, flooding, member severance and missing members Corrosion pits leading to flooding and potential for fatigue initiation	Corrosion of chain link or wire strand, loss of mooring line, drifting units	Spalling and delamination (above water) as result of reinforcement corrosion as well as potential loss of reinforcement area Seepage of corrosion products at cracks under water as result of reinforcement corrosion Chloride ingress above water (Reinforcement corrosion and potential loss of reinforcement area) Corrosion of steel attachments (e.g. pre-tension cable at anchorages, steel support attachments) and potential loss of prestressing capacity
Overload (hurricanes and major storms)	Buckles, dents, holes, cracks at welds and into chords, tears, out-of-plane bowing, severed members, tears, pancake leg severance, punch-through of brace, pull out of brace, conductor torn loose from guide, broken water caissons, broken riser standoffs.	Damage to links, snapping of anchor chains, drifting units, dragging of anchors Drifting units and dragging of anchors may impact other structures and pipelines	To the knowledge of the authors not many instances of major storm damage have been recorded to the concrete structure itself. Run-up of waves resulting in damage to lower decks of topside has been reported.
Overload and accidental damage (vessel impact and dropped object)	Buckling in plates and members Dented and bowed member, member severance and missing member.	Mechanical damage from seabed impact, wear and bending over the fairlead and handling of anchor chains and wire ropes.	Local impact damage to towers resulting in reduction in concrete section area and possible water ingress Local impact damage to tops of storage cells from dropped objects resulting in reduction in concrete section area and possible water ingress
Wear and tear	General loss of thickness	Abrasion/wear of link or rope Lost stud (chain)	Abrasion (local wear) and erosion resulting in minor reduction in concrete section area

(Continued)

Table 18 (Continued)

Cause of damage	Fixed steel and hull structures	Floating platforms - mooring	Concrete platforms
Marine growth	Increased loading and potential overload damage. Possibility for sulphate-reducing bacteria resulting in corrosion.	Increased loading and potential overload damage. Possibility for sulphate-reducing bacteria resulting in corrosion.	Increased loading
Settlement and subsidence	Platform settlement and subsidence may lead to change in loading pattern and exposure condition.	N.A.	Platform settlement and subsidence may lead to change in loading pattern and exposure condition.
Scour and build-up of drill cuttings	Scour around the platform exposing the piles with the potential for pile failure. Drill-cuttings can bury or bear onto lower frames and result in damage.	Loss of holding power to anchors	Scour around the base of the platform possibly leading to instability
Material deterioration	Cracking from hydrogen embrittlement particularly for high strength steel (e.g. Jack-ups)	Brittle fracture of chain links	Loss of the concrete material due to aggressive agents (sulphate, chloride)

Cracking due to shrinkage (after construction)

Cracking, scaling and crumbling due to freeze and thaw (above water)

Strength loss of the concrete material due to sulphate producing bacteria (primarily in oil-storage tanks) |
| Fabrication fault | Lack of penetration or excessive undercutting that could lead to unexpected fatigue cracks, failure to provide vent holes for intended flooded members could lead to implosion, inferior material, incorrect member sizes, incorrect member positions, omissions. | Brittle fracture has been reported in the high strength steel due to manufacturing issues. | Defective construction joints, minor cracking, surface blemishes, remaining metal attachments, patch repairs from construction, low cover to the reinforcement possibly leading to spalling in the splash zone and possible water ingress |
| Under-design | Member buckling, joint failure, tearing, fatigue cracking | Damage to links and failure of anchor chains possibly leading to drifting units. Under-designed anchors (soil strength) could lead to dragging of anchors. | Cracking due to low reinforcement or prestressing, crushing of concrete in rare cases, failure in the junction between shafts and cells. |

transverse brace to column joints, end terminations. It should be noted that floating structures are normally damage tolerant with respect to fatigue cracking in many areas of the structure unless water ingress occurs. Floating structures are also more exposed to corrosion problems due to corrosion in internal tanks. The main damage effects from corrosion include loss of wall thickness, potential loss of watertight integrity leading to buoyancy and stability issues.

3.8.3 Damage Types Specific to Concrete Structures

For concrete structures the main damage types are shown in Table 18. The splash zone is particularly vulnerable due to, for example, the availability of chlorides and oxygen, wave motion and temperature variations.

The degree of erosion and abrasion is very dependent on the quality of the concrete, which is usually high for offshore structures with controlled permeability and a high depth of cover. Cracked sections or areas of low quality are more vulnerable to erosion. Ice abrasion and freeze-thaw cycling can, as mentioned in ISO 19903 (ISO 2019), also lead to early deterioration even with high-quality concrete. The main effect of local erosion of the concrete cover is earlier corrosion of the steel reinforcement with potential loss of strength.

As indicated in ISO 19903 (ISO 2019), debris can cause impact and abrasion damage and can limit the ability to inspect various parts of the structure (e.g. tops of storage cells). Drill cuttings can build up particularly on the cell tops.

The production of acids from sulphate-reducing bacteria (SRB), which attack the concrete, can occur in concrete structures containing both water and oil. Significant loss of material has been shown in laboratory tests to occur when sufficiently acidic conditions exist (Department of Energy 1989). This type of environment can exist in the concrete storage tanks, which are present in several concrete offshore structures. SRBs are known to grow rapidly under certain acidic conditions, which can cause loss of material which reduces the wall thickness. Unfortunately, the storage tanks are very difficult to inspect due to very limited access and hence the level of damage from SRBs is difficult to assess. The thick coating that is expected to exist on the inner walls of the tanks due to the presence of waxes in the oil may be a mitigating factor.

The storage of hot oil in the concrete tanks at the base of many concrete installations can lead to thermal stresses that can produce cracking of the concrete. Concrete is vulnerable to significant temperature differences, which in this case arise from the hot oil on one side of the wall and cold seawater on the other. Tests have shown that temperature differences of up to 45 °C can be sustained with the correct design details (Department of Energy 1989). However, if the coolers fail (the oil is cooled before storage) or unusual conditions occur, oil with temperatures of up to 90 °C can be diverted into the storage cells, with potential cracking of the walls. Over a long period of operation these effects could accumulate, leading to cracking of the concrete and overstress of steelwork, in and around the walls and roofs of the storage cells, including the critical junction with the legs.

3.8.4 Summary Table of Damage to Different Types of Structures

In Table 18 an overview of potential causes of damage for different types of structures is provided, partly based on observed damage in these studies and the authors' experience with offshore structures. It can be seen that some of the categories are common to each type of structure, e.g. corrosion. These damage categories will be developed in more detail later in this book, particularly relating to inspection and repair. For each category there will be particular inspection and repair methods which are also detailed later in the book.

3.9 Bibliographic Notes

Parts of Section 3.1 are based on Ersdal et al. (2019).

References

API (2014), *API RP-2A Recommended Practice for Planning, Design and Constructing Fixed Offshore Platforms*, API Recommended practice 2A, 22nd Edition, American Petroleum Institute, 2014.

API (2019), *API RP-2MIM Mooring Integrity Management*, American Petroleum Institute, 20e19.

Bahadori, A. (2014), *Cathodic Corrosion Protection Systems—Guide for Oil and Gas Industries*, Gulf Professional Publishing, Elsevier Waltham, Massachusetts, US.

Department of Energy (1989a), *Fatigue Correlation Study Semi-Submersible Platforms*, OTH 88 288, HMSO, London, UK, 1989.

Department of Energy (1989b), *OTH 87 248 Concrete in the Ocean Programme—Coordinating Report on the Whole Programme*, HMSO, London, UK.

DNVGL (2019), "Reparasjonsmetoder for bærende konstruksjoner", Report for the Norwegian Petroleum Safety Authority, DNVGL, Høvik, Norway (in Norwegian).

Energo (2006), "Assessment of Fixed Offshore Platform Performance in Hurricanes Andrew, Lili and Ivan", MMS Project no 549, Energo 2006.

Energo (2007), "Assessment of Fixed Offshore Platform Performance in Hurricanes Kartina and Rita", MMS project no 578, Energo 2007.

Energo (2010), "Assessment of Fixed Offshore Platform Performance in Hurricanes Gustav and Ike", MMS project no 642. Energo 2010.

Ersdal, G., Sharp, J.V., Stacey, A. (2019), *Ageing and Life Extension of Offshore Structures*, John Wiley and Sons Ltd., Chichester, West Sussex, UK.

Gabrielson, O., Liengen, T. and Molid S. (2018), "Microbiologically Influenced Corrosion on Seabed Chain in the North Sea", In the Proceedings of OMAE2018, Paper OMAE2018-77460, Madrid, Spain.

HSE (1991), OTH 91 351 *Hydrogen Cracking of Legs and Spudcans on Jack-Up Drilling Rigs—A Summary of Results of an Investigation*, Her Majesty's Stationery Office, London, 1991.

HSE (1995), *MSL Engineering Hurricane Damaged Fixed Platforms and Wellhead Structures*, Report No. C155R002 Rev. 2, July 1995.

HSE (1997), *Durability of Prestressing Components in Offshore Concrete Structures*, Gifford and Partners, OTO 97 053, HMSO, London, UK.

HSE (1998), OTH 555 *A Review of the Effects of Sulphate-Reducing Bacteria in the Marine Environment on the Corrosion Fatigue and Hydrogen Embrittlement of High-Strength Steels*, Prepared by Marine Technology Centre at Cranfield University (M.J. Robinson and P.J. Kilgallon) for the Health and Safety Executive (HSE), HMSO, London, UK.

HSE (1999), *Semi-Submersible Flooding Incident Data, Report OTO 99 016, Health and Safety Executive (HSE)*, HMSO, London, UK.

HSE (2002), OTO 01 010 *Environmental Considerations*, Health and Safety Executive (HSE), HMSO, London, UK.

HSE (2003), *Review of the Performance of High Strength Steel Used Offshore*, Health and Safety Executive (HSE), London, UK.

HSE (2004), HSE RR219 *Design and Integrity Management of Mobile Installation Moorings*, Prepared by Noble Denton Europe Ltd. for the Health and Safety Executive (HSE), HMSO, London, UK.

HSE (2006), HSE RR444 *Floating Production System—JIP FPS Mooring Integrity*, Prepared by Noble Denton Europe Limited for the Health and Safety Executive (HSE), HMSO, London, UK.

HSE (2017a), HSE RR1093 *An Assessment of Proof Load Effect on the Fatigue Life of Mooring Chain for Floating Offshore Installations*—Mooring Integrity Joint Industry Project Phase 2, Prepared by the Joint Industry Project Steering Committee for the Health and Safety Executive (HSE), HMSO, London, UK.

HSE (2017b), *HSE RR1091 Remote Operated Vehicle (ROV) Inspection of Long-Term Mooring Systems for Floating Offshore Installations—Mooring Integrity Joint Industry Project Phase 2, Prepared by the Joint Industry Project Steering Committee for the Health and Safety Executive (HSE)*, HMSO, London, UK.

ISO (2007), ISO 19902:2007 *Petroleum and Natural Gas Industries*—Fixed Steel Offshore Structures, International Organization for Standardization, 1998.

ISO (2013), ISO 19901-7:2013 *Petroleum and Natural Gas Industries—Specific Requirements for Offshore Structures—Part 7:* Stationkeeping Systems for Floating Offshore Structures and Mobile Offshore Units, International Organization for Standardization, 2013.

ISO (2019), ISO 19903 *Petroleum and Natural Gas Industries—Fixed Concrete Offshore Structures*, International Organization for Standardization, 2019.

Kareem, A., Kijewski, T., Smith, C.E. (1999), "Analysis and Performance of Offshore Platforms in Hurricanes", In *Wind and Structures, Vol. 2*, No. 1 (1999), pp. 1–23.

MSL (2000), *Rationalisation and Optimisation of Underwater Inspection Planning Consistent* with API RP2A Section 14, Published as MMS TAP report 345aa.

MTD (1989), *Underwater Inspection of Steel Offshore Installations: Implementation of a New Approach, Report MTD publication 89/104*, London, UK.

MTD (1990), *Design and Operational Guidance on Cathodic Protection of Offshore Structures, Subsea Installations and Pipelines*, Marine Technology Directorate Ltd (MTD), Report 90/102.

MTD (1994), *Review of Repairs to Offshore Structures and Pipelines*, Publication 94/102, MTD, London, UK.

Norges Byggforskningsinstitutt (1999), Bestandige Betongkonstruksjoner—Delprosjekt 2 Konstruksjoner—Rapport 2.4 *Erfaring fra offshore konstruksjoner*, Norges Byggforskningsinstitutt, Blindern, Oslo, Norway (in Norwegian containing an English summary).

Ocean Structures (2009), Ageing of Offshore Concrete Structures—Report for Petroleum Safety Authority Norway, Laurencekirk, Scotland.

Offshore: Risk & Technology Consulting Inc. (2002), "Post Mortem Failure Assessment of MODUs during Hurricane Lilli", Report Published as MMS report 469AA.

PMB (1987), "AIM II Assessment Inspection Maintenance", Published as MMS TAP report 106ak.

PMB (1988), "AIM III Assessment Inspection Maintenance", Published as MMS TAP report 106ap.

PMB (1990), "AIM IV Assessment Inspection Maintenance", Published as MMS TAP report 106as.

PMB Engineering (1993), "Hurricane Andrew—Effects on Offshore Platforms—Joint Industry Project, PMB Engineering, Inc, San Francisco, California, October 1993.

Puskar, F.J., Ku, A.P. (2004), "Hurricane Lili's Impact on Fixed Platforms and Calibration of Platform Performance to API RP 2A", OTC paper 16802, OTC 2004.

Sharp J.V., Stacey A. and Birkinshaw M. (1995), "Review of Data for Structural Damage to Offshore Structures", 4th Intern. ERA Conference, London, Dec.1995.

SSC (1990), "SSC-337 Ship Fracture Mechanisms Investigation", Ship Structure Committee, Washington, DC, US.

SSC (1995), "SSC-381 Residual Strength of Damaged Marine Structures", Ship Structure Committee, Washington, DC, US.

SSC (2000), "SSC-416 Risk-Based Life Cycle Management of Ship Structures", Ship Structure Committee Report *no* 416, 2000.

SSC (2002), "SSC421 Risk-Informed Inspection of Marine Vessels", Ship Structure Committee, Washington, DC, US.

Stacey, A., Sharp, J.V. and Nichols, N.W. (1997), "Fatigue Performance of Single Sided Circumferential and Closure Welds in Offshore Jacket Structures", OMAE Conference, Japan, 1997.

TSCF (1997), *Tanker Structure Cooperative Forum Guidance Manual for Tanker Structures*, Issued by Tanker Structure Co-operative Forum in Association with International Association of Classification Societies, Witherby & Co. Ltd.

UEG (1978), "Underwater Inspection of Offshore Installations—Guidance for Designers", CXJB Underwater Engineers Report UR10, CIRIA Underwater Engineering Group, London, UK.

4

Inspection Methods for Offshore Structures Underwater

The stuff you see beneath the water often seems like a wild parody of the stuff you see above it[1].
—Douglas Adams

When you want to know how things really work, study them when they're coming apart.
—William Gibson

You cannot inspect quality into a product. The quality is there or it isn't by the time it's inspected[2].

—W. Edwards Deming

4.1 Introduction to Underwater Inspection

Many different inspection methods are available to determine the in-service condition of a structure underwater or in the splash zone. They all have their strengths and limitations related to the type of anomaly (degradation, deterioration and damage) that the inspection is aiming at revealing and the availability to access the site. In addition, the choice of inspection method and tool will be dependent on the information required, such as the extent to which an anomaly has to be sized and characterised (e.g. size of cracks).

Underwater inspection commenced in the GoM with the early structures in place there. Busby (1978) reported that up to the date of his review the only requirements for inspection of offshore structures in the GoM were those which the platform operator or owner elected to impose upon themselves. Some statutory requirements were also implemented in 1953 (US Geological Survey) and 1970 (Occupational Safety and Health Administration). The early recommended practices tended to concentrate on design and fabrication. Inspection at the fabrication stage was often included but not inspection in the operational stage.

---

1 *Sources:* Kath Jordan, "Ks3 Success Workbook English 4-7", 2007, Letts and Lonsdale.
2 *Sources:* Rafael Aguayo, "Dr. Deming: The American Who Taught the Japanese About Quality, A Fireside book", 1991, Simon and Schuster.

Underwater Inspection and Repair for Offshore Structures, First Edition.
John V. Sharp and Gerhard Ersdal.

The development of the North Sea offshore industry in the late 1960s and 70s led to the British and Norwegian governments to establish underwater inspection requirements and schedules to which the platform owners had to comply, primarily by referring either to the UK Department of Energy Guidance Notes or to the rules of classification societies such as DNV, Lloyds Register, Germanischer Lloyds, Bureau Veritas (BV) and American Bureau of Shipping (ABS). At the time, only DNV and BV had published requirements for inspection (Busby 1978), typically based on routine intervals as found in ship rules and regulations. As a result, the parts of the structures that were regarded as important in integrity terms were inspected typically every fifth year.

Underwater inspection was mainly undertaken by divers up to the 1980s. Visual inspection using photographic and TV documentation was the primary method (Busby 1978). Inspection included preliminary cleaning of structures which was often found to be more time consuming than the inspection itself. The diving industry was formed in the 1950s and 60s as the need for commercial diving grew as a result of the oil and gas industry moving offshore (Michel 2003). Companies were formed primarily by navy divers utilizing their technology and grew larger and developed new technologies as well as companies involving civilian divers.

While the early offshore structures were small and located in shallow water that allowed for a wide working inspection window, the structures in the mid- to late 1970s were more massive, complex and in significantly more severe environments, especially in the North Sea. The complexity of the structures led to an increased amount of inspection and the demanding environmental conditions led to an increase in fatigue cracking that produced a need for enhanced inspection methods. As a result, many inspection companies started to use NDT instrumentation underwater. These included magnetic particle inspection (MPI), ultrasonic testing (UT), radiography and corrosion potential measurements. These were reported to have been used underwater in 1978 (Busby 1978).

The interest in developing NDT tools for underwater inspection was growing fast in the 1970s, especially in the North Sea region, where the use of such tools was required by regulators. This included research and development in, for example, the following areas (Busby 1978):

- maintenance, inspection and repair techniques (Strathclyde University);
- review of NDT equipment, procedures and operators' qualifications (DNV); and
- underwater inspection (CIRIA 1978; two studies).

Remotely operated vehicles (ROV) started to be developed as an alternative to inspection by divers in the mid-1970s and military ROV technology was introduced to the offshore industry in the Gulf of Mexico (Michel 2003). ROVs were at that time evaluated and found to be capable of producing high-quality video and photographic inspection and could also bring some form of cleaning device (e.g. wire brush, clipping hammers) to the inspection site (Busby 1978). At that stage NDT techniques were becoming available, but as reported by Busby (1978), were not regarded as a suitable tool for operation by an ROV manipulator.

In the late 1970s improved ROV designs became available with more advanced inspection tools, sometimes supported by diving spreads. At this stage, ROVs suffered from reliability issues and time on site was often limited to a few hours each day. With the ever-increasing growth in offshore oil and gas activities in the early 1980s, the workload for underwater inspection increased and this supported the development of larger and more complex ROV units. As a result, the ROV industry grew and ROVs became more reliable and greatly improved their performance.

In the mid-1980s there was a certain downturn in the oil and gas industry due to a lower oil price; development funds became limited and research and development reduced (Michel 2003), particularly in the US. However, the use of ROVs continued to grow. In the 1990s the oil industry encouraged ROV applications to expand, particularly as deeper waters were involved and the safety of divers started to be recognised as an issue. With the increasing reliability and performance of ROVs the industry began to design ROVs for specific project work offshore. More recent developments in ROV technology include larger, more powerful and capable systems, with an increasing array of tools capable of more onerous and difficult tasks. The growth of sub-sea systems for oil production has led to the need for ROVs with a greater capability for sub-sea work such as the operation of sub-sea valves.

In more recent years, autonomous underwater vehicles (AUV) technology was developed for surveying and data-gathering activities and have started to be used for offshore oil and gas inspection and maintenance. A key driver for improvements in underwater inspection and maintenance has been the need to reduce costs, particularly as oil prices have been variable.

Current industry practice now includes a range of non-destructive inspection methods for use underwater such as general and close visual inspection (GVI and CVI), ultrasonic testing (UT), magnetic particle inspection (MPI), eddy current inspection (EC), radiographic testing (RT), leak detection testing (LT) and flooded member detection (FMD). These methods are described in more depth in Section 4.3. In addition, a good overview of most inspection techniques is given in NORSOK N-005 Annex B (Standard Norge 2017).

It should be recognised that there are potential negative consequences from undertaking inspection underwater. Firstly, the cleaning process (brushes or water jetting) may introduce damage which could lead to crack initiations. Further, the use of ROVs and AUVs has led to collisions with the structure itself potentially causing damage such as dents, crack initiators or damage to coatings. These potential consequences need to be addressed in the programme for the execution of the inspection in order to minimise their likelihood.

4.2 Previous Studies on Inspection

4.2.1 Introduction

Since 1953 there have been a number of relevant studies on inspection, particularly in the UK and the US in the 1980s and 90s. The authors' intention of including these reports is to make the reader aware of relevant previous work and thereby avoid unnecessarily repeating research work.

These reports provide a very useful background and insight into the development of this topic. They have also provided data that are now used in current standards. A brief overview of these studies, as summarised in Table 19, is provided in this section.

4.2.2 SSC Survey of Non-Destructive Test Methods

The Ship Structure Committee carried out a study into non-destructive testing (inspection methods) as early as 1953 (SSC 1953), providing an insight into the available inspection methods in the early years of the offshore oil and gas industry. The non-destructive test methods described in this report as being applicable for flaw detection in welds in ship structures were radiography (RT), magnetic particle (MPI), ultrasonic (UT) and fluid penetrants, where radiography was reported to be most extensively used. The ultrasonic method was reported to offer considerable potential and, if developed, to be more expedient than radiography. The magnetic particle method was reported to be established to the point where it serves as a useful inspection tool. The methods were reviewed with respect to their use both in shipbuilding and repair.

Table 19 Overview of Previous Studies on Inspection Methods.

	Section	Inspection methods	Performing inspections	Reliability of inspection methods
Inspection methods (SSC 1953)	4.2.2	X		
Risk-based life cycle management (SSC 2000)	4.2.2	X	X	
Inspection of marine structures (SSC 1996)	4.2.2			X
Underwater inspection / testing / monitoring (Busby 1979)	4.2.3	X	X	
Handbook of underwater inspection (HSE 1988)[a]	4.2.4	(X)	(X)	(X)
Underwater inspection of steel offshore structures (MTD 1989)	4.2.5	X	X	X
Fourth edition Guidance Notes (Department of Energy 1990)	4.2.6	X	X	
Damage to underwater tubulars OTO 99 084 (HSE 1999)	4.2.7	X	X	
Damage to underwater tubulars (Frieze et al 1996)	4.2.7	X	X	
Data for structural damage to offshore structures (Sharp et al 1995)	4.2.7	X	X	
Underwater inspection planning (MSL 2000)	4.2.8	X	X	X
Inspection methods and their reliability (Department of Energy 1984)	4.2.9			X
ICON - Inspection methods and their reliability (Dover and Rudlin 1996)	4.2.9			X
POD/POS Curves for non-destructive examination (HSE 2002)	4.2.9			X
Guidance Notes–type approval of NDT equipment (BV 1998)	4.2.9			X
Concrete in the ocean (Department of Energy 1987a).	4.2.10	X	X	
a) OTI 88 539 is now regarded as out of date				

A later report by the Ship Structure Committee (SSC 2000) provides an overview of the scope for inspection of oil tankers[3], providing a useful overview of the planning requirements for the execution of inspections. The principles outlined in this list are also applicable to offshore structures.

---

3 Although tankers are outside of the scope of this book, the inspection requirements are similar to those for a ship-shaped structure.

- Objective of inspection:
  - detecting defects including fatigue cracks, buckling, corrosion and pitting;
  - reporting present condition of steel plate thickness reduction due to corrosion and the condition of coating and other corrosion protection systems;
  - detecting any other problems such as structural deformation, leakage.
- Types of inspections:
  - mandatory inspections required by classification societies or flag state (annual, intermediate and special surveys); and
  - owner's voluntary inspections.
- Scope of inspection in accordance with IACS Unified Requirements (IACS 2012):
  - locations to be inspected and extent of inspection for structural defects, corrosion, pitting and coating.
- Technical information needed for effective evaluation:
  - main structural plans (drawings);
  - extent of coatings and corrosion protection systems;
  - previous structural survey reports and thickness measurement reports;
  - previous maintenance and repair history;
  - classification society's condition evaluation reports and status;
  - updated information on inspections and actions taken by ship's personnel;
  - critical and high-risk areas for corrosion and structural fractures;
  - survey planning documents; and
  - loading history (cargo and ballast loading and trading route history).
- Preparation for inspection:
  - emptying, cleaning, ventilation (to prevent gas hazard to the inspectors) and general lighting.
- Methods of inspection:
  - access and possible hazards to inspection personnel.
- Recording of inspection data (location, the affected structural member, the type and the size of the defect):
  - structural defects such as crack, buckling and dents;
  - pitting and grooving corrosion including pitting intensity diagram;
  - thickness measurement of steel plates;
  - coating condition including percentage of breakdown, peeling, flaking and blistering;
  - condition of corrosion control systems (sacrificial anode or impressed current);
  - effectiveness of previous repairs;
  - crack growth if previously not repaired; and
  - drawings or photographs to supplement the above data.

The accuracy of inspections and the factors affecting inspection performance is reviewed in more detail in SSC389 (SSC 1996). Knowledge of how likely it is that a flaw will be found during an inspection is important to provide guidance in setting inspection intervals and to compare different inspection technologies. Based on a review of the literature and interviews with inspectors and others involved in tank inspections, a model of the factors that can influence the probability of detection was developed. The following main groups of factors were identified as:

- physical: the structure and the defects in the structure including structural layout, size, coatings, structural details, condition and age and the type of defects (cracks, corrosion or buckling);
- experience of the inspector including his training, workload and motivation and knowledge of the specific structure; and

- inspection environment including level of lighting and cleanliness, temperature, humidity and external weather, degree of ventilation, access and inspection method, crew support and available time.

Many of these factors are relevant in determining the probability of successful detection (see Section 4.2.9).

4.2.3   Underwater Inspection / Testing / Monitoring of Offshore Structures

In 1978 Frank Busby Associates undertook a project for the US Department of Commerce, Department of Energy and the Department of the Interior to study underwater inspection, testing and monitoring of offshore structures (Busby 1979). The purpose of the study was:

- to identify and describe all actual or potential underwater inspection requirements (national and international) for fixed concrete and steel structures promulgated by the governments of offshore oil and gas–producing countries and by the offshore operators themselves;
- to identify and assess the state-of-the-art in underwater non-destructive testing, monitoring and inspection of offshore structures;
- to evaluate the capability of servicing and hardware producers to meet the inspection requirements identified; and
- to describe and establish priorities for specific tasks for technology development that should be undertaken to satisfy current and future requirements.

While this study concentrated on fixed offshore oil and gas structures, the results also reflected the state-of-the-art at the time in underwater inspection and testing for other offshore structures as well as floating power platforms, offshore terminals and deep-water ports.

The data for this study were collected in three stages, a literature review, telephone interviews and personal interviews with individuals and companies active in the industry. It was found that requirements for underwater inspection of these offshore structures and the techniques and tools to conduct such inspections varied widely from country to country. In some instances, periodic inspection was required by law. In a few cases there was no requirement whatever, once the structure had been installed. It was also seen that the instruments to conduct underwater inspections varied and the cost of underwater inspection to the operator was high and would get higher as the water depth and complexity of the structure increased.

It was found that general surveys consisted primarily of visual inspection to identify cracking and pitting, corrosion, broken or bent members, corrosion protection system effectiveness, marine fouling, debris accumulation and scouring at the platform base. Major surveys, as reported in this review, included these items plus a detailed examination of selected parts of the structure (10% was reported as a normal requirement). Cleaning of the structure was seen as a part of a major survey, followed by an examination using MPI or a similar technique to determine the presence of cracking, pitting or corrosion at pre-selected nodes. In practice the annual general surveys were organised such that over a four- to five-year time period, visual inspections were complemented by elements of a major survey. Thus, over the period the requirements of a major survey were met.

If the structure had been subject to ship impact, dropped objects or loading by severe weather, a special survey would be required. Similarly, such surveys would be needed if changes in the condition or operation of the structure had occurred, that might affect its safety or parts of the certification system.

The report notes that five underwater inspection techniques were being used in US waters and the North Sea. These were visual inspection, MPI, UT (thickness and flaw detection), radiography and corrosion potential measurements. Other specialised techniques were also being used including a magnetographic method of crack measurement and acoustic holographic techniques for flaw detection. All of these techniques were discussed with regards to types of defects that can be detected, advantages and limitations. It was further noted that at the time of this review all testing techniques required the structure to be cleaned of marine growth. Although brushing, chipping and scraping would sometimes be sufficient, high-pressure water jetting was reported as often being used but could be potentially dangerous to the operator if used by a diver. Cleaning could constitute the major expenditure of underwater time involved in inspection.

Locating a site to be inspected to conduct the inspection test could be quite difficult, particularly on complex nodes or in the interior of a steel structure. The splash zone was seen as a difficult area to inspect due to sea motions where the periodical rise and fall of the sea surface could prevent the surveyor from maintaining his position at the work site[4]. NDE equipment was reported to be deployed using divers, manned submersibles, ROVs and one-atmosphere diving suits. Each of these deployment methods had strengths and weaknesses for performing NDE. At the time of this review it was claimed that nearly all NDE equipment was designed to be used by a diver. It was claimed that manipulators on ROVs, manned submersibles and one-atmosphere diving suits did not have sufficient capability for managing NDE equipment. However, it was noted that ROV capability was developing for NDE testing.

4.2.4 HSE Handbook for Underwater Inspectors

A handbook was published by the HSE for the Department of Energy in 1988 (HSE 1988) and subsequently taken over by HSE; at the time of writing, it is still available on the HSE website. It was compiled to collate all the information required for an Underwater Inspection Controller training course, based on the 1987 CSWIP syllabus requirements for the 3.3U (Pilot-observer inspector) and 3.4U (Underwater inspection controllers) qualifications (see Section 4.4.5). The text follows the format of a typical training course at the time for candidates wanting to obtain the relevant qualifications. Some parts of this handbook are now significantly out of date, particularly the legislation and regulatory requirements and some aspects of diving practice. However, the material on corrosion monitoring and NDE methods still has some relevance. The text on data recording, inspection planning, intervention techniques and contractual aspects has limitations as these topics have developed significantly with time. Training for CSWIP 3.3U and 3.4U continues using more up-to-date material.

4.2.5 MTD Underwater Inspection of Steel Offshore Structures

As noted in Section 3.3.1, the MTD (1989) report included a review of underwater inspection requirements, techniques available, significance of defects and different approaches to planning inspections. The overall aim was to improve the effectiveness of underwater inspection, particularly through the use of a more rational method of planning inspection operations. The report was produced several years ago and some parts of it are now dated, but it is still useful, as it contains some valuable background material.

The report contains sections on the following:

- damage, deterioration and fouling;
- management of inspection planning;

---

4 This implies that an operator (diver) inspects the splash zone, which is unlikely at the present time.

- inspection and monitoring operations;
- assessment of damage;
- interaction between design and inspection;
- statutory requirements, certification and guidance; and
- principles of probabilistic techniques.

The major outcomes of this review included an overview of damage to structures and an extensive guidance to the evaluation of this damage, based on knowledge at the time. As an example, the report highlighted a crack growth model developed by the Southwest Research Institute as a method for analysing part-though cracks (Hudak et al. 1984). This method has been superseded by improved models, as further discussed in Section 7.5. Another major outcome of the review was a suggested computerised approach for rational planning of inspection operations. At the time, computerised systems to a limited extent were used in data management and inspection planning. However, this revised approach did not appear to find wide acceptance in the industry at the time.

The review also included a chapter on the standard inspection methods available at the time, which are all reviewed in Section 4.3. Cleaning methods were also reviewed, including high-pressure water jetting, brushes and grinders. Intervention methods considered in the report include diving, atmospheric diving suits, manned submersibles and ROVs. The review of ROVs identified that at the time they were only capable of carrying out CP checking and thickness checking by UT. Crack detection equipment available only enabled gross cracks to be detected. Atmospheric diving suits and manned submersibles are no longer in widespread use for offshore inspection.

The section of the report on the interaction between design and inspection has a useful section on designing for ease of inspection and monitoring, particularly concerned with diver inspection access. Similar issues are to a large extent at the time of writing this book included in up-to-date codes and standards.

The report also contains useful information on structural monitoring, which is reviewed in Section 5.2.1.

4.2.6 Department of Energy Fourth Edition Guidance Notes on Surveys

The 4th Edition Guidance Notes (Department of Energy 1990) included a section on surveys, recognising that it related to the certification regime and the certificate of fitness, which applied until 1995. Renewal of the certificate required a major survey to ensure that the installation continued to meet the requirements of the regulations. As part of this it is stated that a schedule of inspections and tests should be drawn up and agreed by the owner and certifying authority.

An installation (platform) with a certificate of fitness was required to have an annual inspection which included a close visual inspection down to and including the splash zone of a fixed structure or to the water line at maximum freeboard of a floating unit. As the certificate of fitness was valid for not more than five years, a major survey was required before it could be renewed. In addition, the certifying authority may require additional surveys in the event of damage, deterioration, repair, alteration or replacement to the installation, where the surveyor could require a close examination to be made of, for example, any underwater repair work undertaken since the last survey. An assessment should also be made of the thickness of marine growth on typical members or areas of the structure.

The object of the major survey for recertification was to ensure that the offshore installation continued to comply with the regulations, that any deterioration was within acceptable limits and that the installation was fit to operate for such a period as the certifying authority may specify, not exceeding five years from completion of the survey. It was noted that fixed structures in deep water presented unacceptable hazards with respect to deep diving. As a result, consideration should be given to using manned or remotely controlled submersibles or tools.

For floating and mobile installations, it was recommended that, where reasonably practicable, they should be moved to a suitable location for a major survey, preferably to a dry dock, or to sheltered waters. Where this was not possible the major survey could be carried out on location, subject to acceptable survey procedures being agreed with the certifying authority.

4.2.7 HSE Detection of Damage to Underwater Tubulars and Its Effect on Strength

The report OTO 99 084 (HSE 1999) reviewed the effect of dent and bow damage to underwater tubulars and the detection of these. The actual work was performed by MSL and the content of this report was preceded by an earlier paper (Frieze et al. 1996) containing much of the same material. The review clearly states that dent and bow damage can, if not detected, lead to significant reduction in member strength (see also Chapter 7).

A survey by interviews of staff was conducted of selected organisations performing jacket inspection in UK waters. The objective was to gather information on inspection methods used for identifying dented and bowed members, gather views on perceived accuracy of techniques and to obtain data on dents and bows. According to the review, two techniques were identified as universally used, namely flooded member detection (FMD) and general visual inspection (GVI), both deployed by ROVs. It was noted that anomalies could be identified by local loss of marine growth, following an incident (impact, extreme storm damage, etc.). After the detection of the anomaly, as much information as possible should be gathered by the ROV to allow for a preliminary evaluation. This included flying over the anomaly to search for signs of damage to other members and to locate a possible dropped object, perform a close visual inspection (CVI), photogrammetry and to take still photos. Accurate dimensions of the damage were in the review described as being needed both to establish the degree of loss of strength and also to plan repair techniques if required. If a work class ROV was available, cleaning and varying the light intensity using two light sources could be used to cause shadow movements to increase detection reliability. If further information was deemed necessary to characterise the damage, divers would be used with tools such as straight edges, gouges (pit dept), callipers (for measuring ovality of small members), Deerman clamps (for measuring ovality and dents in larger-diameter members), photogrammetry (more accurate with divers at the time), taut wire and spot laser ranging.

Although the data from Frieze et al. (1996) is actual data from offshore surveys, it will only give limited information about the frequency of actual damage as the distribution of the damage will be affected by the effectiveness of the inspection. These limits need to be reflected in the understanding of the data but can also be used to give some indication of thresholds for the detection of these types of damage.

Experience indicates that a significant part of this type of damage (dents and bows) to structures will remain either undiscovered or unrepaired. For example, the review by Marine Technology Directorate (MTD 1994) showed that concerning vessel and dropped object damage, 17% was detected during routine inspection or by chance, rather than as a response to an event. Hence, a key aim of periodic inspection is to ensure that any degradation that significantly affects the strength of the structure is detected.

Data on actual offshore damage to offshore platforms were used in the review to establish an overview of dent and bow magnitudes as shown in Figure 29 and Figure 30. The values indicated in the range 0–1 mm relate to magnitudes generally described as negligible or as unknown. As it can be seen from these figures, 44% of bows were reported to be associated with a dent, while only

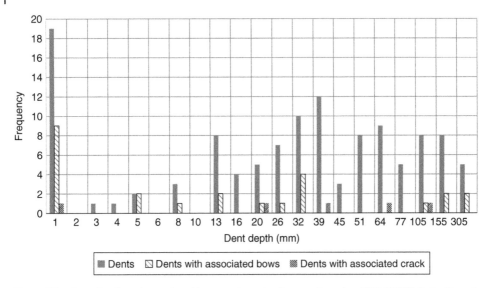

Figure 29 Dent depth and associated bows and cracks. *Sources:* Based on HSE (1999). Detection of damage to underwater tubulars and its effect on strength. HSE Offshore Technology Report OTO 99 084, HMSO, London, UK; Frieze P.A., Nichols N.W., Sharp J.V., Stacey A. (1996). Detection of Damage to Underwater Tubulars and its Effect on Strength, OMAE Conference, Florence 1996.

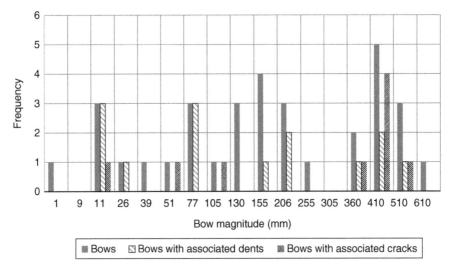

Figure 30 Bows with associated dents and cracks. *Sources:* Based on HSE (1999). Detection of damage to underwater tubulars and its effect on strength. HSE Offshore Technology Report OTO 99 084, HMSO, London, UK; Frieze P.A., Nichols N.W., Sharp J.V., Stacey A. (1996). Detection of Damage to Underwater Tubulars and its Effect on Strength, OMAE Conference, Florence 1996.

23% of dents simultaneously were associated with a bow. The review indicated that this could suggest that dents were easier to detect than bows, which would be consistent with the methods for detection.

Figure 29 shows typical dent depths with associated bows and cracks based on Sharp et al. (1995) and Frieze et al. (1996) from actual offshore data made available for the review from offshore maintenance programmes relating to the southern North Sea. The data show that dents found during inspection can be up to 300 mm in depth but more typically dents are in the range 10–70 mm in depth. Further, the frequency of dent depths is shown and many of the data points are associated with bows resulting from accidental damage. This association adds to the complexity of the evaluation of this type of damage. The figure further shows the frequency of dent depths associated with cracks. There is no obvious correlation between dents and cracks based on these data. However, as shown in Figure 30, cracks are more correlated with bows.

Figure 30 shows the observed frequency of bows with associated dents and cracks. It can be seen that bows of up to 610 mm are registered in this database. It is known that severe collision accidents can create much larger bows. Further, it can be seen in Figure 30 that some very large bows are recorded as having no associated dents. The reason for this is unclear. Figure 30 also shows the frequency of bows with associated cracks and it can be seen that for the largest bows, cracking often occurs and may add to the complexity of the evaluation as further fatigue crack growth may occur. These cracks were reported as usually occurring at member ends associated with a joint. This may require specific inspection and monitoring and possible repair.

Inspection for dent and bow damage will normally follow a known boat collision or dropped object incident, but such damage may also be found during routine inspections. The review indicates that nearly 20% of the dent and bow damage is not detected in inspections following an incident. In light of the possible reduction of strength from such damage, the review indicates monitoring approaches at that time needed reconsideration.

Routine GVI would appear to miss a number of larger dents and particularly bows. Figure 29 and Figure 30 can also provide some indication of inspection thresholds (the minimum damage that can be expected to be found). Figure 29 shows that the first significant set of dent records is around 12.5 mm (imperial measure half an inch) and then there is a second peak around 38 mm. Frieze et al. (1996) suggest that these two peaks in frequency may be a result of two possible thresholds for the visual inspections. The first seems to be around 12.5 mm and may be the threshold for specific inspections following damage when dents are being specifically sought. It is suggested that there is a second threshold of around 38 mm resulting from routine inspections where such damage is less expected. Larger dent sizes have the most significant effect on structural integrity; this emphasises that finding damage after a known incident is important.

Frieze et al. (1996) argues that for bows there are also two thresholds, as can be seen in Figure 30. One of these (possibly around 11 mm) is assumed to be based on that smaller bows seem to be detected only when a dent has also been detected. The second threshold (possibly around 130 mm) is assumed to be the detection limit for the size of the bows when dents haven't been detected.

In ageing structures, accidental damage can accumulate and the combined effect of multiple dents and bows can be significant structurally. This combined effect is not covered in current codes and standards and hence places pressure on operators to detect individual incidents of damage during regular inspections. If these are not repaired, should further damage in a similar location be detected later, appropriate analyses should be undertaken.

Typical repair techniques (as further described in Chapter 8) include grout filling of members and grouted clamps for larger dents. Bows are more difficult to repair, unless by member replacement, which is expensive.

4.2.8 MSL Rationalization and Optimisation of Underwater Inspection Planning Report

The Joint Industry Project titled 'Rationalisation and optimisation of underwater inspection planning consistent with API RP2A section 14' was discussed in Section 3.3.6 (MSL 2000). The database developed as part of the JIP contained platform, inspection and anomaly data from over 2,000 Gulf of Mexico platforms. This project reviewed several different NDE methods available at the time, including visual inspection (GVI or CVI by diver or ROV), non-destructive examination (NDE) focusing on crack detection and sizing (including MPI, EC, UT and ACFM), NDE focusing on identifying members with cracks and other through-thickness damage (FMD) and continuous monitoring methods such as acoustic emission. A brief description of each method was given in the JIP report. Figure 31 shows the relationship between the different techniques, ranging from those suitable for global inspection to those more suitable for local inspections.

The review also included a summary of the uses, limitations and comparative costs for four of the different NDE methods (GVI, CVI, MPI and UT) commonly used in the offshore industry at the time. Limitations were identified and showed that GVI and CVI had limits on what level of damage could be found. The use of divers for GVI would not now be regarded as good practice compared to an ROV, for safety reasons.

The JIP report also addressed the reliability of the different techniques. For visual inspection, the report noted that:

- GVI was a reliable method for detection of damage such as missing or separated members, large dents or bows and large holes or cracks. For smaller types of damage, the reliability of detection for GVI reduced depending on factors such as the degree of marine growth and the ambient lighting level.
- CVI reliability depended on the degree of cleaning and surface preparation, which was determined by the type of cleaning method used. The report states that cleaning with pressure was necessary for reliable visual inspection.

The reliability of UT was also addressed, noting that a rough surface (e.g. severely corroded) would reflect a significant proportion of the input energy making measurements difficult. With a smooth surface UT could be effective in measuring wall thickness. Regions with large pits are also difficult to make accurate measurements as the probe cannot make good contact with the member. Since areas of pitting are those where thickness measurements are more likely to be required, the report suggests that where possible, other methods such as the use of gauges should be preferred.

Further, the reliability of the MPI technique was reviewed in the JIP report. From the feedback from the operators involved in the JIP it was concluded that the mean length of the shortest

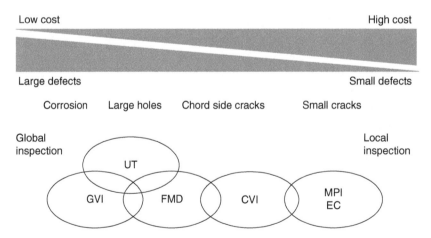

Figure 31 Relationship between the different NDE techniques versus cost and detection ability. *Sources:* Based on MSL (2000). Rationalization and optimisation of underwater inspection planning consistent with API RP2A section 14. Published by MMS, USA as Project Number 345, November 2000.

crack that could reliably be detected was 2.8 cm (1.1″), which is several times the assumed detectable size at the time. A series of studies conducted at the Underwater NDE Centre of University College London (HSE 2002) have shown that a PoD of 90% was obtained for a crack length of 0.2 cm. More recently DNVGL RP-C210 (DNVGL 2019) published PoD curves for underwater use of MPI which are less than those indicated by Visser (HSE 2002). The 90% PoD level for MPI reported by DNVGL is approximately 1.4 cm.

The point was made that before using MPI, CVI is necessary to check for evidence of weld under-cut, inter-bead groves and other fabrication defects which are often mistaken for cracks. It was also noted that successful use of MPI depended on the skill of the operator.

A comparison of reliability and costs for CVI and MPI for four different levels of cleaning is shown in Table 20. The costs relate to the time of carrying out the study (MSL 2000). It was claimed that after cleaning, MPI had the advantage compared to CVI that cracks down to 10 cm (4") long and 0.025 mm (0.001 inches) wide could be detected with reasonable confidence. As shown in Table 20, this approach can be justified on a cost basis since the relative cost of MPI is only about 20% of the total inspection cost including cleaning.

The report also compares gamma ray and UT flooded member detection methods. In accordance with the review the UT method can be used by both divers and an ROV and is the preferred method for GoM inspection. However, it is claimed that there are limitations in using an ROV due to the difficulty of placing the sensors. There are also safety issues in using divers. The gamma ray

Table 20 Reliability and Cost Estimate for CVI and MPI (MSL 2000). *Sources:* MSL (2000). Rationalization and optimisation of underwater inspection planning consistent with API RP2A section 14. Published by MMS, USA as Project Number 345, November 2000.

		CVI Limited cleaning (black oxide)	CVI Complete cleaning (bare metal)	MPI Limited cleaning (black oxide)	MPI Complete cleaning (bare metal)
Detectable crack length		30 cm and greater	30 cm and greater	2.5 cm and greater	2.5 cm and greater
Detectable crack width		0.15 mm and greater	0.05 mm and greater	0.025 mm and greater	0.025 mm and greater
Detectable crack depth		0.75 mm and greater	0.75 mm and greater	0.75 mm and greater	0.75 mm and greater
Cleaning time		32–54 min/m^2	110–322 min/m^2	32–54 min/m^2	110–322 min/m^2
Estimated relative cost per length		1.0	1.8	1.2	1.9
Reliability of detecting crack	L=10 cm W=0.025 mm D=0.75 mm	5%	20%	80%	90%
	L=30 cm W=0.25 mm D=0.75 mm	75%	80%	90%	90%
	L=60 cm W=25 mm D=9.5 mm	90%	90%	90%	90%

technique using an ROV has many advantages, including that cleaning is not normally needed and is now seen as the preferred method.

The review indicated that FMD could be used as part of any underwater inspection as a supplement to other inspection activities and provided general guidance for using FMD. It was recommended that the use of FMD should be combined with a thorough general visual survey (GVI). The combination of FMD and GVI helped to concentrate the visual inspection on the platform nodes. It was shown that FMD could provide a safety net for any unknown damage occurring on the jacket. A SIM engineer could consider rotating the locations for FMD inspections such that all the major framing was inspected over a period of time. For robust structures, such as X-braced framing, FMD was described as a good tool to check for defects that might have been missed by other inspection techniques and could be used as one of the focused and main techniques of the inspection. For less robust structures, such as single diagonal and K-braced framing, FMD was considered an acceptable technique but should not be considered as the main inspection technique. FMD may miss small cracks that have yet caused member flooding but could develop into larger, more significant cracks prior to the next inspection.

If FMD identified a flooded member, a more thorough CVI or NDE inspection was described as required at each end of the flooded member. This would add to the total cost of inspection. However, the criticality of that member should be taken into account before large efforts were spent on identifying the cause of the flooded member. It was reported that FMD should be focussed on the main diagonal framing with less emphasis on the horizontal framing as it contributed less to the overall strength of the platform. The exceptions were the horizontals associated with supporting the conductor guides and tray supports.

As noted above, a good candidate for FMD was the air-filled conductor guide framing. Conductor guide frames are typically located near the water line and at intervals of greater depth. Near the water line the effect of vertical wave motion and buoyancy on some types of conductor guide framing could lead to significant cracking, normally at the 12 o'clock and 6 o'clock joint locations. Flooding may result from cracking both on the brace and chord side. The report also noted that more modern platform designs have improved to minimise this problem. FMD was also claimed to be useful for underwater areas that could be subject to dropped objects or vessel impact. Such areas include critical members near the boat landing and areas used for off-loading.

Further, reliability of the gamma ray FMD method was reviewed based on the ICON FMD trials (Dover and Rudlin 1996). The testing was undertaken on both horizontal and vertical members. The test results were reported as being 100% accurate for both 50% and 100% flooding.

Based on the review by MSL (2000), Figure 32 shows how crack growth can progress and the feasibility of different inspection methods to identify cracks based upon the remaining life. For obvious reasons FMD would not find the crack until the member has flooded. In contrast, MPI and EC would have found the crack well before it reaches the through-thickness stage. The use of FMD needs to take this into account recognising that repair of a through-thickness crack is much more expensive than a repair of a crack detected by MPI or EC. The safety of the structure also needed to be addressed in accordance with the discussion on framing given above.

The reliability of a structure inspected by FMD was further discussed. At the time of the FMD check, a member might have a crack that had not progressed to the through-crack stage, so it would be regarded as not flooded. However, if the member is damaged, depending on the crack growth rate, the damage may propagate to a more hazardous stage prior to the next inspection. In this case FMD would have provided a false negative, giving the impression of a no damage condition.

The finding of the JIP supported the use of FMD as was the industry position at that time with some limitations as discussed above. It was seen to provide an acceptable alternative to CVI and in some cases may be preferable.

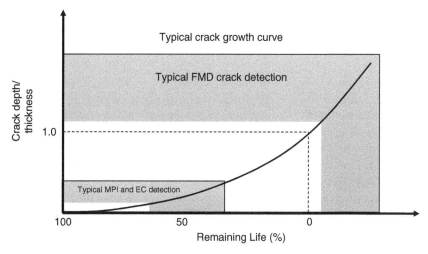

Figure 32 Ability to detect a crack in member for different NDE techniques. *Sources:* Based on MSL (2000). Rationalization and optimisation of underwater inspection planning consistent with API RP2A section 14. Published by MMS, USA as Project Number 345, November 2000.

4.2.9 Projects on Testing of Inspection Methods and Their Reliability

Many trials have been performed to establish the reliability of inspection methods. Many of these are discussed in this section.

Work at the Centre for Non-Destructive Testing at University College London (UCL)

The Centre for Non-Destructive Testing at University College London (UCL), which was opened in 1984 (Department of Energy 1984), was instrumental in developing reliability data for underwater non-destructive testing techniques. Much of the work was carried out with joint sponsorship from the UK Department of Energy, the offshore industry and, in the case of the later ICON (Inter-calibration of offshore non-destructive examination) programme, the European Union. Most of the testing was on fatigue cracks as a result of testing tubular joints of realistic offshore dimensions.

The library of test specimens (Department of Energy 1984) which was developed over a period included unstiffened tubular joints of seven basic types such as simple Y, simple K, complex T, double YT, X and double K joints. In total there were 20 nodes with 80 individual joints. In addition, there were 20 coated specimens with defect indications. The defects varied in length from 2 mm to 600 mm and had depths up to 33 mm. Such an extensive library was necessary in order to establish PODs with any confidence. To obtain a 95% confidence level, it was established that a set of 29 defects with similar crack-lengths was needed.

One of the aims of the UCL NDT Centre (Department of Energy 1984) was to establish a set of pre-cracked specimens, including welded tubular nodes, containing well characterised fatigue cracks. The UCL work concentrated on four different types of techniques for underwater inspection:

- magnetic particle inspection (MPI);
- eddy current methods;
- ultrasonic creeping wave inspection; and
- alternating field measurements (ACFM).

Defect characterisation comprised the process of determining the location, length and maximum depth of cracks in each individual joint.

The NDT facilities at UCL included a diving tank for simulating diver inspection trials and fatigue testing equipment to fabricate the various fatigue defects. The diving tank was 5 m × 4 m and had a maximum depth of 4 m. It had viewing ports for observing the divers in action. A number of divers were used in the trials, most of which had the 3.2U CWIP certification (see Section 4.5.4). A controller was present to supervise the divers' work.

University College London further studied the accuracy of inspection methods in the report OTH 87 263 (Department of Energy 1987). This work included the preparation of 11 pre-cracked tubular joint specimens for trials using several different inspection methods both in air and underwater by divers. The inspection methods used included MPI, EC, UT TOFD and ACPD where for the underwater trials MPI was used for crack detection and ACPD for crack sizing. Following the inspections, the specimens were sectioned so that the size and depth of the defects could be established for comparison. In general, the inspection methods tended to underestimate the crack depths by approximately 10%, while the crack length was both over- and underestimated with an average error of 8%. The diver trials showed acceptable results but not as good as the in-air results. The two divers were able to find all the defect locations although there were problems in detecting the extent of the cracks with up to 30% underprediction reported.

A further set of trials was undertaken by UCL and reported by HSE (2002), where accurate measurements of defect size and length were also obtained by optical measurements and compared with the test results for that defect. These tests showed that MPI only missed surface breaking defects that were less than 3 mm in length and provided a satisfactory measure of length in the majority of cases. The data showed that the eddy current and ACFM inspections gave a much lower number of spurious (false positives) results than MPI. On the other hand, MPI and ACFM inspections were much more reliable in determining defect lengths than the lengths obtained by the eddy current inspection. The data from the UCL programme was later incorporated into the European ICON programme.

These tests were valuable in that they enabled crack measurements from inspection to be compared with true crack sizes from destructive sectioning. The errors reported were relatively small and in addition these results can be used to adjust the crack size measurements obtained from inspections.

The ICON Project

The ICON (Inter-Calibration of Offshore Non-destructive examination) project was a £4.4M major programme funded by the European Union, HSE and several UK and European companies. It provided at the time the most comprehensive database on sub-sea NDT equipment, operating procedures and PoD values (Dover and Rudlin 1996). It incorporated data from the underwater trials at UCL making use of their library of defective nodes. Further, it collected a vast amount of information on NDE of tubular joints in a marine environment. The emphasis was on realistic laboratory trials deployed by ROVs or divers in a laboratory tank. An important part of the project was carried out offshore at a depth of 140 m from the Shell's DSV (diving support vessel). Many variables both in equipment and in the types of test specimens were tested in order to establish PoD and PoS curves for surface breaking and crack-like defects. The tests were carried out at three sites (UCL in UK, Ifremer in France and Technomare in Italy). Bureau Veritas audited the data and provided certified performance.

ICON addressed many different aspects on underwater inspection. The main part of the work was to test eight NDE methods on four different types of samples using both CAT (computer-assisted tele-manipulator) and manual systems. The NDE methods were based on MPI, ACFM and eddy current and the samples were tubular joints, welds between different metals, (corroded) tee butt welds and coated specimens. For many investigations only a subset of the UCL model library of nodes was used. Hence only in a few cases the number of data points is more than 30.

ICON also included tests of flooded member detection (FMD), reporting that at 50% or higher filling of water, the probability of detection (POD) was 100%, but only 70% POD at 10% water fill (HSE 2002, Mijarez 2006).

The final report (Dover and Rudlin 1996) contains much of the concluding results in the form of graphs from this project. An ICON database was also provided which supplied a great deal of information on equipment selection, procedures and PoD type data. ICON software produced by the project included three interactive databases consisting of information on equipment, operating procedures and PoD performance. It provided users with the capacity to compare data and NDT systems for a given offshore task.

Work by Visser

A study by Visser funded by HSE (2002) undertook an assessment of published NDE PoD and PoS data from six major test programmes both from above water and underwater testing. Two of these underwater programmes were from the UCL test centre and the ICON programme. This assessment has been used to develop standard PoD-type curves for different NDE methods. These are now incorporated into codes and standards such as DNVGL RP-C210 (DNVGL 2019).

Work by Bureau Veritas

Bureau Veritas (BV), using ICON data, produced a type approval document, "Type Approval on Non-Destructive Testing Equipment Dedicated to Offshore Structures Underwater Inspection" (BV 1998). This described how manufacturers of NDT equipment could achieve approval for their equipment (2 levels of approvals were issued, namely Type and Basic). Type tests were carried out to determine whether the equipment was capable of meeting BV's requirements for the specified application. These Type tests were to be carried out using the ICON samples to evaluate performance. BV's minimum requirements for crack detection were as shown in Figure 33. For crack sizing, BV indicates for a measured crack depth in the range 10–25 mm the required accuracy was –10% to +20%.

The reliability operating characteristic (ROC) defined by the amount of spurious indications obtained during Type tests was compared to the total amount of characterised effects and the overall POD. For POD values >80% it was stated the ratio of spurious indications to detected defects should not exceed 20%.

This type-approval also applied to flooded member detection (FMD) where tests would be carried out on dedicated samples at various positions (vertical, horizontal and vertical diagonal) at various levels of flooding from 0 to 100%. Results were expressed as a function of the minimum detectable level of flooding (% in volume) by the equipment for each position.

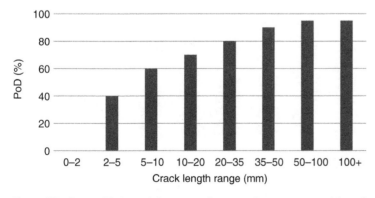

Figure 33 Bureau Veritas minimum requirements for type approval, based on crack detection. *Sources:* BV (1998). Bureau Veritas Guidance Notes Type approval of non-destructive testing equipment dedicated to underwater inspection of offshore structures, NI 422 DTO R00 E, 1998. © 1998, BUREAU VERITAS MARINE & OFFSHORE.

4.2.10 Concrete in the Oceans Programme

The Concrete in the Oceans (CiO) programme included projects on surveys to establish the condition of concrete structures which had been in the sea for long periods and to demonstrate the use of different surveying techniques (Department of Energy 1987). The most relevant of these was the survey of the Tongue Sands fort in 1977 which was built in the Second World War and placed in the Thames estuary for the defence of London (see Figure 34). At the time of writing, it is still in place, situated 7 miles north of Margate in a water depth of 8 m at LAT. Its details are as follows:

- structure: two hollow circular reinforced towers 7.3 m in diameter, 18.3 m in height, on a concrete foundation, supporting a steel deck;
- wall thickness 305 mm, wall reinforcement 25 mm mild steel, vertical 25 mm at 150 mm centres, hoop 25 mm at 305 mm centres (lower 14.6 m, at 280 mm centres for upper 3.7 m);
- cover to reinforcement, 32 mm;
- cement OPC/RHPC, aggregate flint gravel; and
- concrete specification, 1934 code, mix 2.

The structure had been built for short-term use in the Second World War, as noted above, and hence long-term durability was not a main factor in design. Exposed to the weather and marine environment for 35 years (at the time of the survey), it represented a small-scale version of an ageing offshore concrete structure. As the structure was of limited further practical use, destructive examination was acceptable, including the taking of cores. The project involved extensive surveying in all the environmental zones, including tidal, splash and underwater and also inside the hollow shafts. The underwater diver survey showed that the back of the base pontoon to the structure had been severely broken midway between the two towers. This fracture had resulted in the deck splitting away from the top of the towers, causing severe damage to the concrete in these regions. It was concluded that this damage had been caused by scour, exacerbated by storm damage.

The survey involved the use of several different survey methods, both above and below water. These included a detailed visual survey and the taking of 11 cores, of which four were below water, for detailed physical and chemical analysis. Sulphate attack, carbonation and leaching of lime were experienced in the splash zone, but even after 35 years of exposure, this was confined to a depth of only 50 mm.

Figure 34 Tongue Sands Tower. *Source:* John V. Sharp.

In the splash zone the compressive strength of the concrete was found to be high, with the strength varying between 65 and 92 N/mm². The permeability of the concrete varied with an average value of 10^{-10} m/sec, which is a typical value for non-saturated site-produced concrete. The analysis showed that carbonation had occurred but only to a depth of 50 mm. Chloride ingress had occurred to the full depth of the cores examined (330 mm). Corrosion of the reinforcement could have been initiated, but it had not resulted in extensive damage to the concrete cover. The exception was part of the tower which had been cast with a mortar mix; this mix was extensively deteriorated and the reinforcement in areas containing the mortar mix was extensively corroded.

In the tidal zone the concrete was generally found to be in a sound condition. The strength of the concrete was 50 N/mm², which is less than that in the splash zone. There was evidence that corrosion of the reinforcement had been initiated in some areas (indicated by rust stains) but high corrosion rates had not occurred. It was suggested that this was due to the low rate of oxygen ingress into the saturated concrete.

In the underwater zone the strength of the underwater concrete was approximately 61 N/mm². The permeability of this concrete was very low (5×10^{-13} m/sec) compared to the splash zone. Corrosion initiation had deemed to have occurred in all underwater reinforcement (confirmed by chloride levels) although no evidence of steel corrosion was found, which was considered to be due to the slow rate of oxygen ingress, due to the low permeability concrete and also to the low oxygen content in seawater. Localised corrosion was found on one sample which occurred on an exposed steel surface within a void in the concrete.

In the internal zone, the upper section had been severely damaged by the deck splitting away from the top of the towers and corrosion damage was evident. In general, the NDT results showed that the concrete was in a sound state, except for those areas mentioned above. NDT measurements indicated that the reinforcement was in an active state.

Figure 35 shows the variations in chloride levels with height (in relation to the high-water level). It can be seen that the chloride level increases dramatically with height above high water level, reaching 6%

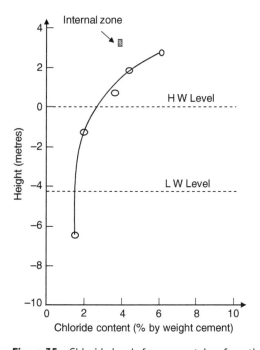

Figure 35 Chloride levels from cores taken from the Tongue Sands Tower vs height above LAT.

by weight of cement, which is six times greater than the surface level in the underwater zone. It was concluded that this increase was likely to be due to the action of sea spray and subsequent water evaporation. High chloride levels are an indication of the potential for corrosion of the reinforcement. However, as noted above, these levels had not generally resulted in actual corrosion, except for the mortar zone.

Several NDT techniques to establish the state of the concrete were deployed. This included Pundit, Gamma backscatter and the Schmidt (rebound) hammer. All three techniques were able to show the presence of the mortar mix. The backscatter technique was most effective in detecting the areas of delamination.

For assessing the state of the reinforcement, potential surveys were deployed. In addition, measurements of resistivity and chloride contents provided data relevant to the potential for corrosion. The several techniques established that the reinforcement was in a corrosively active state. However, only in the areas of the mortar mix had this corrosion led to corrosion and serious delamination to the concrete cover. In other areas, steel corrosion had, at the time of the survey, not damaged the concrete cover. Potential measurements underwater were limited in value as only an overall potential could be measured which was –630 mV compared with a copper sulphate reference electrode. The exposure of significant amounts of steel due to the underwater fracture affected these readings. The potentials in the tidal zone had a maximum value of –460 mV, which was less than the value measured in the splash or internal zones.

Overall the report on the survey of the Tongue Sands tower concluded that deterioration of the concrete and significant corrosion of the embedded reinforcement were restricted to limited areas of the splash zone, where either the specified cover had not been met or the local use of an inferior quality "mortar" mix had resulted in rapid ingress of both oxygen and chlorides and also carbonation of the concrete cover. Despite 35 years of exposure, the concrete and steel throughout the underwater and tidal zones and in most of the splash zone was in remarkably good condition, considering the speed of construction and intended short-term use.

The CiO programme also involved surveys of other structures including a Mulberry Harbour unit still in Portland Harbour and of similar age to Tongue Sands Tower. It was designed to be floated in position on the Normandy beaches and sunk as part of the invasion of France in 1944. It was a hollow rectangular reinforced concrete structure situated in relatively shallow water, only 200 m from the beach. In a second phase of the CiO programme an additional structure was chosen for survey which was only 10 years old at the time of survey but enabled comparison with the older structures in phase 1. This was the Trinity House lighthouse situated in the English Channel 11 km south east of Eastbourne. This structure also had some similarities with an offshore concrete structure as it had a reinforced concrete foundation supporting a hollow concrete tower. This single tower comprised two hollow concrete cylinders 6 and 4.25 m in diameter in telescopic form, which allowed the upper section to be raised into position once the base raft had been placed on the seabed. The use of sulphate-resisting Portland cement in the wave zone was of interest.

The main objectives of the CiO surveys were to assess how the long performance in the sea affected concrete durability, the effect of different exposure zones on the deterioration, how the concrete properties had affected the corrosion protection given to the reinforcement and to determine, if possible, the cause and extent of any reinforcement corrosion.

In terms of corrosion of the reinforcement, the surveys of the two wartime structures can be compared most easily. The Tongue Sands Tower had 25 mm diameter bars at fairly close centres (150 and 300 mm) embedded in concretes with strengths in excess of 50 N/mm^2 (at the time of the survey), with a specified cover of 32 mm. The Mulberry Harbour unit had a similar size of bar at very much wider centres in a concrete of a lower strength (37 N/mm^2 at the time of the survey) with normal covers from 25 to 30 mm. In the splash zone of the Tongue Sands Tower which had the higher strength concrete, surprisingly corrosion had not resulted in any extensive damage to the concrete cover although chloride levels measured at bar depth were such that it was believed that the reinforcement had been activated many years previously. Corrosion of the reinforcement and subsequent spalling were evident in both structures in areas of weak concrete without large aggregates, at voids, and where covers were lower than 10 mm. The above-water area of the Mulberry Harbour unit, with a lower strength concrete and a history of submergence and re-emergence, showed more corrosion, and there was clear evidence of different types of corrosion between the drier west side and the wetter east side. With an assumed plentiful supply of oxygen on the west side and a medium moisture content, the formation of corrosion had led to spalling.

The Royal Sovereign Lighthouse was a much younger structure with a high concrete strength (55 N/mm^2 at 28 days), a low water/cement ratio (0.37) and covers of 50 and 75 mm. As the structure was in use, the cores were taken in such a way as to avoid cutting the reinforcement. No visual evidence of corrosion was found on the surface of the concrete although the potentials were more negative than –280 mV Ag/AgCl in some places. Because of bad weather, the survey programme had to be limited and there was limited opportunity to inspect the reinforcement for corrosion. On the east side, with a high moisture content possibly restricting the access of oxygen, soft or soluble corrosion products were observed as rust stain exudation. In both structures very little corrosion was observed where the concrete was water-saturated; this was assumed to be due to restricted access of oxygen.

Sulphate attack, carbonation and leaching of lime were found in the Tongue Sands Tower, but after 35 years this was confined to a depth of 50 mm. In contrast, little concrete deterioration other than that due to reinforcement corrosion was noted in the Mulberry Harbour structure, in spite of the lower concrete strength. It must be remembered, however, that most of the concrete area examined had previously been submerged for some 18 years. Deterioration was confined to the east side where mechanical abrasion and scaling possibly due to frost damage had eroded the concrete surface. The carbonation depth based on a phenolphthalein indicator was found to be consistently low (less than 3 mm), particularly in areas that had been previously immersed in seawater.

No deterioration of the concrete was observed on the Royal Sovereign Lighthouse (Department of Energy 1987), which was to be expected in this relatively young, high-quality structure. X-ray diffraction of material from core samples taken from a point 8 metres above datum showed that chloride penetration, partial carbonation of lime and a minimal sulphate attack were all confined to the surface of the concrete. In the tidal zone there was a higher proportion of soft marine growth (algae) on the south-west side (exposed to the prevailing wind) than on the north-east side where hard growth (e.g. barnacles) predominated.

Inspection techniques were selected for detection of materials deterioration and for detection of corrosion of the reinforcement and included:

- visual, above and below water;
- ultrasonic pulse velocity (for cover quality), both above and below water;
- gamma ray backscatter (for concrete density);
- rebound hammer (for concrete surface quality); and
- potential measurements (detection of corrosion of the reinforcement).

All the methods were successful in identifying areas of weak concrete and delamination, but the correspondence between the levels (poor, good, etc.) was variable. The gamma ray backscatter method was alone in identifying one area of delamination, subsequently confirmed by coring. However, it was found that the technique had limitations in detecting delamination directly over reinforcement as it led to false readings.

Since corrosion of the reinforcement is an electro-chemical process, its detection can be aided by the measurement of the electrochemical potential of the reinforcement. Measurement of resistivity can also add value by measuring the ionic conductivity of the concrete in the cover zone, indicating moisture content and quality of the concrete.

Potential measurements underwater were taken on the Tongue Sands Tower. However, the results obtained were influenced by the potential of the reinforcement exposed in a major structural crack in the base of the structure and by the immersion of a steel diving cage. Another problem was the need for a connection to the embedded reinforcement underwater. A connection can be made more easily above water, but electrical discontinuities between the reinforcement in the splash zone and that underwater may have given false readings.

4.3 Inspection and Inspection Methods

4.3.1 Introduction

This section presents an overview of methods used for underwater inspection of offshore structures. Primarily, these are non-destructive methods in that the material and structure is not intended to be damaged by the inspection. Such methods are often called non-destructive testing (NDT) or non-destructive evaluation (NDE). A few destructive methods are also commonly in use, particularly chloride ingress tests used for concrete structures which are also discussed in this section. An overview of the different techniques described in this section is given in Table 21.

4.3.2 Visual Inspection

Visual inspection is inspection of a structure using primarily human senses, particularly vision but also in some cases hearing and touch. Visual inspection may include non-specialized inspection equipment such as magnifying lens, callipers, crack measuring ruler, radius gauge, mirrors, endoscopes, borescopes, fibre optics and TV cameras. Visual inspection requires good lighting: 500 lux is required at the surface to be inspected, e.g. by DNVGL-CG-0051 (DNVGL 2015) and hence also a light source and lux meter will be required in many cases.

Experience has shown that many defects can be found visually, also underwater and hence, visual inspection will be the initial inspection method in many cases. Visual inspection is used for the entire structure from the seabed to the top of the flare tower. In addition to the structure itself, visual inspection is also used to look at the seabed for scour and subsidence, the level of marine growth on the structure and the state of cathodic anodes in the corrosion protection system.

A visual inspection of a steel sub-structure will normally include:

- Weld inspections: After cleaning, the welds can be inspected for discontinuities on the weld and in the heat-affected zone. Particularly, signs of cracking will be looked after, but also signs and severity of corrosion and pitting in and around the weld is normally assessed and noted.
- Damage: Along with the inspection of welds, the inspector will normally look for any type of damage to the structure such as buckles, dents, bows, etc., as a result of storms, collisions and dropped objects or other similar events.
- Marine growth: The density and coverage of population, local density variations, types of marine growth, indications of the presence of hard marine growth and possibly size of the average and largest specimen.
- Scour: Any signs and the amount of scour will normally be assessed and if necessary, a profile of the seabed will be established (photogrammetry).

Table 21 Overview of Inspection Methods for Offshore Structures Underwater.

Methods	Damage type that can be identified	Brief description
Visual inspection methods: • general visual inspection (GVI) • detailed visual inspection (DVI) • close visual inspection (CVI)	Any damage that can be seen visually, including missing members, large to medium cracks and damage, dents, bows, buckles, corrosion, corrosion products, spalling of concrete, coating breakdown, depletion in anodes, amount of marine fouling, seabed debris and scour etc. GVI can only detect major defects, DVI can give some more details and CVI can detect visible corrosion, pitting, cracks and other damage.	GVI includes a swim around for general assessment of condition. CVI can be used for more details. Video and still photography are typically used. GVI does not usually require cleaning, and hence the cost of a GVI is minimal. DVI requires minimal cleaning and CVI requires a cleaning process to remove the marine growth completely from a structure. Small defects may be missed by CVI.
Ultrasonic methods: • ultrasonic testing (UT) • time-of-flight-diffraction (TOFD) • phased array ultrasonic (PAUT) • ultrasonic pulse velocity test	UT, TOFD and PAUT can be used for thickness measurements and detection of internal defects in steel structures. The ultrasonic pulse velocity test can be used to evaluate uniformity and quality of concrete, including the presence of voids and cracks.	Ultrasonic methods use high-frequency sound transmitted by a probe that passes through a material to measure distances to reflecting surfaces, such as internal defects and the back wall.
Electromagnetic methods: • magnetic particle Inspection (MPI) • eddy current inspection (EC) • alternating current potential drop (ACPD) • alternating current field measurements (ACFM)	Used to detect surface defects and near surface defects (e.g. cracks). Also capable of determining crack length and location. EC, ACPD and ACFM are also able to measure the depth of defects to varying degrees of accuracy.	MPI: Magnetic particles are used with induced magnetic fields to determine surface defects. Requires cleaning of the surface. EC: A current field is introduced in the material and disturbances in the field are measured. ACPD: Electric current is passed between two probes and the presence of a surface breaking defect between the probes can be detected. ACFM: Electric current is induced in the material, the associated electromagnetic field is measured and the presence of a defect is detected by disturbances in the associated magnetic field. ACFM is a non-contacting method capable of detecting anomalies through coatings.
Radiographic testing (RT)	Voids, thickness losses, e.g. due to corrosion, corrosion under coating, insulation or passive fire protection coating	X-rays or gamma rays are used to determine thickness or density differences.
Flooded member detection (FMD) • radiographic testing • ultrasonic testing	Any damage such as cracking, corrosion, denting, local rupture and anode failure resulting in water ingress to dry members	Check water filling level in members, primarily tubular members

(Continued)

Table 21 (Continued)

Methods	Damage type that can be identified	Brief description
Rebound hammer	Delamination and cavities in concrete structures	A "hammer" that strikes a concrete surface and the extent of the rebound provides a measure of surface quality
Chloride ingress tests • dust samples	Measure the ingress of chloride in concrete, which may give an indication of the likelihood of reinforcement corrosion	Samples of concrete powder are extracted and analysed for chloride ion concentration and cement content (laboratory)
Electro-potential mapping	Corrosion of the reinforcement in concrete structures	Mapping of potential variation on a concrete surface using probes
Cathodic potential measurement	Loss of cathodic potential capacity	Measuring the cathodic potential compared to a standard cell

- Condition of coatings: Where coating is used, signs of breakage or blisters in the coating will normally be reported.
- Cathodic protection system: Anodes will normally be measured to assess the amount of anode material that has been used. If impressed current system is used, cables and insulation will normally be inspected.

Visual inspection is often divided into general visual inspection (GVI), detailed visual inspection (DVI) and close(-up) visual inspection (CVI). No general definition of when a visual inspection is called close, detailed or general exists and it also will depend on the item that is being inspected. However, definitions typically are based on the level of cleaning of the surface of the structure and the nearness of the inspector to the structure. For example, the definition in IACS UR Z 7.1.2.4 (IACS 2002) for close-up visual inspection is "... a survey where the details of structural components are within the close visual inspection range of the surveyor, i.e. normally within reach of hand".

For underwater ROV-based inspection, the following are often used:

- General visual inspection (GVI) is inspection of the structure without any cleaning and is applied to detect large-scale anomalies and to confirm the general condition and configuration of the structure. A benefit of GVI is that large parts of the structure can be inspected in a short time period.
- Detailed visual inspection (DVI) is inspection of the structure after cleaning away the marine growth by brush, water jetting or scraping. DVI will enable the inspector to determine more details of the structure and hence detect smaller anomalies. The cleaning of, for example, a node on a jacket structure for marine growth is reported to take up to 1 or 2 days.
- Close visual inspection (CVI) is inspection of the structure after cleaning and grit blasting the surface. CVI provides a more detailed examination of structural components and is able to determine the existence of cracks, corrosion and damage to the structure. A benefit of CVI is that a certain level of detailed inspection can be achieved for a reasonable time and cost compared to techniques such as MPI and EC.

Visual inspection has for many years made use of photos and videos and are now also using 3D real-time imaging methods, such as photogrammetry and multi-beam sonar. These visual systems can provide a high-resolution 3D image in real time of the surfaces of structural parts underwater, depending on the level of cleaning. Images of cleaned structural parts are highly relevant in detecting damage and anomalies but also for preparing accurate digital models suitable for designing clamp and sleeve repairs.

CVI above water and internally inside dry compartments in floating structures is performed by a qualified inspector. Dye penetration can be used to enhance the presence of cracks under dry conditions.

Visual inspection was responsible for the discovery of 36 of the 40 cases of damage reported in the AIM IV project (PMB 1990); see also Section 3.2.2. Visual inspection was expected to be the primary tool for damage detection because it was the most frequently applied technique. Also, since significant damage was the issue, visual inspection would be very successful in detecting the more obvious damage types (holes, dents, severed or missing members). A review of the detailed reports showed that in at least 28 of the cases no cleaning of the areas was performed for the initial inspection.

Inspection of concrete structures relies heavily on visual inspection (GVI and CVI using an ROV) due to the very large areas on the shafts and cells. There are imperfections on the concrete surface, some of which are important for further investigation. The CiO report (Department of Energy 1987) described a classification of damage types on the surface of a concrete structure and identified a range of different defects which could be found by visual inspection. More information is given in Section 4.8.

4.3.3 Ultrasonic Testing Methods

Ultrasonic testing methods use a transmitter of high-frequency ultrasound that is passed through the material and a receiver records the reflected sound.

Traditional ultrasonic testing (UT) uses a couplant between the transducer and the body of the material being tested increasing the efficiency of the process by reducing losses in the transmission of the ultrasonic energy at the surface. Ultrasound is modified by reflecting from a surface such as a flaw or a back wall. A probe is used to detect the received signal which can provide information on the presence of flaws and the thickness of the specimen.

In the reflection (or pulse-echo) mode, the transducer can act to both send and receive the pulsed ultrasound waves, or two separate transducers are used, as illustrated in Figure 36. Reflected ultrasound comes from an interface, such as the back wall of an object, or from an imperfection within the object. The device displays these results in the form of a signal whose amplitude represents both the intensity of the reflected sound and the distance to the defect or the back wall, characterised by the arrival time of the reflection.

In the through-transmission mode, a transmitter sends ultrasound through one surface, and a separate receiver detects the amount that has reached it on another surface after travelling through

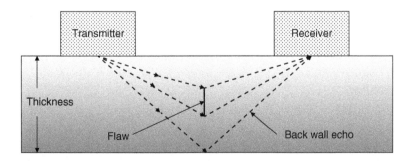

Figure 36 Ultrasonic testing both for thickness and detection of internal flaw.

a medium. Imperfections or other conditions in the space between the transmitter and the receiver are revealed by reducing the amount of sound transmitted. Ultrasonic testing can be used underwater by an ROV but is more usually used internally in hulls and compartments.

Ultrasonic testing (UT) in floating structures is primarily used for thickness measurements and is particularly used for monitoring corrosion on components. In this mode the focus is on measuring the time required to receive a signal reflected from the back wall.

Ultrasonic time-of-flight diffraction (TOFD) testing relies on the measurement of time differences between the signal travelling through known paths and being reflected from a defect. It places little reliance on signal amplitude and so is less sensitive than amplitude techniques to the condition of the steel surface and operator performance than the amplitude techniques (MTD 1989). An ultrasonic pulse is introduced into the steel at one point and diffracted signals are received, recorded and interpreted by a receiver. The technique is particularly useful for the sizing of known defects such as surface-breaking cracks but can also be used with considerable benefit in a "search" mode to locate unknown defects. In a TOFD system, a pair of ultrasonic probes sits on opposite sides of a weld. One of the probes, the transmitter, emits an ultrasonic pulse that is picked up by the probe on the other side, the receiver. In undamaged material, the signals picked up by the receiver probe are from two waves: one that travels along the surface and one that reflects off the far wall. When a crack is present, there is a diffraction of the ultrasonic wave from the tip(s) of the crack. Using the measured time of flight of the pulse, the depth of a crack tips can be calculated automatically by simple trigonometry.

Phased array ultrasonic testing (PAUT) is an advanced method of ultrasonic equipment, creating a beam of ultrasonic pulses from many small ultrasonic elements (array of probes) which can be swept electronically without moving the probe. Each element can be pulsed individually at a computer-calculated timing. This forms a combined view that detects hidden defects or discontinuities inside a structure or weld. Phased array UT is capable of producing precise and reliable images and can effectively be used to inspect large parts of a structure. It is often used as an alternative to radiography.

An ultrasonic pulse velocity test is a non-destructive test to assess the uniformity and quality of concrete. It can indicate the presence of voids and cracks. This test using two probes measures the velocity of an ultrasonic pulse passing through a concrete structure, measuring the time taken by the pulse to traverse the concrete specimen. Good quality concrete is indicated by higher velocities and slower velocities could indicate concrete with cracks and voids. The test can also be used to measure the effectiveness of concrete repairs. The degree of saturation of the concrete affects the pulse velocity, and this factor must be considered when evaluating test results. Typical frequencies used are in the range 40 to 50 kHz. It is also important to note that the pulse-velocity measured in the vicinity of reinforcing steel can be higher than in plain concrete of the same composition. This is because the pulse velocity in steel is up to double that in concrete. It is recommended to avoid measurements close to steel parallel to the direction of pulse propagation. Ultrasonic pulse velocity measurements can be combined with rebound hammer measurements to improve the accuracy of the estimation of compressive strength. A commercial version of this equipment is available, called "Pundit". A standard exists for this test and is provided in ASTM- C597-16 (ASTM 2016).

4.3.4 Electromagnetic Methods

Magnetic particle inspection (MPI) is an electromagnetic method for use on a magnetic material involving induced magnetic fields and magnetic powder to identify the presence of defects. Any cracks or defects in the material will interrupt the flow of current and will alter the magnetic field creating a "flux leakage field" at the site of the damage; see Figure 37. The flux leakage field will

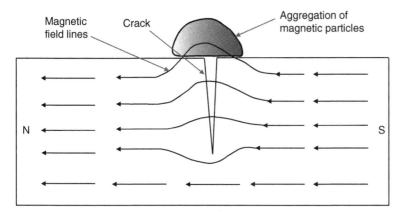

Figure 37 Magnetic particle method illustrated.

draw the magnetic particles to the damage site making it visible. MPI requires very careful surface preparation. However, coatings of up to 50 μm may be accepted.

MPI is specifically useful for determining surface breaking cracks. MPI is capable of detecting quite small cracks and the probability of detecting a crack according to DNVGL-RP-C210 (DNVGL 2019) is similar to that for EC inspection (typically 90% PoD for a crack depth of 12 mm underwater). MPI often requires removal of coatings and it is often difficult to reinstate the original corrosion protection. This may result in future corrosion attack, especially relevant for semi-submersible and ship-shaped structures.

Fluorescent magnetic particles are normally used for underwater MPI and an ultraviolet light is used for illumination. MPI underwater is normally undertaken by divers because of the difficulty of automatization of the inspection process for an ROV on typical structures. As a result of divers being used less for underwater inspection, MPI is at present being replaced by other methods such as ACFM.

Eddy current inspection (EC) is also an electromagnetic method involving the use of magnetic coils to induce eddy currents in members such as steel. The presence of surface breaking cracks distorts the current field, which can be detected by suitable probes. Eddy current inspections are useful for searching for surface cracks and can determine length and depth of the crack.

Eddy current inspection (EC) is capable of detecting quite small cracks, and according to DNVGL-RP-C210 (DNVGL 2019) there is a 90% probability of detecting a crack with a depth of 12 mm underwater and 3 mm in dry conditions. EC can also be used to measure stresses in a material (Lotsberg 2016). EC can be used both in dry conditions and under water. Very careful surface preparation is required for the method to be effective. EC underwater is usually performed by divers, which has cost and safety implications.

The alternating current potential drop (ACPD) technique is where an electric current is passed between two probes placed on the component. The presence of a surface breaking defect between the probes increases the electrical resistance locally. This change is used to detect and size defects. It requires cleaning to bare metal for the probes to be placed and also to be used by a trained diver. Multiple measurements along the crack length may be required if good resolution is required.

Alternating current field measurements (ACFM) is an electromagnetic technique used for the detection and sizing of surface breaking cracks in steel components. It was initially developed to replace ACPD for underwater crack sizing and it was then found that it could provide many benefits compared to ACPD. It combines the ACPD technique and eddy current testing in being able to size defects without calibration and in its ability to work without electrical contact with the surface. The method involves using a probe that introduces an electric current locally into the component and measures the associated electromagnetic field close to the surface. The presence of a defect disturbs the associated magnetic field. The ends of a defect can be identified to provide information on defect location and extent. Accurate sizing of defects up to 25 mm in depth is claimed. ACFM inspection is a non-contacting method and can be performed through paint and coatings and therefore, it is considered to imply less cost and time compared to several other NDT techniques requiring more thorough surface preparation (e.g. MPI).

ACFM inspections can be performed by an ROV. Weld inspection by ROV has been performed in many applications such as structural node welds on jacket structures, welded plates, mooring systems including chains and spud cans.

4.3.5 Radiographic Testing

Radiographic testing (RT) uses either X-rays or gamma rays to examine the structure for any flaws or defects. The test part is placed between the radiation source and film or a digital panel and any thickness differences in the test piece will reduce the amount of radiation penetrating through the specimen. This difference will be visible on the radiographic film or on the digital panel.

Radiographic testing is particularly good at detecting volumetric anomalies and can be used to detect corrosion under coatings, insulation or passive fire protection coating (accuracy approximately ± 0.3 mm). Planar anomalies cannot be detected reliably when the rays are parallel to the anomaly and hence usually the technique is not seen as a good choice for determining fatigue cracks.

X-rays methods are normally viewed to be reliable for structural parts with thickness up to approximately 60–75 mm. Gamma rays based on Cobalt 60 with an energy in excess of 1 MeV are normally seen as the choice for larger sizes. ABS "Guide for Non-Destructive Inspection of Hull Welds" (ABS 2018) indicates that X-rays and Iridium 192 can be used for materials up to 75 mm in thickness and that Cobalt 60 should be used for materials in excess of that.

In general, during a radiographic test the area needs to be closed for access due to the radiation danger. Radiographic testing is used for in-service inspection in special cases on floating structures (semi-submersibles and ship-shaped structures).

Limitations related to this method include that access to both sides of the detail is necessary in order to place the source and the detector opposite to each other. Also, the geometry of details will often limit the use of radiographic testing.

Wet radiography has been used and, for example, subsea digital radiography is a useful method for identifying corrosion and erosion, particularly in thicker sections and also root defects in welds. A digital detector array produces a high-resolution radiographic image which can immediately be relayed to the topside personnel using a fibre optic link. A marinised high-energy X-ray or gamma source is used in an assembly which is used to position the radiation source and detector on the test specimen, using preferably an ROV. Due to the safety hazards, the UK HSE requires notification to be given for the use of offshore site radiography work (HSE 2010).

4.3.6 Flooded Member Detection

Flooded member detection (FMD) can be used to detect the presence of water in initially air-filled submerged hollow members which would indicate a through-thickness crack, damage, corrosion or a weld defect allowing water to enter the hollow member. FMD is primarily used on fixed steel structures where the braces and conductor frames typically are air filled (the legs of a fixed steel structure are normally water filled and FMD is of little use). FMD can only be used underwater and does not necessarily require any cleaning of the member to be inspected. Historically, FMD was initially developed as a diver-based inspection using ultrasonic methods, but at present FMD is primarily performed by an ROV using a radioactive gamma source. A benefit of the method is that it is possible to inspect large parts of the structure in a short time period (as many as 30 components per hour is claimed to be possible by some suppliers). The technique is recognised as providing a relatively rapid inspection tool and, therefore, as cost effective.

Radiographic FMD uses a gamma ray source and a sensitive detector fixed on opposite forks of a changeable yoke system on an ROV and deployed across the diameter of the member under inspection. The method exploits the gamma ray absorption properties of different materials to detect the presence of water in a member, usually resulting from through-thickness cracking or other damage. Typical sources are cobalt-60 with a 1.2 MeV (million electron volts) gamma or caesium-137. Operators of gamma FMD are normally required to be trained in radiation protection, as the gamma source has to be transferred from the protective canister used for transport to the ROV.

The method requires access to both sides of a member with minimal surface preparation. Calibration is essential, using the count rate both in air and in water without a member present. There is a significant difference between the recorded count rate between a dry and water-filled member facilitating the detection of water filling. However, intermediate count rates usually require re-measurement and may indicate partial filling of a member. Vertical or angled members require testing at different locations along the member to check for partial filling. To analyse the results, the member size and wall thickness need to be known. In addition, the presence of any internal ring stiffening or grout fillings would also affect the count rate. Particular problems are tight cracks, cracks covered by coatings or sealed by marine growth and a general difficulty of access to members particularly if drill cuttings are present.

In ultrasonic-based FMD, an ultrasonic pulse is transmitted through the member wall. The presence of a fluid provides an efficient acoustic conduction path, which otherwise does not exist, and the detection of a return echo from the opposite side of the cylinder is used to confirm the presence of water and hence through-wall damage (Hayward et al. 1993). Ultrasonic-based FMD usually requires deployment by divers and an accurate placement of transducers (Mijarez 2006).

The FMD method was evaluated in the framework of the Inter-Calibration of Offshore NDE (ICON) project, which aimed at testing NDE methods and equipment. The FMD method was evaluated using clean 0.4 m and 0.5 m diameter tubes, filled with water to different levels. It was reported that at 50% or higher filling of water the probability of detection (POD) was 100% but was only 70% POD at 10% water fill (HSE 2002, Mijarez 2006).

A typical count rate in air is 3900 counts and via a water path only 47 counts. The FMD gamma rays are typically calibrated both in air and in water to these typical numbers. Typical count rates when performing the FMD test of structural members are (see also Figure 38):

- 862 counts for a dry member;
- 8 counts for a flooded member; and
- 202 counts for partial flooding, as an example.

Hence, there is a clear difference between a dry and a flooded member.

At the time of writing this book, it is understood that developments have been made to allow the frame to be adjusted by the ROV to match different member diameters, thus eliminating the need for the ROV to surface for frame adjustment. However, for special diameters and thicknesses, the radioactive source may need to be changed (Hanson 2020).

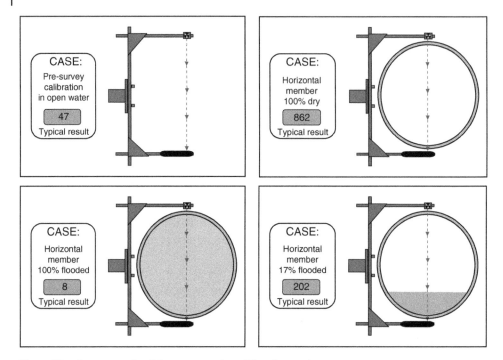

Figure 38 Count rates for different examples of flooding under water.

4.3.7 Rebound Hammer

A rebound hammer (also called a Schmidt hammer) test is a non-destructive testing method for concrete on-site providing a convenient and rapid measure of the compressive strength of the concrete. The compressive strength of concrete depends on various factors such as aggregate size, water-cement ratio, air content in concrete. The test involves the plunger of a rebound hammer being pressed against the surface of concrete when a spring-controlled mass with a constant energy strikes the concrete surface. The extent of the rebound is a measure of surface hardness and is measured on a graduated scale. This value is designated as a Rebound Number (rebound index). The objective of a rebound hammer test is to measure the compressive strength and the uniformity of concrete and can be used to give indication of spalling and delamination, which will absorb more energy, providing a lower rebound value. The test has been standardised by ASTM in their standard ASTM C805 / C805M (ASTM 2018).

4.3.8 Chloride Ingress Test

During immersion of concrete in seawater, ingress of chlorides occurs over time due to the permeability of the concrete. The depth of penetration and the chloride content at depth are important in the context of corrosion of the steel reinforcement. This depth is measured by core drilling at sample locations to prescribed depths. Concrete rebar and prestressed cables are avoided by controlling the location and depth of the drilling. Samples of concrete powder are collected from various depths which are then taken to a laboratory for analysis of the chloride ion concentration and cement content. Once results are obtained, depth profiles can be prepared showing the penetration of the chlorides and hence the susceptibility of the reinforcement to corrosion. A typical example of a chloride profile is shown in Figure 35, as a result of samples being taken from the Tongue Sands fort.

4.3.9 Electro-Potential Mapping

When reinforcing steel in concrete corrodes, coupled anodic and cathodic reactions are taking place simultaneously on its surface. At the corroding site (the anode), iron is dissolved and oxidised to iron ions leaving electrons in the steel. In the cathodic reaction at the steel surface, oxygen is reduced and hydroxyl ions are produced. In localised corrosion, the anodic and cathodic reactions on the surface of the steel are usually spatially separated. Under these circumstances, a significant amount of potential drop separates the anodic and the cathodic sites and therefore the corrosion potential varies with position. Thus, surveys of corrosion potential can be used to locate sites of active corrosion.

The corrosion potential is measured as a potential difference against a reference electrode (half-cell). The numerical value of the measured potential difference between the steel in the concrete and the reference electrode will depend on the corrosion condition of the steel in concrete. A typical reference electrode is the saturated copper-copper sulphate electrode, which is robust for on-site work as well as being sufficiently accurate. However, errors may arise due to contamination of the concrete surface with copper sulphate. The measured values are influenced by concrete cover, concrete resistivity (moisture content) and oxygen availability. Typical potential values measured against the copper-sulphate electrode (CSE) are shown in Table 22.

The corrosion potential of passive steel in concrete depends on the oxygen availability (moisture content) and can vary over a wide range of potentials. Corroding steel in chloride-contaminated concrete (such as in offshore structures) shows potentials ranging from −0.6 V to −0.4 V CSE. The most negative values of the half-cell potentials usually locate the local anode, but the absolute values of the potentials measured are influenced by the cover depth and the resistivity of the concrete.

An electrical connection to the rebar is usually required. This may require drilling a hole into the concrete above the rebar. Potential measurements can be performed with a single electrode (point measurements) or with one or several wheel electrodes (potential mapping). Two techniques for mapping are shown in Figure 39. In the upper part of Figure 39 a reference electrode scans the surface, linked to an electrical connection to the reinforcement. This probe can be replaced with a wheel for easier scanning. In the lower part of Figure 39 two CSE half-cells are used, one fixed and the other to scan the surface. This second method does not require an electrical connection to the reinforcement but only provides comparative information. As noted above, potential mapping is usually performed with a wheel arrangement and a small grid size of approximately 0.15 m.

Table 22 Typical Ranges of Potential of Reinforcing Steel in Concrete (RILEM 2010). *Sources:* Modified from RILEM (2010). Acoustic emission Report TC212-ACD, 2010 International Union of Laboratories, Experts in Construction Materials, systems and structures.

Condition	Range volts CSE
Water saturated concrete without oxygen	−0.9 to −1.0
Wet, chloride contaminated concrete	−0.4 to −0.6
Dry concrete	+0.2

(a) Circuitry for monitoring surface corrosion potentials of steel reinforcement in concrete – Single half-cell method

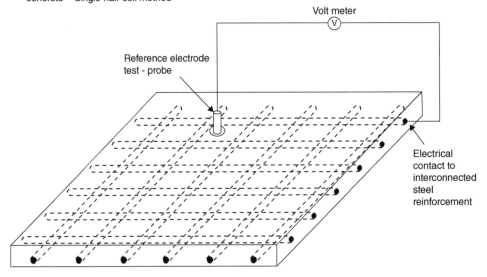

(b) Circuitry for monitoring surface corrosion potential variations across the steel reinforced concrete surface – Two half-cell method

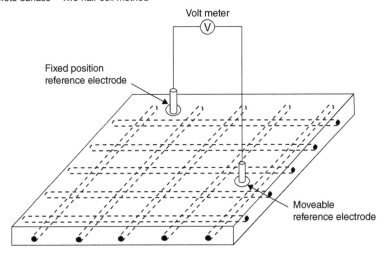

Figure 39 Methods of potential mapping (Department of Energy 1987b).

Representation of half-cell potential data in order to locate areas with different potentials can be performed by colour plots or equipotential contour plots. In an equipotential contour plot, lines of constant potential are calculated and plotted through points of equal or interpolated equal values. As can be seen from Table 22, areas with strongly negative values indicate areas of potential corrosion. However, there are several uncertainties in the readings obtained and further non-destructive or destructive analyses are required to confirm areas of corrosion.

Use of the technique for underwater concrete is more difficult as the seawater provides a low resistance electrical path, making potential measurements more complicated and the results more uncertain. However, the technique is suitable for concrete in the splash zone. A standardised approach to potential mapping is given in the ASTM procedure ASTM C876-15 (ASTM 2015) and RILEM TC 154 (RILEM 2003).

4.3.10 Cathodic Protection Inspection

Cathodic protection systems for offshore steel and concrete platforms require significant survey, monitoring and in some cases mitigation to ensure their effective performance throughout their life.

Corrosion protection systems are designed at the beginning of the life of the installation, with an expected life matching the design life. Life extension may require review of these systems, inspection and consideration of any mitigation required. Sacrificial anodes and impressed current are the two underwater corrosion protection systems in use, with the former dominating.

Sacrificial anodes are designed with typically a 25-year life based on the expected current demand of the system. Regular inspection of the anodes is a formal part of an API level 1 or 2 inspection as well as part of an ISO 19902 level 1 periodic inspection. These inspections include checking the condition of the anodes and measuring the potential usually by an ROV. The potential needs to be measured in a number of locations on the platform using a half cell (typically Ag/AgCl). The measured potentials provide important information on the state of the CP system.

The cathodic protection system for offshore structures includes protection of the structure itself in addition to subsystems depending on the type of the platform such as:

- risers, conductors, pull-tubes and other submerged objects electrically connected to the main structure of fixed offshore steel structures;
- mooring lines, hull and marine systems for floating offshore structures; and
- attached steel work and embedded steel reinforcement for concrete structures.

The initial survey of a platform is likely to be extensive but follow-on periodic surveys should take account of previous results. As an example, for fixed steel structures the follow-on periodic surveys should include:

- each vertical and horizontal frame of the structure;
- representative anodes of the main structure, selected based on previous results; and
- known areas of possible or actual under-protection including conductor bays, pile guides, pile sleeves, shielded areas.

Optimum protection for medium-strength steels requires potentials in the range –850 to –1000 mV. Below –850 mV is considered to be under-protection with the possibility of corrosion increasing as the potential becomes more positive. More negative results than –1000 mV (over-protection) can lead to problems such as hydrogen embrittlement and damage to coatings. Especially for high-strength steels, over-protection is a serious problem particularly for hydrogen embrittlement and potentials are recommended to be in a narrower range from –730 to –830 mV (DNVGL 2017), which is much more difficult to achieve in practice.

Some cathodic potential survey results have been published (MTD 1990). These show out of 38 platforms included in the survey, 6 platforms had individual results more positive than –850 mV and 16 platforms surveyed had individual results more negative than –1000 mV, with 3 with values more negative than –1050 mV, which is likely to indicate a more serious over-protection problem. For the cases of under-protection, remedial actions were noted as a result of the survey, which included retrofit of local anodes, 100% replacement of the anodes, coating of conductors (to limit drain on the CP system) and more regular monitoring. The MDT report (MTD 1990) does not suggest any mitigating actions for the over-protection cases. The problem of over-protection and its effect is more prominent than under-protection and the ensuing corrosion result. As a result, over-protection is often accepted except for high strength steel. Problems of cracking due to hydrogen embrittlement in high-strength steel in jack-ups has been reported (HSE 1993).

Figure 40 shows a series of drawings of sacrificial anodes with increasing loss of material (Ersdal et al. 2019). At some stage, the severely pitted examples result in a less negative protection potential and therefore less efficient cathodic protection. At this stage anode replacement is required to maintain efficient protection.

A decommissioning removal of an installation from the sea allows the state of anodes to be examined in more depth. One example was the recovery of the West Sole WE platform in 1978 after 11 years in the sea. A detailed examination of the several components including the anodes was undertaken, as shown in Figure 40. Weight loss of the anodes was found to be 40% less than the expected value for the period in the sea. Protection potentials were measured before recovery of the platform and were in the range –0.870 to –0.930 mV compared with Ag/AgCl, which is on the more positive level than set at the design stage but nevertheless in an acceptable range. The West Sole platform was also used to trial some structural monitoring systems (see Section 5.2.2).

The splash zone coating on the West Sole platform was a coal tar epoxy and it was found that the protection had broken down from wave and spray action. It was noted that this could have been minimised by more painting during maintenance. Isolated pitting and metal wastage were seen on members in the splash zone, up to a depth of 3 mm in places, which could have become more serious if the platform had been in the sea for a longer period.

4.4 Deployment Methods

4.4.1 Introduction

As previously stated, underwater inspections can be performed by several different deployment methods which are listed in Table 23 including the tasks that each method can perform. These include diving, remotely operated vehicles (ROV) and autonomous underwater vehicles (AUV).

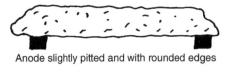

Anode slightly pitted and with rounded edges

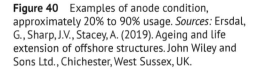

Figure 40 Examples of anode condition, approximately 20% to 90% usage. *Sources:* Ersdal, G., Sharp, J.V., Stacey, A. (2019). Ageing and life extension of offshore structures. John Wiley and Sons Ltd., Chichester, West Sussex, UK.

Anode moderately pitted, approximately 50% missing

Anode severely pitted where frame is visible

Table 23 Deployment Methods for Inspection and Repair.

Tasks	Under water				Splash zone		
	Divers	ROV (observation class)	ROV (work class)	AUV	Rope access	Aerial drones	Rail access ROV
Visual inspection	X[1]	X	X	X	X	X	X
NDT inspection	X[1]		X		X	(X)	X
FMD		X	X	X			
Cleaning	X[1]		X		X		X
CP measurements	X[1]	X	X				
Weld repair	X		X		(X)		X
Reinstate coating	X[1]		X	X			X
Seabed inspection		X	X	X			

[1] Normally not performed by divers but possible

4.4.2 Divers

Divers were used almost exclusively in the early days of the offshore oil and gas exploration, both for air diving (up to a maximum depth of 50 m) and saturation diving for deeper waters. Offshore diving is well regulated by governmental organisations such as HSE in the UK, BSEE in US and PSA in Norway.

As offshore developments moved into deeper waters, saturation diving was required to enable divers to work underwater at depths for longer periods. It requires specialised equipment such as complex and expensive Diving Support Vessels (DSV). Saturation diving involves keeping up to 24 divers under pressure for prolonged periods. It is a diving technique that allows divers to reduce the risk of decompression sickness ("the bends") when they work at great depths for long periods of time. Transfer to and from the pressurised surface living quarters to the equivalent depth is done in a closed, pressurised diving bell. This may be maintained for up to several weeks. The diving bell is linked to the surface vessel by an umbilical. Once saturated, decompression time does not increase with further exposure. Saturation divers typically breathe a helium–oxygen mixture to prevent nitrogen narcosis, but at shallow depths saturation diving has been done on a nitrox mixture. To prevent hypothermia, hot-water suits are commonly used for saturation diving and the breathing gas supply may be heated. Heated water is produced at the surface and piped to the bell through the umbilical and then transferred to the divers through their own umbilical, which is linked to the bell. These umbilicals also have cables for electrical power to the bell and helmet lights and for voice communications and closed-circuit video cameras. Decompression from a saturation dive is a slow process and controlled by the working depth and managed to minimise the risk of decompression sickness.

Diving, particularly saturation diving, is recognised as having safety issues and there have been several accidents and loss of life. For these reasons it is heavily regulated and as a result, ROVs are used for underwater working where possible.

4.4.3 ROV and AUV

Early US ROVs were developed in the 1950s to recover torpedoes and mines. This was followed by continuous improvement and development first for military use and later for offshore use.

ROVs have become more capable in the last decades of undertaking underwater tasks (particularly the work-class ROVs). Similar to diving, ROVs also require support vessels. Most ROVs are tethered to the support vessel using an umbilical (a cable with the necessary communication and control information) and can operate in two different ways, over the side deployment or using a moon pool. The over the side deployment method is generally considered to be most cost effective and most frequently used. An A-frame is normally used to deploy the ROV into the water. The moon pool technique allows an ROV to be deployed directly into the water. The stability of this technique is better in comparison to over the side deployment, as the ROV enters the sea in a more controlled way which reduces the likelihood of damaging the vehicle. The moon pool method is also less weather dependant.

ROV's are usually classified into the following (Michel 2003 and Christ and Wernli 2014):

- observation class (from micro-ROVs to vehicles of around 100 kg which are depth limited to approximately 300 m, minimal to no payload, power typically less than 15 kW, often used as backup to divers or other ROVs or for shallow water inspection tasks);
- light work class (ROVs from approximately 100 kg to 1000 kg which can be used at depth up to 1000 meters with moderate payload, power up to 55 kW);
- work class (depth up to 3000 meters with heavy lift and payload for construction work, power up to 75 kW); and
- heavy work class (depth up to 5000 meters, ultra-heavy lift and payload, power more than 100 kW).

An example of an ROV (FCV 4000) with a seabed device in its manipulators about to enter the water is shown in Figure 41.

ROVs may be instrumented with the necessary equipment to perform inspection work. The ability of ROVs to perform NDE inspections is steadily improving, but there remains a difficulty in inspection of welded nodes because of their complex geometry. Automatic controlling of an inspection probe requires pre-programmed setup for each individual joint. In addition, the

Figure 41 ROV being deployed. *Source:* Courtesy of Fugro – www.fugro.com.

ROVs need to be fixed to the structure in order for the manipulator to move the head of the NDE equipment over the weld with the necessary accuracy.

ROVs are also used to accommodate magnetic crawlers or robotised inspection tools as shown in Figure 42. A magnetic crawler can move around on the structure on wheels that are magnetic enabling attachment to a steel member or plate. These can crawl along horizontal and vertical surfaces to inspect the integrity of a structure underwater. The unit is typically controlled from the topside or a ship through a tether connection. Crawlers can be equipped with multiple operational attachments such as pressure washer to remove debris, rust, scale or fouling in addition to various NDT equipment and cameras. The crawlers can also be deployed by the use of rail systems as shown in Figure 45 and further discussed in Section 4.4.4. For tubular members, specialized tools that grab around and are able to crawl along or around the member are an option as shown in Figure 43.

Figure 42 ROV installation of Robotised Inspection Tool. *Source:* Courtesy of OceanTech, www.oceantech.no.

Figure 43 RIT: Robotised inspection tool. *Source:* Courtesy of OceanTech, www.oceantech.no.

Figure 44 Two examples of AUVs. *Source:* Courtesy of Stinger — http://www.stinger.no/.

Autonomous Underwater Vehicles (AUV) systems, as illustrated for example in Figure 44, are suitable for inspection and are operating remotely from a vessel without an umbilical and can run either a preprogrammed or logic-driven course (Christ and Wernli 2014). AUV systems usually have a dedicated docking station on the seabed which is connected to the topside facility. Such a system can be continuously available and while docked, the AUV can be charged, data can be downloaded from the AUV and future mission information can be uploaded to the AUV. Where tether length is a limitation in undertaking inspection using an ROV, the solution could be to use an AUV.

To undertake CVI-type inspections, the AUV requires enhanced capabilities to be able to position its cameras and lights close to the inspection target without damaging the structure. This requires a manoeuvrability similar to that of an ROV. This hovering ability is a key capability which is often used to distinguish bathymetry-class AUVs (used for monitoring the seabed) from inspection-class AUVs.

4.4.4 Splash Zone Access

Carrying out inspection and repair in the splash zone is recognised as difficult and specialised access techniques have been developed to enable this to take place. These include rope access and more recently vertical access tools and drones. The main types of access methods are rope access and more recently aerial drone access and various types of rail systems.

Rope access has been the accepted deployment method for splash zone inspection with personnel requiring specialized training and work while suspended by a harness. The tasks typically include riser, conductor, caisson, leg and brace inspection. Expertise includes the ability to take high-quality photographic records and videos as well as being able to undertake ultrasonic testing and NDT.

Specialised companies have developed various splash inspection tools for cleaning and inspection. This includes the introduction of rail systems to deploy AUVs and ROVs where the rails are attached to the structure in the splash zone. OceanTech's Vertical Access Tool (VAT), shown in Figure 45, is an example of such splash zone deployment by a rail system. Access to the splash zone is dependent on weather conditions including wave and total states.

With the advancement in aerial drones and related technologies, drones are increasingly finding their way into industries such as oil and gas. Companies are now relying on drones as a cheaper and more effective alternative in assisting them with day-to-day operations. Visual inspection by an unmanned drone is a quick and cost-effective inspection method of critical areas such as the splash zone.

In floating structures, inspection of internal structures and tanks can be performed using a method applicable to structures above water. Cleaning and venting tanks are a significant prerequisite. However, this type of "underwater" inspection is not the main topic of this book. For some floating offshore structures re-ballasting to transit mode can allow for easier access to outer surfaces for inspection personnel.

Figure 45 Rail installation (Vertical Access Tool) of Robotised Inspection Tool. *Source:* Courtesy of OceanTech.

4.4.5   Summary of Inspection Methods and Their Deployment

As has been discussed in Section 4.3, a wide range of different inspection methods are available for detecting, locating and characterising defects and anomalies. Further, their deployment methods have been discussed in this section. These are summarised in relation to the various inspection tasks and the relevant inspection methods as given in Table 24 with indications of cleaning needs and deployment methods.

4.5   Competency of Inspection Personnel and Organisations

4.5.1   Introduction

Demonstration of both individual and organisational competency is increasingly being recognised as very important for managing safety by minimising gross errors where humans and organisations are involved. Competency of organisations can have a significant effect on the ability of individuals to control hazards to structural safety. The competence of individuals at all levels in an organisation is vital, from the inspector, the inspection supervisor and the operator's site representative to the onshore structural engineer and the management.

The area of competency is recognized to be part of the wider topic of human factors. HSE's report HSG 48 "Reducing Error and Influencing Behaviour" (HSE 1999b) provides guidance aimed at managers with health and safety responsibilities, health and safety professionals and employee safety representatives. The report's message is that proper consideration of 'human factors' is a key ingredient of effective health and safety management. The guidance in the report provides practical help on how to tackle some of the important issues, e.g.:

- human error and behaviour can impact health and safety;
- human behaviour and other factors in the workplace can affect the physical and mental health of workers;

Table 24 Methods for Inspection with Indication of Cleaning Needs and Deployment Method.

Inspection task (damage type to identify)	Description	Methods used	Cleaning required	Deployment method
General assessment of structural condition	Swim around to identify general damage	General visual inspection (GVI) Video and still photography	NO	ROV
	Debris survey to note and record debris on the installation			
	Seabed survey to measure and record degree of scour and debris			
	Measure distribution and thickness of marine growth			
CP and coating survey	Measure CP potentials, assess depletion in anodes, check condition of any coatings	CP checking using reference cell, Visual inspection	NO	ROV
Detailed inspection of selected areas	Inspect for visible damage, such as cracking, denting, bows, pitting etc.	Close visual inspection (CVI) using photography, video, taut wire measurement, profile gauging, photogrammetry, etc.	Yes	ROV
Identification of member flooding	Check tubular members for water ingress due to cracking, corrosion, denting, local rupture, anode failure	FMD method (usually gamma ray)	No	ROV
Crack detection	Detailed NDT methods to detect cracks	ACFM	No / Yes	ROV / Diver
		MPI, Eddy current, radiography, ACPD	Yes	Diver / ROV
Measure length and depth of identified cracks	Specialised NDT	MPI (length), eddy current, ACPD (length and depth)	Yes	Diver / ROV
Condition of concrete material in splash zone	Detection of reinforcement corrosion	Electro potential testing	Yes	Rope access
	Visual inspection for corrosion products and spalling	General visual inspection (GVI) and Close visual inspection (CVI)	Yes	Rope access
	Delamination	Rebound hammer	Yes	Rope access

- practical ideas on what you can do to identify, assess and control hazards arising from the human factor; and
- illustrative case studies to show how other organisations have tackled different human problems at work.

4.5.2 Regulatory Requirements on Competency

In the UK the Health and Safety at Work Act (HSE 1974) includes general requirements for competency. HSE's report HSG 65 "Managing for Health and Safety" (HSE 2013) states that the competence of individuals is vital at all levels in an organisation but especially those with safety-critical roles. It ensures they recognise the hazards in their activities and can apply the right measures to control and manage the risk associated with these hazards. Further, it also provides a checklist of factors which apply to activities when done effectively, badly or not at all (HSE 2013 Table 2). In Norway the Petroleum Safety Authority (PSA) has issued similar requirements in their Activities Regulation concerning competency.

The HSE's Provision and Use of Work Equipment Regulation (HSE 1998) specifically states that it is necessary to ensure that persons involved in all areas of inspection need sufficient competency. This includes those who determine the nature of inspections required and those who carry out inspections. It is also stated that every employer (duty-holder, contractor, etc.) should ensure that all persons have received adequate training for relevant equipment. This includes training in the methods which may be adopted when using the equipment and any hazards involved and the precautions to be taken.

4.5.3 Requirements on Competency in Standards

For steel jacket structures, ISO 19902 (ISO 2007) covers in-service inspection and structural integrity management which includes a separate section on personnel qualifications covering several functions (including SIM engineer and the inspection team). This part of ISO 19902 (ISO 2007) has recently been transferred to a new ISO 19901-9 (ISO 2019a) keeping more or less the same competency requirements.

For concrete structures, ISO 19903 states that personnel involved in inspection planning and condition monitoring, as well as in assessment of the findings, shall have relevant competence with respect to:

- design of marine concrete structures,
- concrete materials technology;
- fabrication (execution) of concrete structures; as well as
- specific experience in the application of inspection techniques and the use of inspection instrumentation and equipment.

These four requirements listed in ISO 19903 (ISO 2019b) do not incorporate important areas in structural integrity management, for example, as listed in Table 25. Particularly experience in evaluation of findings and inspection planning which are normally regarded as important areas requiring competent personnel are missing from this list.

Further, in ISO 19903 (ISO 2019a) it is recognised that because each offshore concrete structure is unique, inspectors should familiarize themselves with the primary design and operational aspects of that structure before conducting an inspection. In addition, inspectors should have adequate training appropriate for supervisors, divers or ROV operators as specified in accordance with national requirements where applicable.

NORSOK N-005 (Standard Norge 2017) is a Norwegian standard for integrity management of offshore structure and marine systems. It should be noted that while ISO 19902 (ISO 2007) and ISO 19901-9 (ISO 2019a) only cover fixed jacket structures, NORSOK N-005 covers all types of offshore structures, including concrete structures, semi-submersibles and ship-shaped structures.

Table 25 Competency Requirements (Based on the SIM Standards Mentioned in This Section).

Function	Application	Competency requirement
Integrity strategy	Technical authority for structural integrity	<ul><li>Competent in offshore structural and marine engineering;</li><li>Understand the hazards and failure modes of structures and marine systems;</li><li>Understand how the structures and marine systems are designed, operated and maintained to withstand these hazards;</li><li>Be familiar with relevant information about the facility;</li><li>Have knowledge of ageing effects for structures and marine systems;</li><li>Be aware of general in-service inspection findings in the offshore industry.</li></ul>
Evaluation	Engineer or technicians	<ul><li>Familiar with relevant information about the specific structures under consideration;</li><li>Experienced in evaluation of inspection findings;</li><li>Knowledgeable about underwater corrosion processes and control;</li><li>Knowledgeable in offshore inspection planning, inspection tools and techniques and their limitations;</li><li>Cognisant of general inspection findings in the offshore industry.</li></ul>
Data collection and update of database	Engineer or technicians	<ul><li>Competent in inspection recording protocols, database administration and requirements for recovery of data;</li><li>Familiar with the typical data to be archived such as structural arrangement, inspection techniques, analysis techniques and interpretation of analysis results, material specifications, certificates and welding specifications and qualification practices.</li></ul>
(Long-term) Inspection programme	Engineer or technicians	<ul><li>Knowledgeable of the content of the overall integrity strategy;</li><li>Experienced in offshore inspection and maintenance planning of structures and marine systems;</li><li>Experience and technical expertise commensurate with inspection tasks to be performed;</li><li>Knowledgeable of the structural arrangement, inspection techniques, analysis techniques and interpretation of analysis results, requirements for inspector competency.</li></ul>
Offshore inspection execution team	Supervisors, divers, ROV operators, etc.	<ul><li>Qualified and accredited to international or equivalent regional standards (underwater pre-qualification trials are usually required for divers);</li><li>Competency in inspection methods and practice;</li><li>Competency in applying acceptance criteria during their inspection, e.g. reporting of major damage and potentially serious hazards.</li><li>Knowledgeable of how and where to look for damage and situations that could lead to damage;</li><li>Familiarity with the structure owner's data validation, recording and quality requirements.</li></ul>

The guidance given for competency for the team of SIM engineers is, however, similar to the guidance given in the ISO standards.

API RP2-SIM, 2FSIM and 2MIM (API 2014a, 2019a, 2019b) also detail competency requirements associated with structural integrity management similar to those given in the ISO and NORSOK standards.

In the commentary section to ISO 19902 (ISO 2007) additional information is given on the competency requirements for the offshore inspection execution. This includes accreditation to recognised schemes as discussed in the next section.

4.5.4   Certification and Training of Inspectors

Certification of inspectors should be according to recognised schemes such as the UK Certification Scheme for Welding Inspection Personnel (CSWIP) (https://www.cswip.com/) or equivalent such as ISO 9712:2012 (ISO 2012). Training is a required part of the preparation to achieve such certification and also to retain the certification after the initial approval.

Regarding the execution of the inspection programme the UK CSWIP scheme is available for the examination and certification of individuals to demonstrate their knowledge and competence. The scheme covers both underwater and topsides inspection personnel. The different underwater inspection categories are shown in Table 26.

CSWIP accreditation requires both written tests and for commercial divers, tests on suitable specimens in an underwater habitat. The CSWIP 3.1U course includes theoretical instruction to an CSWIP approved syllabus, which includes general underwater and close visual inspection, recording by video and photography, cathodic protection measurements, ultrasonic digital thickness measurements followed by an end-of-course assessment. The next level CWIP 3.2U course adds a capability in advanced underwater NDT techniques such as magnetic particle inspection, weld toe grinding and an overview of electromagnetic techniques. The ROV inspector is a further development of the 3.1U and 3.2U courses focussing on quality assurance, closed circuit television, calibration of equipment, cathodic protection systems, interpretation and recording methods. The underwater inspector controller is a follow-on to the ROV inspector with teaching in advanced NDT techniques, recording and processing data, interpretation and recording methods, quality assurance, intervention techniques, inspection, planning and briefing. The course material is available and provides a useful and detailed source of background information to inspection methods (TWI 2006).

4.5.5   Trials to Study Inspector Competency

In inspection, operator competency is recognised as having a significant effect on performance. Human factors are a recognised aspect for all manual inspection methods, but even mechanised

Table 26   CSWIP Categories.

Scheme	Description
3.1U	Underwater (diver) inspector* (visual, cathodic protection, and ultrasonics)
3.2U	Underwater (diver) inspector* (as 3.1U plus MPI, weld toe grinding)
3.3U	ROV inspector
3.4U	Underwater inspector controller

*Note: a concrete endorsement is available for grades 3.1U and 3.2U to cover inspection of concrete structures.

UT systems require interpretation and thus operator performance should be regularly checked. The PISC (Programme for Inspection of Steel Components) was established in the 1980s. PISC-II was set up to examine in more detail which techniques could provide the desired level of capability in detection and sizing of defects in nuclear pressure vessel components. One aspect investigated was human reliability. The work concentrated on Round Robin Testing (having the same specimen tested by several different inspectors or organisations) of four thick plates of approximately 250 mm thickness. In order to check human capability, manual UT inspectors with relevant experience were observed by skilled observers. The Transportable Environmental Laboratory (TEL) enabled different environmental conditions to be set for inspectors to work in. It was found a clear difference was demonstrated between skills, knowledge and working practices in these tests. The main conclusions were (Nichols and Crutzen 1988):

- variability of calibration was acceptably small;
- flaw detection performance varied between 65% and 100% between inspectors;
- variability for single inspectors was also found to be due to tiredness (a factor 2 is quoted); and
- there were a significant number of reporting errors (left for right, etc.) when using ultrasonic testing.

Nichols and Crutzen (1988) made suggestions to increase the NDE effectiveness from human errors which included:

- some form of indication warning device from the NDE tool;
- use of a UT simulator and training;
- avoid long working periods, since they have an effect on performance; and
- use of automated UT tools.

4.5.6 Organisational Competency

Measuring the capability of an organisation to perform an important process is a key factor in ensuring that the objective of, for example, managing structural integrity is met. If used correctly it can also be a relevant performance indicator. One such model, the Capability Maturity Model (CMM) was developed by The Software Engineering Institute as a method for evaluating the maturity of software companies. This model was used to enable their software projects to be carried out on time, on budget and with the expected quality (Paulk et al. 1993). The CMM model focuses on the processes that an organisation needs to carry out successfully to meet the project goals and indicates which processes exist in successful organisations. The Capability Maturity concept is widely used to measure organisational maturity in several industries and has been incorporated into ISO 9004 (ISO 2000) as a measure of the maturity in quality assurance.

Generally, the following elements have been identified as important for the capability of an organisation:

- management (organization, materials, management documents, performance indicators);
- leadership and involvement;
- competence and technical knowledge;
- communication;
- quality assurance;
- learning in the organization;
- cultural aspects (participation, shared perception, trust, job satisfaction, training);

- productivity versus safety; and
- resource allocation.

It is important to notice that the maturity model is evaluating the organisations' capability to perform and not the product itself. For example, it is a method that looks at the structural integrity management process rather than the integrity of the structure per se. It is based on an assumption that if the integrity is managed properly, then the capability maturity model result can be used as a leading indicator for the actual integrity.

The creation of the CMM descriptions allows organisations or auditors to establish the current level of maturity for each of the particular characteristics associated with an activity. The five basic levels used in CMM to describe an organization are given in Table 27. Further, it is used to identify what steps are necessary to enable the organisation to progress to a higher level, building on their strengths and improving on their weaknesses.

CMM-type models have been developed for several different offshore applications. These include the design process (Sharp et al. 2002), asset integrity management (Energy Institute 2007), structural integrity (Poseidon 2007 and Sharp et al. 2011) and the management of ageing safety critical elements (Energy Institute 2009).

Poseidon (2007) introduced a rating system for the assessment of structural integrity management activities based on the CMM model. Key aspects of this model are inspection and repair, which are the subject of this book. The seven main processes associated with managing structural integrity identified by Poseidon are shown in Figure 46. These can be assessed using the above described CMM model.

It is generally accepted that the minimum maturity level for an efficient organisation is level 3, a level which is regarded as only just meeting the minimum regulatory requirements. However, most regulators also require a certain level of continuous improvement, which indicates that a level 4 would be preferred. Actual audits undertaken with several offshore operators (OMAE 49335 2011) showed only a range of maturity levels from 2 to 3, which indicates that the industry at that time did not perform in an optimal way.

Table 27 Five Basic Maturity Levels Used in CMM.

Level	Description of organisational capability at this level
1	At the *initial* level, processes are disorganised, even chaotic. Success is likely to depend on individual efforts and is not considered to be repeatable because processes would not be sufficiently defined and documented to allow them to be replicated.
2	At the *repeatable* level, basic project management techniques are established, and successes could be repeated because the requisite processes would have been made established, defined and documented.
3	At the *defined* level, an organisation has developed its own standard processes through greater attention to documentation, standardisation and integration.
4	At the *managed* level, an organisation monitors and controls its own processes through data collection and analysis.
5	At the *optimising* level, processes are constantly being improved through monitoring feedback from current processes and introducing innovative processes to better serve the organisation's particular needs.

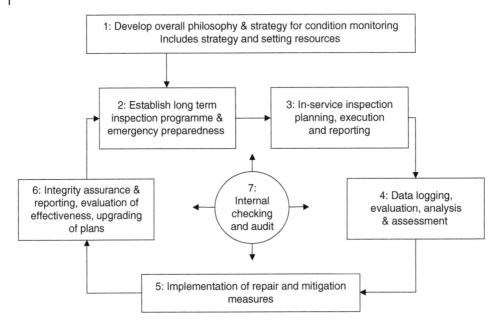

Figure 46 Seven CMM processes for maturity model for managing structural integrity.

4.6 Reliability of Different Inspection Methods Underwater

Accurate detection and characterisation of anomalies (such as fatigue cracks) is important in determining the safety of an existing structure. This accuracy is normally measured in the method's probability of detection (PoD[5]) and sizing (PoS[6]) and these have impact on:

- the choice of the inspection method; and
- the inspection planning.

As an example, when predicting the remaining fatigue life of a cracked welded tubular joint, an accurate sizing of the existing crack is the key parameter to be determined, while in most cases where no known defects exist, the probability of detection is the key parameter.

The probability of detection (PoD) is often illustrated as a function of the size of the defect as a PoD-curve. For fatigue defects this is typically provided as a function of crack length and depth. PoD is often seen as consisting of three parts, which are:

- defects of a size below which they cannot be detected;
- large defects which are always detected; and
- a transition region.

---

5 Probability of detection (PoD) is defined as total number of successful inspections divided by the total number of defects present. This definition is however not sensitive to the size of the database used. This is overcome by defining a PoD with a confidence level which can be based on the number of test specimens. A typical confidence level is 95% which, using a binomial distribution, arises when the tested database consists of 29 specimens. This is usually defined as a PoD with a 95% level of confidence which is often used in the statistical analyses of inspection data. PoD can be based on crack length or depth.

6 Probability of sizing (PoS) is the probability of correct sizing of a defect for acceptance or rejection. Although this term is often used, it will reflect, in general, the accuracy of estimating the size of a defect.

The shape of the transition region is important for inspection planning. This type of information is useful in being incorporated in probability-based inspection planning (see Section 6.2.5).

Probability of detection and sizing curves for different inspection methods have been determined by trials using specimens with known defects in both offshore conditions and in a controlled laboratory environment. Several projects on the probability of detection for various inspection methods have been initiated. As earlier mentioned, examples of such projects include SINTAP (Structural INTegrity Assessment Procedures) and for offshore structures the ICON project (Dover and Rudlin 1996, HSE 1996) and the TIP project (Rudlin and Austin 1996, Rudlin et al. 1996). For inspection methods underwater the tests undertaken by University College London and later in the European part funded ICON project (see Section 4.2.9) provided the basis for establishing POD curves for a number of inspection methods. The NDT facilities at UCL included a diving tank for simulating diver inspection trials. The HSE study (HSE 2002) collected and analysed data from the most relevant projects and developed POD and POS curves both for inspection in air and underwater. These were later used as the basis for POD-type curves in recommended practices such as DNVGL RP C210 (DNVGL 2019).

Figure 47 shows the POD curve for EC, MPI and ACFM based on DNVGL RP C210 (DNVGL 2019). It can be seen that 90% probability of detection for all these inspection methods is achieved for a defect depth of approximately 12 mm. Figure 48 shows the POD curve for CVI for two levels of access difficulty, based on DNVGL-RP C210 (DNVGL 2019). For the moderate level of access, 90% probability of detection is achieved for a crack length of 400 mm, whereas for the difficult level of access, which is applicable underwater, the probability level of detection for a 400 mm length crack is only approximately 83%. These cracks are significant and the depth of such a crack would typically be in the range of 30 mm. Hence, in many cases these would be close to or through thickness, which demonstrates the difficulty on relying on CVI for fatigue crack detection.

The MSL study (MSL 2000) showed that CVI reliability depended on the degree of cleaning and surface preparation, which was determined by the type of cleaning method used. The report states that cleaning with pressure was necessary for reliable visual inspection. Complete cleaning to bare metal allowed a crack of 300 mm in length to be detected (without providing a level of probability).

The reliability of flooded member detection was assessed in the ICON trials (Dover and Rudlin 1996) and the reliability of the gamma ray FMD method was reported as being 100%

Figure 47 PoD curves for different NDT inspection methods based on DNVGL RP-C210 (DNVGL 2019).

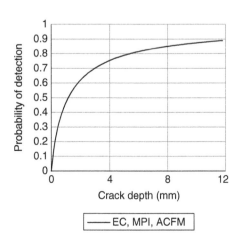

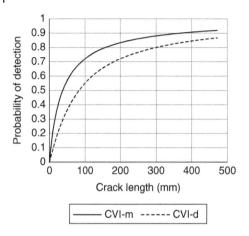

Figure 48 PoD curves CVI inspection for two levels of access difficulty based on DNVGL RP-C210 (DNVGL 2019).

accurate for both 50% and 100% flooding. The testing was undertaken on both horizontal and vertical members. However, the reliability of detection of a through-thickness crack was only 70% at 10% water fill (HSE 2002). The DNVGL RP-C210 (DNVGL 2019) indicates that the reliability of FMD is considered good and a probability of detection should be assumed to be 95%.

As already mentioned, the probability of sizing (PoS) is another important factor. However, very few data appear to have been reported and no PoS curves are provided in recognised offshore standards. Visser (HSE 2002) reported that the length accuracy could be summarised by the following two statements:

- the length accuracy for MPI is 20%; and
- the length accuracy for EC and ACFM is 40%.

However, the data reported by Visser (HSE 2002) indicates that, not surprisingly, the sizing accuracy is much worse for small defects (less than 15 mm).

So-called "spurious indications" are indications obtained during the inspection which do not correspond to actual defects (false positives). Such indications can be analysed in different ways, normally as a percentage of the number of found defects. These false indications are a limitation when carrying out inspections and can lead to unnecessary and expensive re-inspections and in the worst cases repairs. Visser (HSE 2002) states that the rate of spurious indications should preferably not exceed 20%, which to the authors seems to be on the high side particularly for high values of PoD.

4.7 Inspection of Fixed Steel Structures

Fixed steel structures are normally inspected underwater, in the splash zone and above the splash zone. This book is primarily about the inspection of the underwater and splash zone areas. Typical inspection tasks on a fixed steel structure would include:

- welded connections between legs and brace members and between connecting brace members;
- member flooding;
- leg and brace members for dents and bows;
- conductor frames for cracks, dents and bows;
- CP and the condition of the anodes;

- caisson, j-tubes and risers and the status of their supports;
- pile sleeves and the status of the grouting between pile sleeve and pile;
- marine growth thickness; and
- scour and potential build-up of debris (e.g. drill cuttings).

The underwater part of the fixed steel structure can be inspected by either divers or ROVs and recently the emphasis is on ROVs because of the safety issues with using divers. Inspection of the splash zone is more difficult and can be done by rope access or more recently by aerial drones or by using fixed rails for deployment.

The current approach to underwater inspection of fixed steel structure is general visual inspection (GVI), cathodic potential measurement, close visual inspection (CVI) and flooded member detection (FMD). These have the key advantages of being relatively quick and hence less expensive and can cover most of the installation below water in a short period. They are normally undertaken from an ROV. NDT methods and wall thickness measurements may be required for further investigation if the presence of damage or anomalies are identified.

FMD has the significant disadvantage of being able to detect only through-thickness cracks, since it relies on the detection of water penetrating the member. As noted in Chapter 7, the remaining life is short after detection of flooding. Hence, the efficient use of FMD relies on understanding the criticality of individual joints. CVI and GVI also have limitations and as indicated earlier, the probability of detection of crack lengths less than 350–400 mm by CVI underwater is limited. GVI will typically be expected to identify member severance or severe denting. In most cases jacket structures have sufficient redundancy to allow these inspection methods to be used without compromising safety. However, this redundancy should be properly established for each individual fixed steel structure. There are also concerns about relying on FMD for ageing structures where cracking is more likely to occur. This may require more frequent inspections or the use of more detailed NDE methods. Redundancy may also be reduced for ageing fixed steel structures as, for example, widespread damage might be expected at some stage making these inspection methods less acceptable. The topic of system strength is discussed in more detail in Ersdal et al. (2019).

The structural implications of the dent and bow damage are described in Chapter 7. Frieze et al. (1996) indicated thresholds for the identification of dents based on visual inspection which were determined to be 12.5 mm and 38 mm. Such dents led to typical strength reductions assessed to be up to 6% and 17% for legs and braces respectively, based on the strength calculation formulae for dented member available at the time. Corresponding values for strength reduction due to bow sizes of 130 mm and 350 mm were as high as 37% and 68% respectively (Frieze et al. 1996). Such reduction in strength indicates that the structure could be compromised particularly in the cases where the damage is not found following an incident. Frieze et al. (1996) concludes that for detection of dents and bows, current North Sea inspection practice is not sufficiently effective.

The inspection of guide frames and support structures for risers, conductors and caissons has historically been included within the scope of structural inspection. In the early years a large amount of fatigue cracking was found in the guide frames due to inadequate design. For some of these older structures this is still a problem area requiring inspections and possibly repair but less of a problem for newer structures. For guide frames in or above the splash zone, corrosion will also be a major issue. These are to some extent protected by different types of coating, and the integrity of these coatings needs to be one of the items in an inspection programme.

Inspection of the corrosion protection system involving anodes attached to the structure is undertaken by measuring local corrosion protection potentials, with respect to a standard electrode using an ROV. More positive potentials may indicate high usage of the anodes, which can be assessed visually and replacement may be needed.

Pile sleeve connections are a critical part of a structure. The foundation piles are usually grout connected to the pile-sleeves. Recently there have been indications of such grouted connections failing, especially related to wind turbines. Improvements to the design method for grouted connections, as a result of these failures, have indicated that pile-sleeve connections also on fixed steel structure are vulnerable. It is recognised that inspection of these grouted pile-sleeve connections is difficult, but methods for monitoring and inspection of such grouted connection are being developed and tested at the time of writing this book. The piles themselves are also not easily accessible for inspection, but in a few cases the piles may be inspected internally by specialized methods.

Underwater inspection will normally also include surveillance for scour around piles, location of build-up of seabed materials, drill cutting and debris. Scour can reduce the pile efficiency as it exposes the upper parts of the piles (introducing buckling and corrosion as potential failure modes). Build-up of materials around the base of the structure reduces access to inspection of the buried members and this may also introduce unexpected loads into these members. The inspection of the degree of scour and build-up of materials around the base of the structure is normally performed by a GVI using an ROV. Removal of built-up materials may be required if critical structural members have been buried.

Standards for inspection of fixed structures have been developed to document the inspection and integrity management processes, including ISO 19901-9 (ISO 2019), API RP 2A (API 2014a), NORSOK N-005 (Standard Norge 2017) and earlier versions of ISO 19902 (ISO 2007) as discussed in more detail in Chapter 2.

It should be recognised that some structural components are difficult or impossible to inspect but may be vulnerable to degradation processes. This presents a particular problem as their current condition is difficult to determine. Such components could be in very deep water or with difficult access for an inspection tool. Examples are piled foundations where the circumferential butt welds are very difficult to inspect and ring-stiffened joints where the internal stiffeners are also difficult to inspect using conventional equipment. This aspect needs to be taken into account in the fatigue assessment of older structures and the inspection of adjacent members and joints preferably with correlated loading is used to achieve some level of information about an un-inspectable detail; see, for example, NORSOK N-006 (Standard Norge 2015) for guidance.

Inspection methods appropriate to a range of damage causes and types specific to fixed steel structures are shown in Table 28.

Structural monitoring methods are also available providing real-time data on the condition of the structure (see Chapter 5).

4.8   Inspection of Concrete Structures

Concrete gravity-based structures (GBS) are massive compared to steel jackets. As noted earlier, see Section 1.3.3, concrete gravity-based structures consist of one to four shafts (legs or towers), caissons (storage cells), a steel-to-concrete transition between the concrete base and topsides, a shaft-to-base junction (junction between a shaft and the base), prestressing including anchorages, foundations and a cathodic protection (CP) system.

The probability of failure of concrete structures compared to steel structures is more related to corrosion and less to fatigue. Below the water line the development of failure is effectively limited by the presence of a corrosion protection system (anodes). Hence large areas of the underwater concrete structure (caisson walls, upper dome surfaces and the walls of shafts) have a relatively

Table 28 Damage Causes and Suggested Inspection Methods for Fixed Steel Structures.

Cause of damage	Damage or anomaly type	Inspection method
Fatigue	Surface cracks	CVI, MPI, EC, ACFM
	Internal cracks (e.g. full pen welds)	Ultrasonic testing
	Embedded defects (full penetration welds)	Ultrasonic testing, radiography, ACFM
	Through-thickness cracks	FMD, leak detection
	Severance and missing members	GVI (and FMD)
Corrosion	General corrosion on members resulting in reduced wall thickness	GVI. Ultrasonic tests to determine remaining wall thickness
	Patch corrosion	GVI, DVI and CVI. Ultrasonic tests to determine remaining wall thickness
	Corrosion pits	GVI, CVI, EC
	Holes and severance (separation)	FMD, GVI
	Loss of corrosion protection system	CP readings
Overload (hurricanes and major storms)	Buckles, dents, out-of-plane bowing, severed members, pancake leg severance, punch-through of brace, pull out of brace, conductor torn loose from guide, broken water caissons, broken riser standoffs	GVI and DVI possibly using straight-edge and mechanical measurements
	Holes and tears	FMD, GVI
	Cracks at welds and into chords	FMD, MPI and UT
Overload and accidental damage (vessel impact and dropped object)	Dented, bowed and buckled members	GVI and DVI possibly using straight-edge and mechanical measurements
	Dented and bowed members with cracks (fracture)	FMD, MPI and UT
	Severance of members	GVI
Wear and abrasion	Loss of material	GVI, DVI and CVI possibly with mechanical measurements. Ultrasonic to determine remaining wall thickness
Marine growth	Increased loading and potential overload damage. Possibility for corrosion damage from sulphate-reducing bacteria	GVI
Settlement and subsidence	Platform settlement and subsidence may lead to change in loading pattern and exposure condition	Global positioning systems (GPS)
Scour and build-up of drill cuttings	Scour around the platform exposing the piles with the potential for pile failure. Drill-cuttings can bury or bear onto lower frames and result in damage	GVI

(Continued)

Table 28 (Continued)

Cause of damage	Damage or anomaly type	Inspection method
Material deterioration	Cracking from hydrogen embrittlement particularly for high-strength steel (e.g. jack-ups)	CVI and NDT methods.
Fabrication fault	Lack of penetration or excessive undercutting that could lead to unexpected fatigue cracks	FMD, MPI and UT
	Failure to provide vent holes for intended flooded members could lead to implosion, incorrect member sizes, incorrect member positions, omissions	GVI and DVI
Under-design	Member buckling, joint failure	GVI and DVI
	Tearing, fatigue cracking	GVI, DVI, CVI and FMD

low probability of corrosion failure unless the CP system is ineffective. However, accidents can cause damage to shafts and the domes of caissons. Ageing concrete structures will experience increasing amounts of chloride ingress in the splash zone that may lead to corrosion of the reinforcement in areas of poor cover.

Typical inspection requirements for the GBS type of platforms include the:

- concrete shafts (towers or legs) both externally and internally;
- caissons for oil storage;
- steel concrete transition;
- seabed scour, settlement and subsidence;
- construction joints;
- CP and anodes; and
- marine growth.

The shafts and cell walls are characterised by very large general featureless areas. Typical damage arises from corrosion of the reinforcement in the splash zone, impact damage from ships or dropped objects and possibly chemical deterioration.

General Visual Inspection (GVI) is the normal inspection method, usually from an ROV. However, many shafts can be inspected internally and GVI could identify evidence of water ingress or possibly reinforcement corrosion. Delamination may not be visible on the concrete surface and hammer tapping may detect such areas. The Concrete in the Oceans programme (Department of Energy 1987a) reviewed typical features which could be found by inspection of shafts by an ROV. These defect types on concrete structures are listed in six groups (numbers in brackets indicate number in each group):

- Group 1: defects from the construction process (34);
- Group 2: redundant built-in items (15);
- Group 3: concrete materials (12);
- Group 4: repairs and surface treatments (13);
- Group 5: superficial marks and stains (15); and
- Group 6: physical damage and deterioration (26).

More details of the defects in the more significant groups 4, 5 and 6 are shown in Table 29.

Table 29 Examples of Defects on Concrete Structures in groups 4, 5 and 6 (Department of Energy 1987a). *Sources:* Department of Energy (1987a). Offshore Technology Report OTH 87 248 Concrete in the Ocean programme – coordinating report on the whole programme, HMSO, London, UK. © 1987, H.M. Stationery Office.

Group	Examples (numbers in brackets)
Group 4 (repairs and surface treatments)	Making good (4)
	Resin systems (2)
	Epoxy/pitch epoxy paints (11)
	Sprayed concrete (2)
Group 5 (superficial marks and stains)	Stains (4)
	Exudation (4)
	Surface cleaning marks (1)
Group 6 (physical damage and deterioration)	Impact damage (9)
	Abrasion (5)
	Structural cracks (9)
	Deterioration of surface treatments (9)
	Spalling (1)
	Erosion (4)
	Deterioration of repairs (2)
	Deterioration of Jarlan-type breakwater wall (4)
	Corroding reinforcement (2)

The 45 examples in group 6 are the more important and may particularly require follow-up, depending on severity. Typical examples of impact damage and structural cracks are shown in Figure 49.

For storage cells the most likely form of damage is spalling of the concrete on the tops of the cells from dropped objects, as has been observed. This can be detected by GVI from an ROV. However, some cleaning may be required as the tops of storage cells often accumulate debris or drill cuttings. An example of this type of impact damage is shown in Figure 94.

In the splash zone the components of a concrete platform are the towers. Inspection of these using GVI requires access, which can be by rope access or more recently by drone. The features of interest are the same as those in Table 30. However, because of the higher oxygen content in the splash zone, corrosion of the reinforcement is more likely and subsequent spalling of the concrete cover. In addition, significant chloride ingress possibly resulting in reinforcement corrosion will occur. As noted earlier, internal inspection of a tower and shaft could indicate evidence of water ingress or reinforcement corrosion.

Most offshore concrete have massive foundations with skirts which penetrate the seabed as they are gravity-type structures. At the installation stage the foundation is usually grouted to the seabed, providing a solid base. As structures age, this grouting may be damaged by scour, which can be detected by GVI using an ROV. The presence of debris may make this inspection difficult and cleaning may be needed.

In most offshore concrete structures, the reinforcement is electrically connected to the cathodic protection (CP) system which is present to protect the attached steelwork. In the early structures there were attempts at the construction stage to isolate the reinforcement from this system. In most cases this failed as there was unintended electrical connectivity to flowlines

Figure 49 Two examples of inspection of concrete structures underwater. The left-hand picture shows a mild steel rod with a type of washer plate and wedge assembly for restraining shutter during construction. The right-hand picture shows concrete cracking with a gap more than 2 mm on line of fracture probably caused by structural movement (Department of Energy 1986). *Source:* Offshore Research Focus, No 51, January 1986, ISSN: 0309-4189. Published by CIRIA for the Department of Energy, London, UK.

Table 30 Damage Types and Inspection Methods Specific for Concrete Structures.

Damage cause	Damage type	Inspection method
Corrosion at construction joints (above water)	Spalling and delamination (above water) as result of reinforcement corrosion as well as potential loss of reinforcement area	GVI, CVI Rebound hammer (compressive strength and uniformity of concrete) Level of chloride ingress (see below)
	Seepage of corrosion products at cracks under water as result of reinforcement corrosion	GVI for evidence of corrosion products associated with cracking Cathodic protection levels Rebound hammer (compressive strength and uniformity of concrete)
	Chloride ingress above water (reinforcement corrosion and potential loss of reinforcement area)	Drilled dust sampling to measure chloride ingress Core extraction
	Corrosion of steel attachments (e.g. pre-tension cable at anchorages, steel support attachments) and potential loss of prestressing capacity	GVI for evidence of corrosion products and spalling of concrete cover to anchorage
	Corrosion at construction joints (above water)	GVI for evidence of corrosion products and spalling
Accidental damage and overload	Local impact damage to towers resulting in reduction in concrete section area and possible water ingress	GVI (damage to concrete and reinforcement) Ultrasonic testing to evaluate degree of penetration
	Local impact damage to tops of storage cells from dropped objects resulting in reduction in concrete section area and possible water ingress	GVI (damage to concrete and reinforcement) Ultrasonic testing to evaluate degree of penetration
Loss of drawdown	Drawdown provides compression in towers, resisting extreme weather, maintained by pumping	Monitoring of pumping system

Table 30 (Continued)

Damage cause	Damage type	Inspection method
Wear and tear	Abrasion (local wear) and erosion resulting in minor reduction in concrete section area	GVI, Ultrasonic pulse velocity test (thickness measure)
Marine growth	Increased loading	GVI (usually ROV) including measurement of thickness
Settlement and subsidence	Platform settlement and subsidence may lead to change in loading pattern and exposure condition	GVI (usually ROV)
Scour	Scour around the base of the platform possibly leading to instability	GVI (usually ROV)
Material deterioration	Loss of the concrete material due to aggressive agents (sulphate, chloride)	GVI (usually ROV)
	Cracking due to shrinkage (after construction)	GVI (usually ROV)
	Cracking, scaling and crumbling due to freeze and thaw (above water)	GVI (usually ROV), rope access, drones
	Loss of strength of the concrete material due to sulphate-producing bacteria (primarily in oil storage tanks)	Difficult to inspect
Fabrication fault	Defective construction joints, minor cracking, surface blemishes, remaining metal attachments, patch repairs from construction, low cover to the reinforcement possibly leading to spalling in the splash zone and possible water ingress	GVI, DVI and CVI Rebound hammer for delamination inspection
Under-design	Cracking due to low reinforcement or prestressing, crushing of concrete in rare cases, failure in the junction between shafts and cells	GVI

and pipelines as well as to other external attachments. This led to a higher than planned drain on the sacrificial anodes and in some cases, these had to be replaced.

The availability of cathodic protection has the advantage that it protects the reinforcement to some extent, minimising the level of corrosion in situations where chlorides have permeated to the steel. Current design criteria for CP systems recommend or require a minimum of 1 mA/m^2 for the reinforcing steel. Inspection and maintenance of the CP system is therefore a basic requirement to minimise the corrosion reaction. This is usually carried out both visually and by monitoring the CP potentials using a probe operated from an ROV.

Table 30 shows the different inspection methods in use for the different types of damage for concrete structures.

4.9 Inspection of Floating Structures and Mooring Systems

Parts of floating structures are inspected underwater, including the mooring, external skin of the pontoons, ship hulls, sea-chests (sea-water intake) and the CP system. However, in most cases the structures are brought into a transit mode to allow for above water access. Much of the inspection can also be performed internally on floating structures, for example, pontoons, columns and braces on semi-submersibles and hull, transverse and longitudinal bulkheads on ship shaped structures.

A major component of floating structures is the mooring system, which to a large degree dominates the underwater inspection programme. This needs to be monitored regularly to detect any deterioration in its condition, typically inspected utilising a work class ROV which performs a fly-by of the lines. Mooring lines and anchors for MODUs are on a regular basis recovered when they move from one location to another which provides periodic opportunities to undertake a more detailed in-air inspection. In contrast, floating units permanently used for production are in a fixed position for a significant period of time without dry docking, which makes inspection and repair more difficult. For these units it is possible to recover a mooring line part-way through a field life for detailed inspection. However, the lines may be damaged either during recovery or re-installation and such inspection is rather costly.

There has been considerable effort to develop in-water inspection methods particularly for the sections of moorings subject to the greatest deterioration forces (HSE 2006). These sections particularly include the seabed touchdown zone (impact zone) and vessel interface zone (e.g. around fairlead); see Figure 21.

All lines need to be inspected, but a particular check for wear should be undertaken on the leeward lines as these have been found to be most vulnerable to damage. Further, when inspecting mooring lines at the touchdown zone, attention should include the potential hazards from rocks or debris on the seabed as they can cause mooring line abrasion.

Underwater inspection of mooring lines has several issues. Mooring lines which have been in sea water for extended periods accumulate varying levels of marine growth. It can typically be heavy, depending on geography, water depth and season. In general, this growth needs to be removed so that the underlying mooring components can be inspected. Cleaning options include ROV-deployed high-pressure water or grit-entrained high-pressure water. Once marine growth is removed, it is possible to conduct various levels of inspection including visual inspection (DVI and CVI), dimensional measurement and assessment of mechanical fitness. Unfortunately, cleaning off marine growth and scaling by high-pressure water jetting may accelerate corrosion by exposing fresh steel to the corrosive effects of saltwater.

A number of in-water mooring chain size measurement systems for wear and elongation have been developed with varying success. These range from simple diver-deployed manual callipers to a prototype stand-alone robotic system and ROV-deployed systems. The most established ROV-deployable chain measurement system is effectively an 'optical calliper'. It comprises multiple high-resolution video cameras and lights on a deployment frame which is equipped with scale bars in pre-assigned orientations and at set distances from each other and the cameras. The system measures the chain parameters by calibrating from the tool scale bars and resolving dimensions and optical distortions using offline image analysis software.

As well as checking of chain dimensions, it is also important to assess link integrity and condition. The overall condition of mooring components can give an indication of the types of deterioration processes that are present. For example, surface pitting may indicate pitting corrosion, 'scalloping' or indentations of wear (HSE 2006).

Detection of loose studs is another important part of the mooring inspection and these have been implicated in crack propagation and fatigue. Accordingly, studded chain inspection requires the assessment of the numbers of loose studs and degrees of 'looseness'. However, there is no agreed industry opinion with respect to loose stud rejection criteria. Traditionally, chains have had to be recovered for detailed loose stud determinations and have relied on a manual test on land, either moving the stud by hand or using a hammer to hit the studs. The resulting resonance (a 'ping' or 'thud') is used to assess whether a stud is loose or not. A company has developed an ROV-deployable loose stud detection system. The system uses an electronically activated hammer to impact the stud and uses a hydrophone and a micro-accelerometer as sensors. A software program is used to distinguish between 'loose' and 'tight' responses (HSE 2006).

As part of any in-water mooring line inspection, it would be prudent to identify and inspect all kenters (joining links) and shackles to confirm that they appear to be intact and that all split pins are present.

Underwater visual condition assessment by ROV is particularly difficult because of the inherent flatness of video images from standard 2D inspection cameras. With these cameras it is very difficult to distinguish whether a visual artefact on a surface is merely a mark or a region from which material has been lost (e.g. a pit). A number of 3D visualization systems have been implemented and have proven very effective for the assessment of the surface condition and general geometry of mooring components.

API RP-2MIM (API 2019b) provides accuracies of the different techniques relevant for mooring systems; see Table 31.

Wire rope is particularly difficult to inspect and the results can be somewhat subjective. Sheathed spiral strand wire is even more difficult to inspect since it is obscured by the sheathing. In particular, it is difficult to assess if the sheathing has becoming damaged during installation, for example going over the stern roller. If the sheathing is damaged letting in seawater, this could result, over time, in accelerated undetected corrosion. In addition, abrasion could occur on the seabed and this would also be difficult to determine. An in air–based wire rope inspection unit has been developed using a magnetic head which can detect damage and has been tested in practice on the Buchan FPS (HSE 2006).

Synthetic ropes are also difficult to inspect, relying mainly on GVI or CVI from an ROV for detection of damage to the rope. The use of short sections of rope (inserts) as part of the mooring lines for offshore installations, with periodic removal (e.g. every 2.5 years) for laboratory testing

Table 31 Typical Accuracies for Inspection Techniques (API 2019b). *Sources:* Modified from API (2019b), API RP-2MIM Mooring Integrity Management, American Petroleum Institute, 2019.

Component	Technique	Accuracy
Anchor (scour survey)	GVI	±500 mm (extent/depth)
Anchor	GVI	
CP system	Potential monitoring	±10 mV
Anodes	GVI	Visual consumption within ±10% of volume
Chain link (wear, corrosion)	High resolution photography	±0.1 mm
Chain elongation	GVI	±5 mm
Coating survey	GVI	±10% of local area being surveyed
Marine growth	GVI	±10 mm on thickness, 15% of area being examined
Structure	Corrosion, pitting depth, cracks	Within ±2 mm typically
Trenching survey	GVI	Trench depth within ±1 m.

has also been in use. ABS (2011) outlines inspections requirements for synthetic ropes, which include an annual survey, limited to above water and a periodical survey on a five-year basis. This includes a checking of the terminations. Synthetic ropes are subject to creep and this survey requires a review of the records of anchor leg re-tensioning caused by non-recoverable elongation. It is also necessary to check that adequate lengths of chain or wire rope segments are available for further re-tensioning, such that the fibre rope does not come into contact with the fairlead and stays below the water surface. Specific guidance on the scope and frequency of inspections for mooring lines is given in IACS (2010); see Section 2.4.3.

Offshore mooring failure detection systems are normally required by regulation to provide real-time monitoring information in order to keep the line tension within operational limits and alert on-board personnel to mooring line failures. The tension in mooring lines can be measured directly by load cells or indirectly by changes in properties of the mooring line. Such properties may be the eigenfrequency of the mooring line, typically between the fairlead and the gypsy wheel, the elastic deformation of chain links in the mooring line and angle measurement by using an inclinometer fitted on the mooring line.

A load cell will typically include the use of compression cells under chain stopper plates or winch footings. The basic system to measure elastic deformation is usually an instrumented pin in a shackle or other connecting link or it may be a specifically designed link so that a strain gauge is measuring axially along the mooring line in the area of maximum strain. A number of load measuring devices within the one unit can be included to allow for redundancy.

A hull or seabed-based sonar system can monitor the continued presence of all mooring lines. A sonar head array on the seabed can provide simultaneous images of all mooring lines and also provide information on catenary shape and location. As noted above, this can then be used to calculate the mooring line tension as well as confirming that all the mooring lines are present. In addition, deployment of cameras from time to time can provide a high confidence method of checking that all the moorings are in place assuming a suitable deployment tube for the camera is available.

An overview of inspection methods is given in Table 32.

Table 32 Damage Causes and Suggested Inspection Methods for Floating Structures.

Cause of damage	Fixed steel and hull structures	Underwater inspection method
Fatigue	Fatigue cracks in hull structure	CVI, MPI, EC, ACFM, but primarily inspection would be performed from inside in dry conditions
	Fatigue crack in chain link, link failure, loss of mooring line, drifting unit	CVI, MPI, EC, ACFM Failure detection systems
Corrosion	General corrosion will result in reduced wall thickness and in some cases patch corrosion with potential for holes and flooding Corrosion pits leading to flooding and potential for fatigue initiation	GVI, DVI and CVI Ultrasonic tests to determine remaining wall thickness. Internal leak detection would also register through wall corrosion
	Corrosion of chain link or wire strand, loss of mooring line, drifting unit	Chain link size measurement (possibly by ROV). Video surveillance Failure detection systems

Table 32 (Continued)

Cause of damage	Fixed steel and hull structures	Underwater inspection method
Overload (hurricanes and major storms)	Buckles, dents, holes, cracks at welds, tears in hull structure	GVI, DVI and CVI, but primarily inspection would be performed from inside in dry conditions
	Damage to links, snapping of anchor chains, drifting units, dragging of anchors	GVI, DVI and CVI Loose stud detection system. Failure detection systems
	Drifting units and dragging of anchors may impact other structures and pipelines	Failure detection systems. Operation. Station keeping system monitoring (GPS).
Overload and accidental damage (vessel impact and dropped object)	Dents and buckles in hull structure	GVI, DVI and CVI, but primarily inspection would be performed from inside in dry conditions
	Mechanical damage to mooring lines from seabed impact, wear and bending over the fairlead and handling of anchor chains and wire ropes	GVI, DVI and CVI
Wear and tear	Abrasion/wear of link or rope	GVI, DVI and CVI with high resolution photography
	Lost stud (chain)	GVI.
Marine growth	Increased loading and potential overload damage	GVI
	Possibility for sulphate-reducing bacteria resulting in corrosion	
Scour and build-up of drill cuttings	Loss of holding power to anchors	Failure detection systems Operation. Station keeping system monitoring (GPS).
Material deterioration	Brittle fracture of chain links	GVI, DVI and CVI
Fabrication fault	Lack of penetration, excessive undercutting and misalignment that could lead to unexpected fatigue cracks, incorrect member sizes, incorrect member positions, omissions.	GVI, but primarily inspection would be performed from inside in dry conditions
Under-design	Buckling, tearing, fatigue cracking	GVI, DVI and CVI, but primarily inspection would be performed from inside in dry conditions
	Damage to links and failure of anchor chains possibly leading to drifting units Under-designed anchors (soil strength) could lead to dragging of anchors	Failure detection systems Operation. Station keeping system monitoring (GPS)

References

ABS (2011), "Guidance Notes on the Application Fiber Role for Offshore Mooring", American Bureau of Shipping, Houston, Texas, US.

ABS (2018), "Guide for Non-Destructive Inspection of Hull Welds", American Bureau of Shipping, Houston, Texas, US.

API (2014a), *API RP-2A Recommended Practice for Planning, Design and Constructing Fixed Offshore Platforms*, API Recommended practice 2A, 22nd Edition, American Petroleum Institute, 2014.

API (2014b), *API RP-2SIM Recommended Practice for Structural Integrity Management of Fixed Offshore Structures*, American Petroleum Institute, 2014.

API (2019a), *API RP-2FSIM Floating System Integrity Management*, American Petroleum Institute, 2019.

API (2019b), *API RP-2MIM Mooring Integrity Management*, American Petroleum Institute, 2019.

ASTM (2015), *ASTM C876 Standard Test Method for Corrosion Potentials of Uncoated Reinforcing Steel in Concrete*, ASTM International, West Conshohocken, Pennsylvania, US, 2016 (www.astm.org).

ASTM (2016), *ASTM C597-16 Standard Test Method for Pulse Velocity Through Concrete*, ASTM International, West Conshohocken, Pennsylvania, US, 2016 (www.astm.org).

ASTM (2018), *ASTM C805 / C805M Standard Test Method for Rebound Number of Hardened Concrete*, ASTM International, West Conshohocken, Pennsylvania, US, 2016 (www.astm.org).

Busby, F (1979), "Underwater Inspection/Testing/Monitoring of Offshore Structures", Conducted by R. Frank Busby Associates under Department of Commerce Contract No. 7-35336, NOAA/Office of Ocean Engineering, Rockville, Maryland, US (also published as BSEE/MMS TAP report 001).

BV (1998), *Bureau Veritas Guidance Notes Type: Approval of Non-Destructive Testing Equipment Dedicated to Underwater Inspection of Offshore Structures*, NI 422 DTO R00 E, 1998.

Christ, R.D. and Wernli, R.L (2014), *The ROV Manual—A User Guide for Remotely Operated Vehicles*, Second Edition, Elsevier / Butterworth-Heinemann, Waltham, Massachusetts, US.

CIRIA (1978), *Underwater Inspection of Offshore Installations, Guidance for Designers*, Report URIO CIRIA Underwater Engineering Group, February 1978.

Department of Energy (1984), "Offshore Research Focus, No. 42, 1984", Published by CIRIA for the Department of Energy, London, UK.

Department of Energy (1986), "Offshore Research Focus, No 51, 1986", ISSN: 0309-4189, Published by CIRIA for the Department of Energy, London, UK.

Department of Energy (1987a), *Offshore Technology Report OTH 87 248: Concrete in the Ocean Programme—Coordinating Report on the Whole Programme*, HMSO, London, UK.

Department of Energy (1987b), *Offshore Technology Report OTH 87 250: Repair of Major Damage to the Prestressed Concrete Towers of Offshore Structures. A report by Wimpey Laboratories for the Department of Energy*, HMSO, London. UK.

Department of Energy (1987c), *Offshore Technology Report OTH 87 263: Study of Calibration Procedure for Accurately Quantifying Crack Sizes in Welded Tubular Joints*, Prepared by University College London (V.S. Davey) for the Department of Energy, HMSO, London, UK.

Department of Energy (1990), *Offshore Installations: Guidance on Design, Construction and Certification*, Fourth Edition, HMSO, London, UK.

DNVGL (2015), *DNVGL-CG-0051 Non-Destructive Testing*, DNVGL, Høvik, Norway.

DNVGL (2017), *DNVGL-RP-B401 Cathodic Protection Design*, DNVGL, Høvik, Norway.

DNVGL (2019), *DNVGL RP-C210 Probabilistic Methods for Planning of Inspection for Fatigue Cracks in Offshore Structures*, DNVGL, Høvik, Norway.

Dover, W.J. and Rudlin, J.R. (1996), "Defect Characterisation and Classification for the ICON Inspection Reliability Trials", Proceedings of the Conference on Offshore Mechanics and Arctic Engineering, ASME.

Energy Institute (2007), *Capability Maturity Model for Maintenance Management,* (ISBN 978-0-85293-487-6), Energy Institute, London, UK.

Energy Institute (2009), *Research Report: A Framework for Monitoring the Management of Ageing Effects on Safety Critical Elements,* (ISBN 978-0-85293-545-3), Energy Institute, London, UK.

Ersdal, G., Sharp, J.V., Stacey, A. (2019), *Ageing and Life Extension of Offshore Structures*, John Wiley and Sons Ltd., Chichester, West Sussex, UK.

Frieze P.A., Nichols N.W., Sharp J.V., Stacey A. (1996), "Detection of Damage to Underwater Tubulars and its Effect on Strength", OMAE Conference 1996, Florence, Italy.

Hanson, R (2020), Private communication with Gerhard Ersdal.

Hayward, G., Pearson, J. and Stirling, G. (1993), "An Intelligent Ultrasonic Inspection System for Flooded Member Detection in Offshore Structures", IEEE TUFFC, 40(5), 512–521.

HSE (1974), *Health and Safety at Work etc Act*, HMSO, London, UK.

HSE (1988), *A Handbook for Underwater Inspectors*, Edited by L.K. Porter, Published as HSE Offshore Technology Information OTI 88 539, HMSO, London, UK.

HSE (1993), *Offshore Technology Report OTH 91 351: Hydrogen Cracking of Legs and Spudcans on Jack-up Drilling Rigs—A Summary of Results of an Investigation*, Prepared by Techword Services by K. Abertheny, C M Fowler, R Jacob and V.S. Davey for the Health and Safety Executive (HSE), HSE Books, Sudbury, Suffolk, UK.

HSE (1996), *Offshore Technology Report OTN-96-150: Intercalibration of Offshore NDT (ICON)*, Commercial in confidence PEN/S/2736, HSE, UK.

HSE (1998), *Provision and Use of Work Equipment Regulation 1998" (PU WER)*, HSE, UK.

HSE (1999), *Detection of Damage to Underwater Tubulars and Its Effect on Strength, HSE Offshore Technology Report OTO 99 084*, HMSO, London, UK.

HSE (1999b), *Health and Safety Guidance HSG 48 Reducing Error and Influencing Behaviour*, HMSO, London, UK.

HSE (2002), *HSE Offshore Technology Report 2000/018 2002 POD/POS Curves for Non-destructive Examination*, Prepared by Visser Consultancy Limited for the Health and Safety Executive (HSE). (ISBN 0717622975), HSE Books, Sudbury, UK.

HSE (2006), *HSE RR444 Floating Production System—JIP FPS Mooring Integrity*, Prepared by Noble Denton Europe Limited for the Health and Safety Executive (HSE), HMSO, London, UK.

HSE (2010), *Ionising Radiations Regulations 1999—Notification of Offshore Site Radiography Work*, Operations notice 34, HMSO, London, UK.

HSE (2013), *Health and Safety Guidance HSG 65: Managing for Health and Safety*, HMSO, London, UK.

Hudak, S.J., Burnside, O. H. and Chan, K. S. (1984), "Analysis of Corrosion Fatigue Crack Growth in Welded Tubular Joints, OTC 4771", Houston, US.

IACS (2002), *IACS UR Z 7.1: Hull Surveys for General Dry Cargo Ships*, International Association of Classification Societies, London, UK.

IACS (2010), *Guidelines for the Survey of Offshore Chain Cable in Use*, No. 38, 2010, International Association of Classification Societies (IACS).

IACS (2012), *Unified Requirements,* International Association of Classification Societies and International Association of Classification Societies Limited, London, UK.

ISO (2000), *ISO 9004 Quality Management Systems—Guidelines for Performance Improvement*, International Standardisation Organisation.

ISO (2007), *ISO 19902:2007 Petroleum and Natural Gas Industries—Fixed Steel Offshore Structures*, International Standardisation Organisation.

ISO (2012), *ISO 9712:2012 Non-Destructive Testing—Qualification and Certification of NDT Personnel*, International Standardisation Organisation.

ISO (2019), *ISO 19901-9 Petroleum and Natural Gas Industries—Specific Requirements for Offshore Structures—Part 9: Structural Integrity Management*, International Standardisation Organisation.

ISO (2019b), *ISO 19903 Petroleum and Natural Gas Industries—Fixed Concrete Offshore Structures*, International Standardisation Organisation, 2019.

Lotsberg, I. (2016), *Fatigue Design of Marine Structure*, Cambridge University Press, 2016.

Michel, D. (2003), "A Short History and Overview of the Commercial ROV and AUV Industry", In Randall, R. and Ward, E.G. ROV/AUV "Capabilities—Proceedings and Final Workshop Report Prepared for the Minerals Management Service (MMS)", MMS Project Number 446, September 2003.

Mijarez, R. (2006), "A Remote and Autonomous Continuous Monitoring Ultra-Sonic System for Flood Detection in Sub-Sea Members for Offshore Steel Oil Rigs", Ph.D. Thesis, University of Manchester, UK.

MSL (2000), "Rationalization and Optimisation of Underwater Inspection Planning Consistent with API RP2A Section 14", Published by MMS, US, as Project Number 345, November 2000.

MTD (1989), *Underwater Inspection of Steel Offshore Installations: Implementation of a New Approach*, Report MTD publication 89/104, London, UK.

MTD (1990), *Design and Operational Guidance on Cathodic Protection of Offshore Structures Subsea Installations and Pipelines*, Marine Technology Directorate Ltd., London, UK.

MTD (1994), *Review of Repairs to Offshore Structures and Pipelines,* Publication 94/102, MTD, London, UK.

Nichols, R.W. and Crutzen, S. (1988), *Ultrasonic Inspection of Heavy Section Steel Components: The PISC II Final Report*, (ISBN 1-85166-155-7), Elsevier Applied Science, Barking, UK, 1988.

Paulk, M.C., Curtis, B., Chrissis, M.B. and Weber, C.V. (1993), "Capability Maturity Model SM for Software version 1.1", CMU/SEI-93-TR-24.

PMB (1990), "AIM IV Assessment Inspection Maintenance", Published as MMS TAP report 106as.

Poseidon (2007), "Revised Structural Integrity Management Capability Maturity Model Incorporating Sub-Processes for Life Extension", Poseidon Group AS, POS-DK07-138-R01, December 2007.

RILEM (2003), "TC 154 Materials and Structures / Matériaux et Constructions", Vol. 36, August–September 2003, pp. 461–471.

RILEM (2010), "Acoustic Emission Report TC212-ACD", 2010 International Union of Laboratories, Experts in Construction Materials, Systems and Structures.

Rudlin, J.R. and Austin, J., "Topside Inspection Project: Phase I Final Report", Offshore Technology Report OTN 96 169 Nov. 1996.

Rudlin, J.R, Myers, P. and Etube, L., "Topside Inspection Project: Phase II Final Report", Offshore Technology Report OTN 96 169 Nov. 1996.

Sharp J.V., Stacey A., and Birkinshaw M. (1995), "Review of Data for Structural Damage to Offshore Structures", 4th Intern. ERA Conference, Dec.1995, London, UK.

Sharp J.V., Strutt J.E., Busby J., Terry E. (2002), "Measurement of Organizational Maturity in Designing Safe Offshore Installations", OMAE 2002-28421.

Sharp J.V., Wintle J. and Terry E. (2011), "A Framework for Managing Ageing Safety Critical Elements Offshore", OMAE 2011-49203.

SSC (1953), "The Present Status of Non-Destructive Tests Methods for Inspection of Welded Joints in Ship Structures", by R.J Krieger, S.A. Wenk and R.C. McMaster for Ship Structure Committee, Ship Structure Committee report no 72, Washington, DC, US.

SSC (1996), "Inspection of Marine Structures", Ship Structure Committee report no 389.

SSC (2000), "SSC-416 Risk-Based Life Cycle Management of Ship Structures", Ship Structure Committee report no 416, 2000.

Standard Norge (2015), *NORSOK N-006 Assessment of Structural Integrity for Existing Offshore Load-Bearing Structures,* Edition1; March 2009, Standard Norge, Lysaker, Norway.

Standard Norge (2017), *NORSOK N-005 In-Service Integrity Management of Structures and Maritime Systems,* Edition 2, 2017, Standard Norge, Lysaker, Norway.

TWI (2006), "Tuition Note for 3.2U Course (DIS 2)", TWI, Cambridge, UK.

5

Structural Monitoring Methods

Strength and growth come only through continuous effort and struggle[1].

—Napoleon Hill

What gets measured, gets managed[2].

—Peter Drucker

By failing to prepare, you are preparing to fail[3].

—Benjamin Franklin

5.1 Introduction

5.1.1 General

Structural monitoring (also referred to as on-line monitoring or part of structural heath monitoring) is the observation of a structure using sensors that continuously or periodically measure structural behaviour. This enables changes to material, geometric properties and boundary conditions of a structure to be identified. Structural monitoring also includes use of analysis, such as statistical pattern recognition of the data, to identify changes that indicate damage or other changes to the structure.

The most used methods of structural monitoring include:

- natural frequency response;
- air gap;
- global positioning system (GPS);
- mooring chain tension;
- acoustic emission;
- acoustic fingerprinting;

---

1 *Source:* Napoleon Hill.
2 *Source:* Peter Ferdinand Drucker, "The Changing World of the Executive Drucker Library", 2010, Harvard Business Press.
3 *Source:* Benjamin Franklin.

Underwater Inspection and Repair for Offshore Structures, First Edition.
John V. Sharp and Gerhard Ersdal.
© 2021 John Wiley & Sons Ltd. Published 2021 by John Wiley & Sons Ltd.

- strain monitoring (e.g. strain gauges or optical fibres); and
- leak detection.

An important aspect of structural monitoring is that it can complement existing inspection techniques to provide greater confidence in structural integrity. In addition, in some instances it can potentially reduce inspection costs by reducing inspection frequency. It can also assist in demonstrating compliance with regulatory requirements. In combination with other methods it can be used to demonstrate continued safe operation of a structure beyond its original design life. Other typical applications include the monitoring of known local damage or a more critical part of a structure.

On-line instrumentation has also been used to benefit the design of offshore structures. Projects such as Shell's Tern and BP's Magnus have provided a much better understanding of the loading on these platforms, enabling more efficient designs to be developed for future structures (HSE 2009). Monitoring of cracks on offshore structures, such as using acoustic emission on the Ninian Southern platform (Mitchel and Rodgers 1992), provided confidence that operations could continue whilst repairs were being planned and undertaken.

5.1.2 Historical Background

As a result of the research strategy developed by the Department of Energy's Offshore Energy Technology Board in 1977, which identified inspection and integrity monitoring as a priority area, the Department funded a set of programmes on the vibration analysis of steel platforms with three companies involved at a cost (at the time) of ~£1M. The platforms involved were Amoco's Montrose platform, Occidental's Claymore platform and BP's Forties field (Department of Energy 1977a). The technique involved monitoring the structural vibrations caused by wave excitation with accelerometers to detect changes in the vibration signature; the pattern being made up from vibrations from individual members and any changes could indicate component failure. The three projects utilised the same monitoring technique but used different methods of analysis. The research involved placing sensors on the structure, both topside and in some cases up to 30 m below the surface. On Amoco's Montrose platform six horizontal accelerometers were mounted level with the spider deck. Two additional horizontal accelerometers were mounted on a major node at a depth of 11 m and, two vertical accelerometers, above and below the splash zone, monitoring the vibrations of horizontal braces. Occidentals' Claymore platform had 12 high sensitivity accelerometers at 21 m above the waterline and a number of lower sensitivity detectors above the waterline, 12 m and 30 m below the waterline. BP Forties platform had four pairs of orthogonal, horizontal accelerometers mounted at two levels, 13 m and 24 m above the waterline.

In addition, the Department commissioned Lloyds Register to assess the effectiveness of the three systems using vibration measurements (Department of Energy 1977b). The aim was to decide on the extent to which vibration monitoring could be relied upon to give warning of underwater structural damage, thus limiting the requirement of divers to undertake underwater inspection.

Identification of the various natural frequencies was achieved by comparison with a finite element model of each platform (Department of Energy 1982). The long-term stability of the natural frequencies was studied in detail. It was found that fundamental frequencies showed a variation of up to 2% of their mean. Major changes in platform operating conditions were detected by the monitoring systems. As an example, the addition of a significant mass of drilling equipment on one platform lowered the sway frequency by up to 5%. It was found both from the tests on the platforms and from modelling that for a four-legged platform failure of a single member produced a frequency drop in one or more modes which could be detected by the monitoring system. Analyses of an eight-legged structure showed that detection of member loss was dependant on its location. For example, loss of a diagonal member caused significant changes in natural frequencies. Loss of a horizontal brace caused smaller changes which were less detectable. It was concluded that to distinguish between changes in natural frequencies due to degradation of the structure or other effects it was necessary to have continuous information on the operational changes on the platform.

The removal of the West Sole WE platform, shown in Figure 50, in 1978 proved an opportunity for trials of vibration monitoring. Structural Monitoring Ltd performed tests before the platform was removed, using wave loading as the excitation force (Department of Energy 1980). Two sets of eight accelerometers were installed on the structure, one set above sea level, the other below. The platform response was measured both before and after damage was inflicted purposely. A pre-selected diagonal member was progressively cut until it was completely severed and then a horizontal member was flooded. It was found that the accelerometers placed above the water could only detect a completely severed member. However, its location could be located using a theoretical model of the structure. Those accelerometers located below the surface were able to detect every stage of the damage sequence in the member severance. It is important to note that West WE platform was a "minimum" structure with only four legs and hence more likely to be accessible to the benefits of structural monitoring.

Figure 50   West Sole WE platform.
Source: John V. Sharp.

Overall this early government-funded research showed that meaningful results could be obtained for low redundancy platforms (e.g. four-legged structures) but for larger eight-legged structures any changes due to damage were likely to be within the natural day-to-day variation in frequencies. However, improvements in accelerometers and computing analysis methods have made it possible to detect damage more easily in structures.

Recognising the potential of acoustic emission as an inspection monitoring technique for offshore platforms, the UK Department of Energy funded several projects in the late 1970s to develop the technique (Department of Energy 1979). Propagating cracks in metal structures were known to generate ultrasonic stress waves which could be detected and characterised to provide information on the nature and severity of the defects present. The initial work was funded at the Admiralty Marine Technology Establishment at Holton Heath (AMTE) where work on the method had been underway for over 10 years. As part of the work some nodes being tested as part of the UKOSRP steel research programme (Department of Energy 1987) were monitored using the AMTE technique. It was found that fatigue cracks could be detected and located much earlier than they could be seen. Further tests comparing the acoustic emission results with those from eddy current testing showed promising results. Two additional projects were funded by the UK Department of Energy aimed at further developing the equipment initially developed at AMTE. These involved EMI Electronics and the Unit Inspection Company. Both of these developers mounted their own development tests on the large-scale fatigue specimens being tested as part of the UK Offshore Steels Research Project (Department of Energy 1987). The EMI technique used a minimum of four transducers for good location of a growing crack. Tests on nodes at the National Engineering Laboratory (NEL) showed that the equipment provided advance warning of node failure and that there was good correlation between the observed emission characteristics and crack growth. At the time EMI considered that the technique had application for offshore structures with transducers mounted around nodes likely to be subjected to high stresses or where existing cracks were present.

The Unit Inspection Company (UIC), which was part of British Steel, also tested their version of the equipment on a large-scale node with a 1.8 m diameter chord. Two arrays of acoustic sensors were used around the node. The first sign of acoustic emissions occurred after 100,000 cycles, but there was no evidence of cracking at the surface. However, ultrasonic examination revealed a crack 110 mm long which was the source of the acoustic emissions. Further testing of the large node showed that crack extension occurred for relatively short periods of time when the acoustic signals were strong and regular and the amplitude of the emissions increased with increasing crack depth. UIC claimed that the results showed useful potential for monitoring offshore platforms. They also claimed that a four-sensor array was adequate for monitoring purposes and that to monitor a complete platform satisfactorily 40 such arrays would be sufficient.

Later, the Department of Energy funded AVT (Acoustic and Vibration Technology) Engineering Services Ltd (Department of Energy 1986) to develop further their acoustic emission system called Vulcan 8, developed for both onshore and offshore. The Vulcan system had an array of up to eight transducers capable of detecting acoustic emissions. For subsea applications the transducer cables were gathered into a unit clamped or strapped to the node being monitored. The Vulcan system has been evaluated in laboratory testing and in offshore trials. The former involved monitoring cracks growing in full-scale nodes, comparing the results with those from other NDE methods such as the ACPD technique. It was shown that correlation between the AE and NDE results was good. Offshore experience involved monitoring welds in a column of a semi-submersible drilling unit and a complex subsea node joint on a production platform. The developers claimed that the AE method had proved capable of detecting and locating fatigue cracks at a stage when low-cost repairs such as weld dressing were appropriate. The penalty of not detecting cracks at this early stage is that much more expensive repairs such as clamps or hyperbaric welding would be required.

The first wave of interest in structural monitoring lasted up to the late 1980s. It was recognised that the techniques at that time were only capable of detecting major damage and therefore the interest in this methodology diminished. In more recent years the techniques and analysis methods to identify damage have improved, and the need for online monitoring has become greater, particularly for ageing structures, leading to a significant increase in its use.

5.1.3   Requirements for Monitoring in Standards

A number of offshore standards and recommended practices make reference to structural condition monitoring. These include ISO 16587 (ISO 2004), ISO 19902 (2007), API RP2-SIM (API 2014), NORSOK N-005 (Standard Norge 2017) and DNV-OSS-102 (DNV 2010). The references indicate that structural condition monitoring in most cases can be used as an addition to normal inspection.

In API RP-2FSIM the monitoring of floating systems is reviewed and could involve purpose-built systems such as:

- tension monitoring;
- inclinometers and underwater video;
- environmental monitoring (wind, wave and current);
- motion monitoring (heave, roll and accelerations);
- measurement of rotation of turret;
- vibration monitoring; and
- global or local stress monitoring.

Given the recognised importance of structural monitoring, particularly for ageing structures, there is a minimal level of reference to it in standards and recommended practices. The few mentions of structural monitoring are discussed below.

Air gap measurement is included in ISO 19902 (ISO 2007) in the section on in-service inspection and structural integrity management. The commentary lists a range of inspection methods and on measuring the air gap states: "Where air gap measurement devices are correctly set up, calibrated and maintained, continuous records of wave heights and tide can provide very useful information on environmental conditions. Where this can be combined with directionality data and ideally some method of estimating actions (e.g. strain gauges), the data can be used in analyses and assessment of defects and of remaining life, possibly reducing conservatisms. Satellite surveying techniques can often be used to determine levels".

For fatigue sensitive joints API RP2-SIM (API 2014) indicates that the monitoring of these joints and reported crack-like indications may be an acceptable alternative to analytical verification.

ISO 16587 (ISO 2004) describes the performance parameters for assessing the condition of structures, including types of measurement, factors for setting acceptable performance limits, data acquisition parameters for constructing uniform databases and internationally accepted measurement guidance (e.g. terminology, transducer calibration, transducer mounting and approved transfer function techniques).

NORSOK N-005 (Standard Norge 2017) includes in-service inspection methods. In the section on selection of methods it is stated that monitoring can be used as a supplement to conventional inspection methods, particularly for areas with limited accessibility. Monitoring may also be used to verify structural behaviour, for example for novel design solutions. Particular examples of applications listed in NORSOK N-005 are:

- strain monitoring of jacket structures, local and global hull structural components, risers and mooring systems;

- accelerometers to measure motion behaviour to detect any anomaly or survey any trends in, for example, natural frequencies;
- vibration sensors on shafts and pumps;
- leak detection to determine leakage into brace-to-column transitions on semi-submersibles, tendons of TLPs, tanks and compartments of floaters;
- load cells to measure tension in tendons and mooring lines; and
- pressure gauges to measure tank and foundation pressure.

The DNV-OSS-102 rules (DNV 2010) contain a section relating to extended fatigue life. It states that a fatigue utilisation index (FUI) should be calculated (for mobile offshore units). An index above 1.0 requires further steps to be applied in relation to thickness measurement, increased inspection regularity and post crack-detection procedures. Included in these steps is the installation of a permanent flood detection system in critical sections.

5.2 Previous Studies on Structural Monitoring Methods

5.2.1 MTD Underwater Inspection of Steel Offshore Installations

In 1989 the Marine Technology Directorate (MTD) undertook a review of underwater inspection of steel offshore installations, including the implementation of a new approach which included the introduction of structural monitoring (MTD 1989). In Section 4.2.5 the background to this review was described.

In terms of structural monitoring there are three significant sections in the report, which are:

- Outline of Structural Monitoring; This details the role of monitoring, in relation to NDE inspection, for maintaining structural integrity. The sensitivity of the process is discussed, together with planned coverage of a platform and how the technique might be used in practice (continuous or occasional use).
- Monitoring methods; This section covers vibration-monitoring methods, acoustic emission methods, fibre-optic crack monitoring, performance-based monitoring, crack propagation and flooded member monitoring. In addition, the components of monitoring systems are reviewed, such as data acquisition, transmission and processing. Examples are given of the mode shapes expected from vibration monitoring, together with expected accuracies of measuring frequency changes (1% for frequencies, 10% for deformed shapes). Results from monitoring a North Sea platform with a severed member were included, showing both the predicted and measured frequency changes, with generally good agreement between these.
- Design and Fabrication to incorporate Structural Monitoring; This short section outlined the requirements to fit equipment for monitoring during fabrication, including techniques such as vibration analysis, flexibility monitoring, acoustic emission, strain and foundation monitoring.

5.2.2 HSE Review of Structural Monitoring

HSE commissioned a review of structural monitoring of fixed steel structures with Fugro which was published in 1998 (HSE 1998) covering the previous 10 years of operating experience. It reviewed previous work, mainly sponsored by the UK Department of Energy. The main method reviewed in this report is based on monitoring of natural frequencies which are excited by waves.

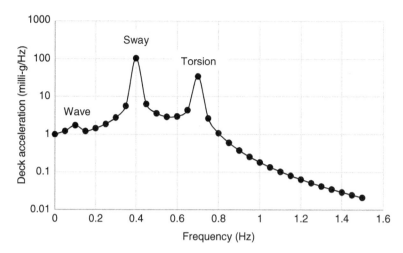

Figure 51 Typical fixed steel platform response spectrum. *Source:* Based on HSE (1998). Offshore Technology Report OTO 97 040 Review of structural monitoring Prepared by Fugro for the Health and Safety Executive (HSE), HSE Books, London, UK.

These include north-south sway, east-west sway and a torsion mode as shown in Figure 51. Platform natural frequencies are normally measured using sensitive accelerometers.

A key input was an assessment of the stability of natural frequencies which form the basis for the monitoring. It was noted that there were natural frequency variations over a year which were typically 1.1% for north-south sway, 1.2% for east-west sway and 0.8% for the ratio of these. These variations in natural frequencies limit the ability of the monitoring method to detect damage as any change in frequency due to minor structural changes could be masked.

Previous studies were analysed to note the reductions in frequency for member severance, as shown in Table 33. For the larger Ninian Northern platform, the change analysed for member severance was only 0.5–1.8%, within the observed natural frequency variation. For the small West Sole platform, the changes were much larger, up to 23% in torsion for member severance in the lower level main diagonal.

The report also reviewed technological developments of the equipment for structural monitoring and concluded that the method was a low-cost and low-labour activity. It was also noted that extension of platform life was becoming an important area which could benefit from structural monitoring. The report also reviewed developments in structural analysis which are relevant to analysis of the effects of various levels of damage on a structure. In addition, the report reviewed data collection and processing and noted that improvements in signal processing were making monitoring techniques more useful.

In Figure 52 the natural frequency variation is indicated with a band within which it would be difficult to observe a structural change. The effects of member severance on the north-south frequency are also shown indicating a clearly detectable reduction in frequency.

The review concludes that a change of at least 3% in natural frequency is needed to enable damage to be confidently detected, which is likely to be the case for low redundancy structures (e.g. West Sole). Hence, jackets with low redundancy configurations are more amenable to monitoring, whilst for a redundant structure such as Forties A, two members or more need to fail in the same frame before the failure could be detected.

For low redundancy, structural monitoring may be essential to give a warning of a member severance as these structures will often not be structurally sound with such damage present. In

Table 33 Sensitivity of Natural Frequencies to Members Severance Based on Analysis. *Source:* HSE (1998). Offshore Technology Report OTO 97 040 Review of structural monitoring. Prepared by Fugro for the Health and Safety Executive (HSE), HSE Books, London, UK.

Platform and description	Member severed	Reduction in frequency	
		Sway	Torsion
West Sole WE	Main diagonal mid-level	3%	10%
Southern UK Basin	Main diagonal lower level	12%	23%
4-legged K-braced jacket	Horizontal	1%	3%
3 bays			
Forties A	One-member failure	1–2% in at least one fundamental mode and 7–16% in higher natural frequencies	
Central UK North Sea			
4-legged X-framed jacket			
6 bays			
	Two-member failure	1.5% in at least two fundamental modes and more than 4.5% in some cases	
	Deck mass changes	Up to 3%	
Ninian Northern	Upper level braces	Below 0.5% for X-braces and between 0.5 and 1.8 for diagonal braces	
Northern UK North Sea			
8-legged diagonal and X-braced jacket			
7 bays	Horizontal braces	No change in frequency	
Ninian Southern	Main diagonal	9.5–11.5%	
Northern UK North Sea	Main horizontal	2.5–4%	
4-legged K-framed jacket			
7 bays			

contrast, for high-redundancy structures, a single member severance is unlikely to be detected but will also not be likely to reduce structural capacity to a level which is unsafe.

5.2.3 HSE Updated Review of Structural Monitoring

The purpose of the HSE RR685 study on structural integrity monitoring (HSE 2009) was to identify current monitoring capabilities for offshore structures at the time. The review included techniques for fixed steel structures (jackets) and semi-submersible structures but excluded ship-shaped floating structures (FPSO). This review identified limitations in approaches and identified areas for further development and opportunities for technology transfer from other industries.

The key structural integrity monitoring methods relevant to offshore structures were reviewed. The review found that offshore experience of structural integrity monitoring was limited to date and that all systems were for bespoke applications. However, several of these techniques are in more regular use at the time of writing this book, such as air gap monitoring. An overview of several different uses of structural monitoring included in this review is given in Table 34.

The review further indicated that codes and standards referred in a limited way to the use of structural integrity monitoring and this may have influenced the limited take up to date of this

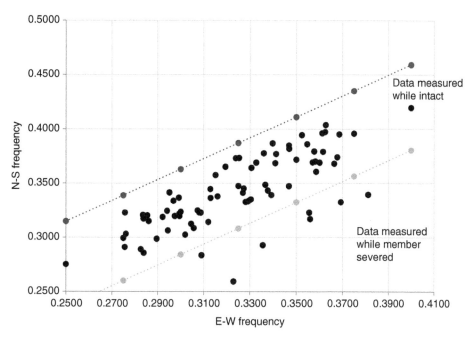

Figure 52   Effects of member severance on NS natural frequency. *Source:* Based on HSE (1998). Offshore Technology Report OTO 97 040 Review of structural monitoring. Prepared by Fugro for the Health and Safety Executive (HSE), HSE Books, London, UK.

Table 34   Examples of the Use of Structural Monitoring Methods (HSE 2006).

Location	Issue	Technology	Outcome
Ninian southern	Fatigue crack growing at one end of an access window	Acoustic emission	Three-year monitoring period
Forties A, Heather and Piper Alpha	These studies used accelerometers both subsea and at deck level to develop the technology	Natural frequency monitoring	Understanding of the senitivity of the technique to levels of damage
Ninian Southern	The system was installed on Ninian Southern following a member severance	Natural frequency monitoring	Following repair the platform's frequency was detected as increasing
FOINAVON Umbilical Monitoring System	Vortex-induced vibrations (VIV) occurred on the Foinavon flexible risers at up to 600m depth The problem was to measure the bending and tensile stress at the vessel	Natural frequency monitoring	It was concluded that VIV can be reasonably modelled—both current and vessel motions—to reduce VIV

technology. However, the review stated that there was a regulatory requirement for leak detection in ageing semi-subs which was introduced by DNV in 2007 in their classification rules DNV-OSS-101 (DNV 2007).

Structural integrity monitoring methods that were reviewed and discussed included (HSE 2009):

- acoustic emission monitoring;
- leak detection;
- air gap monitoring;
- global positioning system monitoring;
- fatigue gauges;
- mooring chain monitoring;
- continuous flooded member detection monitoring;
- natural frequency response monitoring;
- acoustic fingerprinting; and
- strain monitoring.

A detailed description of these techniques as given in the review is included in Section 5.3.

The review concluded that it was clear that the offshore industry had implemented several techniques during the previous decades in order to continuously monitor the integrity of offshore structures. As such, to that date continuous structural monitoring had not been widely embraced by the industry. However, as structures were increasingly called upon to operate beyond their original design life, continuous structural monitoring could have an important role to play in providing a back-up system to conventional inspection.

There existed a wide range of available monitoring systems, from those which concentrated on detecting global damage to those that concentrated on detecting the onset of local crack growth. Each of these had strengths and weaknesses that would make them more suited to certain circumstances. However, there was little guidance at that time on when and how the use of continuous monitoring techniques should be recommended.

5.2.4 SIMoNET

Structural Integrity Monitoring Network (SIMoNET) was a joint venture between industry and government organisations initiated in 1997 and managed by Cranfield University and University College London. The aim was to facilitate communication on structural monitoring between researchers and those interested in the application of the method. The main activity was a web-based source (Simonet 2020) and the holding of up to two seminars each year to promote knowledge on the method across all relevant industry sectors including offshore. The main objectives were to:

- establish the appropriateness of the economic and technical aspects of monitoring techniques for the management of the structural integrity of large structures and plant safety;
- identify the latest available technology, experience in use and methods of retrieving and interpreting data;
- encourage harmonisation of data storage and retrieval, by identifying incompatibility within systems; and
- propose general guidelines for the practice and application of structural monitoring, and to identify priorities for further development.

At the time of writing this book the website provides details of 18 seminars, with the last one undertaken in 2010. Many of these have useful information on applications of structural monitoring in several industry sectors including offshore; several case studies were also included. The website also provides links to a number of relevant articles.

5.3 Structural Monitoring Techniques

5.3.1 Introduction

The offshore industry has, as already mentioned, implemented several techniques during the last few decades to continuously monitor the condition of offshore structures. There is good evidence that the existence of damage on a structure and in some cases the location of this damage can be detected by the latest structural monitoring methods. The most advanced monitoring methods combined with improved analysis methods have shown to be capable to detect both the type of damage and its severity (May et al. 2008).

As noted in Ersdal et al. (2019), where there is uncertainty about the condition of an ageing structure, continuous structural monitoring provides an opportunity to decrease this uncertainty more than could be achieved using periodic inspection. This requires that the most probable and critical failure modes of the structure are well understood and that the structural monitoring system is designed to be capable of identifying such failures.

5.3.2 Acoustic Emission Technique

The principle of acoustic emission (AE) monitoring techniques is to use an arrangement of sensors to detect characteristic patterns which arise from acoustic stress waves that occur when a material undergoes changes in its internal structure. These changes could signal the presence of anomalies locally in the structure. AE has applications to offshore structures and has been used for the monitoring of damage mechanisms such as cracks. The system can be used in locations which are known to be at a high likelihood of fatigue cracking and where normal inspection is difficult, unreliable or costly.

The system can effectively monitor for crack initiation (with pre-installed sensors at an expected crack site) and crack growth. The technique can be used in conjunction with strain gauges to correlate AE signals with structural stress levels.

AE provides real-time information on fatigue crack initiation and growth. It can be used to detect fatigue cracks at an early stage before conventional NDT methods would be able to detect them. However, due to the attenuation of the AE signal, this method is only suitable for local monitoring (i.e. over a few metres in the structure). AE sensors need to be located close to the crack site which could be externally on an offshore jacket or internally to the structure in, for example, dry semi-submersible hull members.

AE systems have been under development for some time and have subsequently been used for a long period in the offshore industry, and the technique is generally regarded as being suitable for the structural monitoring of safety-critical members.

AE systems are generally considered to be reliable with a high probability of detection. When there is a large amount of background noise, problems may arise that cannot be effectively filtered out. This noise can mask the presence of activity due to defects. The system relies on interpretation of the data by an operator to enable the structural anomaly to be detected.

An example of offshore use is the monitoring of cracks on the Ninian Southern platform in the period 1989–92 (Mitchell and Rogers 1992). The decision to use AE monitoring was taken following the severance of a horizontal brace as a result of a fatigue crack growing at one end of an access window. Detailed NDE inspection of other access windows had shown a large number of similar defects of various sizes. Several were repaired using saturation diving. It was decided to monitor some of the remaining ones by using conventional NDE methods, such as ultrasonics. One particular unrepaired defect had a predicted remaining life of only 5 years and it was decided to monitor this using AE methods. It was found that measured defect sizes from the AE results were in good agreement with previous data from other NDE methods. Over the three-year monitoring period, defect growth was found to be minimal, and it was concluded that a repair was not necessary at that stage.

5.3.3 Leak Detection

Leak detection involves a sensor to detect water ingress in dry compartments and structural members and then raises an alarm via an audible or visual unit. Leak detection is particularly important for managing watertight integrity in floating structures, for which regulators and classification societies to a large extent require an approved leak detection system.

Several detection methods have been used ranging from simple buoyancy-based systems, pipe systems to monitor the presence of water in compartments to specialist systems involving inundation alarms to field effect electronic cells that detect slight disruptions caused by the presence of water. Such systems have been used for a number of decades in the offshore industry for ship-shaped structures and semi-submersible platforms. A secondary system such as a CCTV system can be added to identify the cause of the water ingress in, for example, breach of hull integrity.

5.3.4 Global Positioning Systems and Radar

The methodology for calculating the position, tilt and air gap (subsidence) has changed considerably with current use of GPS or downward looking radar, enabling repeatability and reducing inaccuracies between surveys.

The position of a floating structure (station keeping) can be monitored using a Global Positioning System (GPS). The basic positioning GPS technology is mature and has real-time accuracy in the region of 3–10 m horizontally.

To determine a reliable GPS reading of the position on a fixed structure, the equipment is usually left in place for 24 hrs. A value for the air gap and tilt can be calculated from measured values which are recorded and averaged over a given time period. A reference point is established from early readings which will form the basis for comparison with all repeat readings. For such a fixed location, the accuracy can be within millimetres horizontally. The general rule of thumb is that the vertical error is 3 times the horizontal error.

Monitoring of tilt and subsidence of fixed offshore structure is important, and this technique is one of the most important methods for these. In addition, moored floating platforms can experience severe mooring loads in harsh sea conditions. A combined mooring and thruster positioning system can be used to minimise these loads which rely on a thruster system being positioned accurately by the GPS signals.

The air gap is defined as the positive difference between the lowest point of the underside of the cellar deck and the crest height of an extreme wave for a given return period (often 100 or 10,000 years). Reduction in the air gap can lead to wave impact on the lower deck sections with structural implications. The purpose of monitoring the air gap is to establish whether the platform foundation is suffering from subsidence which is an important input along with the maximum wave crest height to establish the air gap.

5.3.5 Fatigue Gauge

A fatigue gauge is a method for measuring fatigue damage from a period of cyclic loading of a welded component in a structure. One such method is TWI's CrackFirst™ which was patented in 1990 and is currently available from Strainstall. The fatigue gauge sensor consists of a thin steel shim which is attached to the structure near the fatigue sensitive area. A fatigue crack will then grow in the gauge under cyclic loading. The gauge should be calibrated so that the fatigue crack growth measured in the gauge relates to the amount of fatigue damage in the detail under observation. This could provide a representative record of cumulative fatigue damage occurring on a structure under variable amplitude loading. There is limited information to date on the reliability of this method (HSE 2009). However, at the time of writing this book the CrackFirst™ gauge is in the process of being further developed to be fit for marine use (Wintle 2020).

5.3.6 Continuous Flooded Member Detection

Continuous flooded member detection can be used both internally and externally to detect internal flooding, typically for fixed steel structures. Internally a device is placed inside a nominally dry member, and if the member becomes flooded, a warning signal is transmitted to a central recording unit. The preferred method (HSE 2009) is where the salinity of the solution activates a galvanic cell which generates an acoustic signal. Internal continuous flooded member detection can only realistically be installed during the construction phase to newly built structures. Leak detection on floating structures as mentioned above is essentially a variant of this internal method.

For external use, a device is clamped to the outside of a nominally dry member. The device transmits a high-frequency acoustic signal through the member (similar to early flooded member detection) and detects any changes to the signal and hence the presence of water inside the member. The device could be set up to send an alarm signal to the control room.

Continuous flooded member detection is an untried method in the offshore industry (HSE 2009). One of the controlling limitations for external use is battery life. However, periodic flooded member detection used to detect through-thickness cracking continues to be in regular use in the offshore industry; see Section 4.3.6.

5.3.7 Natural Frequency Monitoring

Natural frequency response monitoring is the instrumented observation of a structure using sensors that continuously or periodically measure structural behaviour to identify changes such as material degradation, geometric properties and boundary conditions of the structure (e.g. foundation stiffness). This technique can continuously monitor for a change in platform frequency and then identify the significant damage such as failed members. It can be used to detect changes due to incidents such as an extreme storm, an earthquake, a ship collision or a dropped object.

Accelerometers are normally placed on the topsides of an installation, although subsea accelerometers have also been used. The response of a structure to wave loading is continuously monitored, and the measured accelerations are transformed into the frequency domain which can identify the natural modes of the structure. Any significant structural damage to the platform will result in a change in the stiffness in one or more modes, thereby reducing its natural frequency, which can be detected. Examples of such damage include:

- damage to a fixed steel platform member altering the frequency in the sway direction where the member contributes to the stiffness;
- changes in soil stiffness reflected by changes to frequencies in both platform directions; and
- failure of pile sleeve grouted connections similarly changing stiffness and frequency.

Finite element analysis is often performed alongside the measurement to help in identifying which mode shape corresponds to which frequency. This information is of value when a change in frequency is reported, as it is then possible to identify which member is the most likely to have failed.

As noted earlier, the method is only sensitive enough to detect significant changes in frequency because of the background variations, meaning that it cannot detect minor damage such as small defects in redundant structural members. However, the most likely damage to be detected is the severance or significant damage to members that result in a large reduction in global structural strength. Members which cause small changes in frequency on severance or significant damage tend not to be structurally significant.

A Joint Industry Project (JIP) in which HSE participated showed that in low redundancy jackets with single diagonal bracing systems, an 80% circumferential through-wall defect is detectable using this method (EQE 2000). The method is less useful for heavier jackets with high redundancy bracing schemes.

The technique has been in operation since the late 1970s on a number of platforms. A system was installed on Ninian South following a member severance (Mitchel and Rodgers 1992).

5.3.8 Strain Monitoring

Strain monitoring is carried out to measure a change in strain (and hence the stress) as a result of a change in loading on the structure. There are a number of well-established strain measurement techniques available including conventional strain gauging, fibre optics and stress probes. It is widely applied across many industries including the offshore industry. Calibration and appropriate surface preparation are required.

Fibre optic methods have been developed to create a strain sensor where the optical fibre is inscribed during production with a so-called Fibre Bragg Grating (FBG), with a length of typically 5 mm, which reflects the laser light differently from the rest of the fibre. When the optical fibre is attached to a material, it will be strained along with the material. The reflected light takes a little longer or shorter time to travel back when the FBG is under strain, which can be accurately measured. In addition, the wavelength of the reflected light is changed due to the grating. This enables a fibre to have several different FBGs enabling strain to be measured in different locations using the same fibre. FBGs optical fibre sensors are very susceptible to temperature, as the fibre expands when the temperature rises and contracts when the temperature drops. The refractive index changes as well. These changes require compensation such as installing a temperature sensor next to the strain sensor.

For all methods, the measured strain will in turn allow an analysis of the mechanical stress in the material, which is the aim of most strain measurements. It can also be used to verify design calculations, for example, associated with wave loading. It is particularly useful to gain information about the long-term stress range distribution needed for fatigue calculations. It is sensitive to the positioning of the gauges in areas where there are steep stress gradients. Installation of strain monitoring systems in an existing structure is rather cumbersome and expensive.

Strain monitoring and load sensors may be used to provide relevant information on the tension of mooring lines; see more in Section 4.9.

5.3.9   Riser and Anchor Chain Monitoring

The purpose of this technique is to monitor the position of risers and anchor chains and alert the operator to any displacements outside predefined limits. Systems for riser and anchor chain monitoring consist of a sonar array located beneath the platform, which emits signals in and around the horizontal plane. The system then detects and analyses reflected signals from items that are within the monitoring region. This data can then be used to determine the accurate location of these items (mooring lines and risers) and to continuously monitor their position. If there is a problem, the system will alert the operator via an audio or visual alarm.

5.3.10   Acoustic Fingerprinting

Acoustic fingerprinting involves the transmission of acoustic signals into the structure at locations above the water level. The method is to the authors' knowledge so far only been tried on a representative jacket structure. In practice, receivers would be placed above the water level and monitor the signal that is returned, measuring any changes which could indicate damage. The method depends on the complex behaviour of the acoustic signals passing through the steel, with reflections and refractions occurring at section changes and nodes. The technique relies upon measuring any changes between the transmitted signal and the received signal. The technique is at early development stage, and research has been carried out by using time-of-flight computation on the received signals to diagnose failure in a laboratory-based structure using a polycarbonate material to simulate steel. (HSE 2005). If the technique were to be proven, it is likely that it would enable continuous structural monitoring but would only be sensitive enough to detect fully severed members.

5.3.11   Monitoring with Guided Waves

For inspection of pipes or large areas of plate-like structures, ultrasonic waves can be employed to detect changes in the specimen's thickness. Guided ultrasonic wave testing utilises a lower frequency region (typically several hundred kHz) than standard ultrasonic testing (see Section 4.3). The excited wave can propagate over large distances and pass through the thickness of a structure. Corrosion damage can be detected by use of this this technique. Successful measurements have been performed on pipelines where propagation distances of several hundred metres have been achieved. For the monitoring of structural integrity, a permanently attached guided ultrasonic wave array type is required. The array consists of a ring of piezoelectric transducers for excitation and reception of the guided wave. Most of the experience to date has been with pipelines, although there is the potential for use on plated structures (see e.g. www.simonet.org.uk).

5.4   Structural Monitoring Case Study

A four-legged jacket installed in 150 m water depth in the North Sea with inverted K brace structure was used as an example for a monitoring trial (Killbourn 2008). The K-bracing together with it being a four-legged structure implied that it had low redundancy. The platform was ageing and had previously suffered member failures. The installed monitoring system comprised eight Fugro accelerometers located in four horizontal bi-axial pairs at each corner of the topside deck. The signals were transmitted through cabling to a central computer and stored. In addition, a wave radar unit was installed to record wave profiles and tidal changes. The information from the radar unit was synchronised with the data from the accelerometers.

A routine inspection of the platform using FMD showed that one of the horizontal brace members was flooded indicating a through-thickness crack. Detailed visual inspection showed that there was a crack with a length of 10% of the brace circumference. Following inspections showed that the crack was growing and as a result, holes were drilled through the brace wall at each end of the crack, which proved unsuccessful in stopping the crack growing. Therefore, severance of the brace was expected.

Approximately six months after the detection of the crack and in a moderate winter storm, the monitoring system detected a significant change to the response characteristics of the jacket. Checking the on-board monitoring system indicated that there had been a member failure. The plot in Figure 53 shows a significant decrease in the north-south sway frequency below a level set as an alarm level. There was little change in the east-west sway frequency, but the torsion frequency also showed a significant drop. The alarm level is set at just below the normal day-to-day variation in natural frequencies. The changes in the response characteristic were analysed to identify potential braces that could have failed. The jacket member that best fitted the results was the brace previously identified by FMD as being flooded. A subsequent visual examination by an ROV confirmed that the cracked brace had severed.

Repair of this severed brace was required. To allow welding of the brace, a pressurised habitat was constructed over the severed section of the brace. A portion of the brace was removed at either side of the severance, allowing a pup piece to be inserted and welded to one side of the brace. Four axial tension rods were welded to the brace and the pup piece allowing the original tension in the brace to be restored, once welding of the pup piece to the other end of the original brace was achieved. The internal volume of the brace was emptied of water before being welded up allowing FMD to be used for subsequent monitoring.

The monitoring system was operational during the repair, and the results following the repair indicated that the observed reductions in the characteristic responses in the north-south and torsion modes following the severance were restored by the repair.

The case study demonstrated the effectiveness of vibrational structural monitoring for detecting severance of a brace enabling a rapid response. However, at the time of writing, the industry has not taken up the use of monitoring to a great extent, which could provide a useful warning system for structural damage.

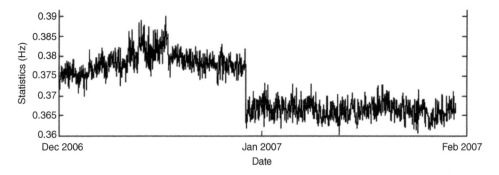

Figure 53 Change in natural frequencies at, for example, a member failure. *Source:* HSE (2009), Health and Safety Executive research report RR685 Structural integrity monitoring—Review and appraisal of current technologies for offshore applications, Prepared by Atkins Limited, HSE Books, 2009. Licensed under Open Government License.

5.5 Summary on Structural Monitoring

Structural monitoring in most cases is a useful adjunct to conventional inspection to provide forewarning of significant changes in the structural condition. Furthermore, structural monitoring can complement existing inspection techniques to provide more confidence in structural integrity, thus potentially reducing inspection frequency and therefore inspection costs.

The first wave of interest in structural monitoring, as discussed in Section 5.1.2, lasted up to the late 1980s and was aimed at creating a UK capability in this field. However, since the techniques at that time were only capable of detecting major damage, interest in further commercial development of this technology waned.

In 2008 it was reported (May et al. 2008) that monitoring was installed on at least twenty platforms world-wide, with half in European waters and the remainder between Mexico, Canada, the Caspian Sea and West Africa. In addition, the company Pulse has indicated use of structural monitoring for the ageing structures in Malaysian waters (https://www.pulse-monitoring.com/), being used both to monitor the condition of the structure and to calibrate design models for future use. Monitoring has included strain, wave height, currents and platform motions. May et al. (2008) also added that a further eight systems had been ordered at that time. More recently, the techniques and methods to analyse the data (like system identification and machine learning algorithms) have improved. These improvements have coincided with a greater need for online monitoring, particularly for ageing structures and this has led to an increase in its use.

At the time of writing, so-called structural health monitoring (SHM) is in widespread use in many industry sectors for monitoring structural integrity. This has led to further development of technologies and the use of digital information methods including such as Internet of Things, artificial intelligence, robotics, swarm robotics and digital twins, which are applicable to offshore use. These emerging technologies are further discussed in Chapter 9.

Table 35 provides an overview of the capabilities of the different techniques that are available, which have been discussed in this chapter; it is based on HSE RR685 (HSE 2009).

Table 35 Structural Condition Monitoring Techniques.

Technique	Example techniques	Monitoring capability	Maturity	Limitations
Air gap monitoring	GPS, laser	Loss of air gap	Widespread use	Requires specialist contractor to carry out measurements
Positioning system	GPS	Loss of station keeping Loss of air gap	Widespread use	Few
Acoustic emission	Acoustic stress wave monitoring devices	Fatigue crack growth Corrosion	Has been used for specific applications, not as a general monitoring method	Requires specialist installation with sensors hard-wired back to a control unit Data can be difficult to interpret. Background noise, cabling

(Continued)

Table 35 (Continued)

Technique	Example techniques	Monitoring capability	Maturity	Limitations
Leak detection	Pipe systems to monitor the presence of water in compartments or buoyancy-based leak detection	Breach of water-tight integrity Member leak detection Through-thickness cracks Through-thickness corrosion	In regular use in floating structures	May be affected by condensation or humid environment and necessary pipes often corrode
Continuous flooded member detection	Electrical detector cells activated by water (internal) Strapping acoustic sensors to a member (external)	Member leak detection	Unproven	Internal detectors are only suitable for new build Battery life is a key limitation for external systems Localised detection only
Natural frequency response monitoring	Accelerometers	Member severance Significant damage to members and joints Damping	Mature technique, which has been in operation for many years on a number of platforms	More effective in detecting damage in low redundancy structures Limited information on heavily redundant structures
Riser and anchor chain monitoring	Load cells, natural frequency monitoring, angle of mooring from fairlead	Loss of mooring line tension	This technique is widely used in the offshore industry	It may not always be feasible to locate a system in certain offshore environments
Fatigue	Fatigue gauges	Fatigue cracking	Regular use in some industries but limited in the offshore industry	Attachment to the structure (close to the weld toe) is often difficult particularly underwater The sensor may prove vulnerable to the offshore environment
Acoustic finger printing	Techniques are under development but not in use	Through-thickness cracks	Untested on real structures	Only likely to detect fully severed members
Strain monitoring	Strain gauges	Local stresses and loading	Widely applied across many industries including the offshore industry	Appropriate surface preparation required. Installation of gauges on existing structures can be difficult Equipment can be sensitive to damage by the environment especially cabling

5.6 Bibliographic Notes

Section 5.3 is to a large extent based on HSE (2009), where John Sharp was one of the co-authors.

References

API (2014), *API RP-2SIM Recommended Practice for Structural Integrity Management of Fixed Offshore Structures*, American Petroleum Institute, 2014.

Department of Energy (1977a), "Offshore Research Focus, No.3", 1977.

Department of Energy (1977b), "Offshore Research Focus, No.5", 1977.

Department of Energy (1979), "Offshore Research Focus, No. 13", June 1979.

Department of Energy (1980), "Offshore Research Focus, No. 17", 1980.

Department of Energy (1982), "Offshore Research Focus, No.34", 1982.

Department of Energy (1986), "Offshore Research Focus, No. 51", 1986.

Department of Energy (1987), "United Kingdom Offshore Research Project - Phase II (UKOSRP II) Summary Report", HMSO, OTH-87-265, 1987.

DNV (2007), "DNV Offshore Service Specification DNV-OSS-101 Rules for Classification of Offshore Drilling and Support Units, Special Provisions for Ageing Mobile Offshore and Self-Elevating Structures, 2007", Høvik. Norway.

DNV (2010), "DNV Offshore Service Specification DNV-OSS-102 Rules for Classification of Floating Production, Storage and Loading Units," October 2010, Høvik, Norway.

Ersdal, G., Sharp, J.V., Stacey, A. (2019), *Ageing and Life Extension of Offshore Structures*, John Wiley and Sons Ltd., Chichester, West Sussex, UK.

EQE (2000), "JIP Guidance on the Use of Flooded Member Detection for Assuring the Integrity of Offshore Platform Structures", EQE Report N°179-03-R-07, Issue 1, 6th June 2000.

HSE (1998), *Offshore Technology Report OTO 97 040 Review of Structural Monitoring*, Prepared by Fugro for the Health and Safety Executive (HSE), HSE Books, London, UK.

HSE (2005), *Cost Effective Structural Monitoring, Research Report RR326*, Prepared by Mecon Ltd for HSE, HSE Books, London, UK.

HSE (2009), *Health and Safety Executive Research Report RR685 Structural Integrity Monitoring—Review and Appraisal of Current Technologies for Offshore Applications*, Prepared by Atkins Limited, HSE Books, 2009.

ISO (2004), "ISO 16587:2004 Mechanical Vibration and Shock—Performance Parameters for Condition Monitoring of Structures", International Standardisation Organisation, 2004.

ISO (2007), "ISO 19902:2007 Petroleum and Natural Gas Industries—Fixed Steel Offshore Structures", International Standardisation Organisation, 1998.

Killbourn, S.D (2008), "Detection of Jacket Member Severance Using Structural Instrumentation", *OMAE* 2008 57391.

May P., Sanderson D., Sharp J.V., Stacey A. (2008), "Structural Integrity Monitoring—Review and Appraisal of Current Technologies for Offshore Applications", OMAE2008-57425, OMAE Conference 2008, Estoril, Portugal.

Mitchell, J.S. and Rogers, L.M. (1992), "Monitoring Structural Integrity of North Sea Production Platforms by Acoustic Emission," Offshore Technology Conference, OTC 6957, 1992.

MTD (1989), "Underwater Inspection of Steel Offshore Installations: Implementation of a New Approach", Report number 89/104. MTD, London, UK.

Simonet (2020), www.Simonet.org.uk, accessed 19 July 2020.

Standard Norge (2017), "NORSOK N-005 In-Service Integrity Management of Structures and Maritime Systems", Edition 2, 2017, Standard Norge, Lysaker, Norway.

Wintle, J. (2020). Private communication.

6

Inspection Planning, Programme and Data Management

To the optimist, the glass is half full. To the pessimist, the glass is half empty.
To the engineer, the glass is twice as big as it needs to be.

—Unknown

Give me six hours to chop down a tree and I will spend the first four sharpening the axe.

—Abraham Lincoln

6.1 Introduction

6.1.1 General

The long-term planning[1] of inspection, the inspection programme and inspection data management are key elements in structural integrity management. In essence, a survey is performed with the aim of knowing all the different changes that have occurred to a structure after it was built, storing these change data in a suitable system for easy retrieval and as a result, achieve an overview of the as-is state of the structure (see Figure 54). Inspection is a part of this survey process, primarily determining the current structural condition and configuration and hence any changes that have occurred since fabrication or previous in-service inspections.

Underwater inspection is often the most time-consuming and costly part of the survey process and has safety implications, particularly if divers are used. Hence, it is usually impracticable and often unreasonable to inspect all members, components and areas of a structure. As a result, the selection of inspection locations, methods, their deployment and the scheduling of inspections are key factors in planning an effective and cost-efficient inspection programme that provides the necessary background information to evaluate the safety of a structure.

The layout of the structure is determined in the design process and has implications for the ability to inspect and repair while in service. Unfortunately, this is not given the priority it deserves. As a

---

1 The long-term inspection plan is often called strategy, inspection strategy, surveillance strategy or even integrity strategy. The word strategy in these contexts imply a high-level long-term plan for how to gather the necessary information about what is needed to be known about the structure in order to make sure that the structure is sufficiently safe. Hence, in this book it is decided to use the term long-term inspection planning to be representative of the above.

Underwater Inspection and Repair for Offshore Structures, First Edition.
John V. Sharp and Gerhard Ersdal.
© 2021 John Wiley & Sons Ltd. Published 2021 by John Wiley & Sons Ltd.

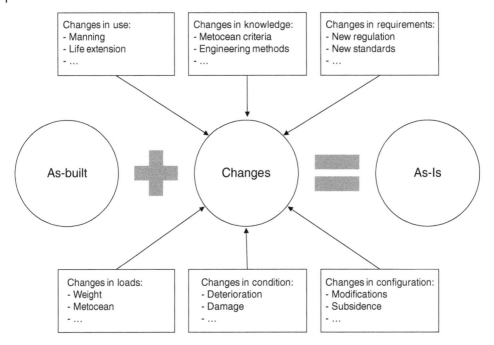

Figure 54 Transition from the as-built to the as-is condition of the structure.

result, the condition of the parts that are difficult or impossible to inspect need to be assessed by other means. An assessment of fatigue cracking in these cases can utilize inspection results from similar or nearby components with significantly lower fatigue life in the structure; see for example, NORSOK N-006 (Standard Norge 2015). The resulting layout from the design process will also determine the accessibility for ROV inspection and failure to properly plan for such may also leave parts of the structure uninspected or possibly require the use of divers to perform the inspection. These issues highlight the importance of bringing in inspection awareness at the design stage of a new structure and the life of a structure with uninspectable and unrepairable components will often be limited (Ersdal et al. 2019).

6.1.2 Long-Term Inspection Plan

The long-term inspection plan should aim to:

- detect in good time any change that may cause unacceptable structural safety, risk to personnel, risk to the environment, interruption in operation or—if these changes are left unattended—a more significant damage and more costly repairs;
- provide information that enables the structural integrity management to be planned on a rational basis in a systematic manner;
- ensure that information obtained about anomalies and changes relevant for the structure is reliable—this will primarily require that inspection is undertaken by suitably experienced and competent staff using appropriate qualified inspection methods and procedures; and
- determine how to obtain information that can be used to indicate the condition of items that are uninspectable.

The reliability of the selected inspection method as mentioned above needs to be taken into account in developing the long-term plan. This reliability is often expressed in terms of a probability of detection (PoD) curve or value as discussed in Section 4.6.

Further, the long-term inspection plan should specify:

- where and the extent to inspect (depending on the approach to inspection planning, e.g. areas of high likelihood of failure and large consequence of failure);
- what type of change is to be detected (e.g. crack or corrosion);
- when inspections should be performed (e.g. specific year, time-interval, frequency);
- how to inspect (the appropriate inspection method and procedure for the given location and the type of change to be detected);
- who should inspect (e.g. diver, ROV-controller) and the required personnel qualifications and competence.

As already mentioned, detecting any change relevant to the structure in good time can be an overwhelming, time-consuming and expensive task. As a result, the integrity management engineer will seek to develop a plan that achieves sufficient safety at a reasonable cost. With this in mind, the long-term plan should optimise the where, what, when, how and who, as mentioned above.

The long-term inspection plan should be based on an evaluation (see Chapter 7) of all available data about the structure, including the likelihood and consequence of failure. The long-term plan must also consider a number of other factors such as:

- regulatory requirements;
- operational requirements (e.g. safety requirements, bed space, other simultaneous activities);
- the known condition of the structure including robustness, damage tolerance, fatigue life, age;
- previous accidental events and inspection findings;
- the number of similar details on the same structure (a representative number may only need inspection or, if defects are found, all similar details need to be inspected);
- known problems on similar types of structures; and
- the availability and effectiveness of inspection techniques (application and reliability).

This book focuses on underwater inspection and as a result, the changes that are discussed are primarily those relevant to the condition of the underwater part of the structure. This also includes changes to features such as scour, seabed condition and marine growth as they are important for the safety and functionality of the whole structure.

An example of a 10-year long-term inspection plan for a fixed steel structure assumed to be installed in 2020 with a baseline inspection performed in the same year, is illustrated in Table 36 covering the splash zone and the underwater structure.

Several international standards, such as ISO 2394 (1998), ISO 19901-9 (ISO 2019) and ISO 19902 (ISO 2007), specify requirements for structural integrity management including the long-term planning of inspections. In addition to these international standards, several regional standards such as API RP-2SIM (API 2014), API RP2-FSIM (API 2019a) and API RP-2MIM (API 2019b) and NORSOK N-005 (Standard Norge 2017) also provide guidance that is relevant.

6.1.3 Approaches for Long-Term Inspection Planning

Initially regulators and operators relied on time or calendar-based inspections, as required by the Certificate of Fitness (CoF) regime in the UK sector and the NPD regulations from 1976. These calendar-based inspection plans indicated a fixed inspection interval, typically annually and up to

Table 36 Example of a Long-Term Inspection Plan for a Jacket-Type Structure.

	Previous inspection			Member description				Ten-year underwater inspection plan									
Component	Year	Inspection method	Anomaly description	Fatigue life	Strength utilisation	Redundancy	Comments including previous events, findings and known problems	2021	2022	2023	2024	2025	2026	2027	2028	2029	2030
Jacket legs	2020	GVI	None	>35y	<1.0	0.5	Water filled				GVI/CP CVI	GVI/CP CVI				GVI/CP CVI	GVI/CP CVI
Diagonal bracing in splash zone	2020	GVI	Coating	>50y	<1.0	0.8					GVI/CP FMD	GVI/CP FMD				GVI/CP FMD	GVI/CP FMD
Diagonal bracing below splash zone	2020	GVI	None	>50y	<1.0	0.8					GVI/CP FMD	GVI/CP FMD				GVI/CP FMD	GVI/CP FMD
Horizontal bracing below splash zone	2020	GVI	None	>50y	<1.0	0.95					GVI/CP FMD	GVI/CP FMD				GVI/CP FMD	GVI/CP FMD
Mudmats, pile sleeve, grouting, etc.	2020	GVI	None	>200y	<1.0	0.7					GVI/CP CVI	GVI/CP CVI				GVI/CP CVI	GVI/CP CVI
Anodes	2020	GVI	None	>20y							GVI					GVI	
Air gap	2020	GVI	None									GPS	GPS	GPS	GPS	GPS	GPS
Marine growth	2020	GVI	None								GVI					GVI	
Debris	2020	GVI	None								GVI					GVI	
Scour	2020	GVI	None								GVI					GVI	

5-year intervals depending on the application. It was, however, regarded as too prescriptive in some cases and not sufficient in other cases. These calendar-based inspections found a certain number of anomalies but little damage or deterioration that had a significant impact on the integrity of these relatively new structures.

This led to a condition-based approach to inspection planning, where the interval between inspections was determined by the state of the structure identified in previous inspections. However, it was found to be difficult to prioritise between inspections of the various details due to the complexity of the structures. As there was a continuing need for the operator to optimise the inspection process and make it more cost-efficient, the criticality of welded details was introduced into the condition-based inspection planning in the 1990s. This led to the probabilistic or risk-based approach, in which inspections were focused on areas of the structure considered to be at the highest risk of failure, where the meaning of risk in this context is the probability and consequence of failure of a member, detail or area of a structure. An example of such a high-risk area of the structure would be a welded detail:

- which is likely to fail (in terms of a probabilistic fatigue analysis or in terms of a short calculated fatigue life relative to the design life); and
- where failure would be critical for the structure (its failure would result in a significant reduction in the strength or robustness of the structure that would reduce its ability to survive an extreme loading event without collapse and loss of life).

At the time of writing this book emerging technologies are being introduced, including the use of robotics in performing inspection together with artificial intelligence and digital twins (see Chapter 9 for more on emerging technologies). These changes in the approach to inspection planning are shown in Figure 55.

Risk-based inspection planning is currently the main approach used in the industry. Several models have been developed, with either a quantitative or qualitative assessment of the probability and the consequences of failure of a specific component in the structure. However, many uses of these inspection planning approaches will be somewhere in the middle between quantitative and qualitative.

An example of a *qualitative approach* is the use of risk matrices. A risk matrix is produced to highlight the components with the highest risk scores based on: (1) a qualitative evaluation of the consequence if this component should fail and (2) the probability of failure for this component. Based on this risk score the components are prioritised for inspection. An example of a typical 5 × 5 risk matrix for risk-based inspection planning is shown in Figure 56. Other configurations are possible and larger matrices are known to be used by the offshore industry to provide a more refined risk assessment.

Severity levels A–E in Figure 56 are typically determined based on qualitative or quantitative evaluation of the consequence of failure of the member. Severity level A typically indicates very low consequence of failure of the member, while severity level E would typically indicate that the structure would collapse if the member, component or area of the structure should fail. Accumulated probability would similarly be determined by a qualitative evaluation indicating how likely the component is to fail. Experience and databases of failure rates are helpful in this qualitative evaluation. However, data on failure rates on structural components is limited and is often difficult to use in practice. A rigorous structural reliability analysis may be used to determine a more quantitative likelihood of failure, as described below. The major benefit of a risk matrix is its simplicity and transparency. Further, the risk matrix can be used for any type of anomaly as long as a qualitative assessment of the consequence and the likelihood of an anomaly can be executed.

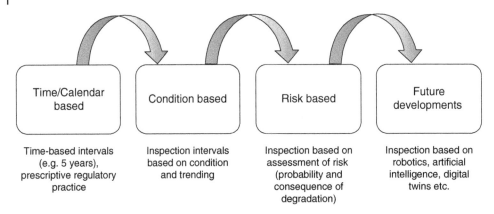

Time/Calendar based	Condition based	Risk based	Future developments
Time-based intervals (e.g. 5 years), prescriptive regulatory practice	Inspection intervals based on condition and trending	Inspection based on assessment of risk (probability and consequence of degradation)	Inspection based on robotics, artificial intelligence, digital twins etc.

Figure 55 Approaches to planning offshore structural underwater inspections.

Consequence				Accumulated Probability				
				1 Rarely been observed	2 Has been observed	3 Observed often	4 Almost certain	5 Certain
People	Environment	Cost	Severity Level	$<10^{-3}$	$10^{-3}-10^{-2}$	$10^{-2}-10^{-1}$	$10^{-1}-1$	~1
Multiple fatalities	Massive effect	Very high	E	M	H	H	H	H
Single fatality or permanent disability	Major effect	High	D	M	M	H	H	H
Major injury	Localised effect	Medium	C	L	M	M	H	H
Slight injury	Minor effect	Low	B	L	L	M	M	H
Superficial injury	Slight effect	Negligible	A	L	L	L	M	M

Figure 56 Example of risk matrix for risk-based inspection planning. L indicates low risk, M indicates medium risk and H indicates high risk.

One of the most useful *quantitative approaches* is the application of probabilistic structural analysis (also called structural reliability analysis). The analysis will then typically be based on:

- a calculation of the probability of a single member or component failure (e.g. through wall degradation or severance) due to one or more degradation mechanisms (e.g. fatigue and corrosion) at different points in time, combined with
- an assessment of the likelihood that the structure in the damaged condition will experience total collapse (e.g. serious leaning or toppling). This is typically assessed by:
  - calculating the probability of system failure for the damaged structure
  - assessment of the remaining capacity of the damaged structure by a non-linear pushover analysis.

Many types of degradation mechanisms may occur at several different locations, so in practice a sum of the probabilities of the component failure as a result of all these different degradation mechanisms need to be combined with the probability of system failure in a damaged condition. The method can be expanded to include the combined probability of multiple failures, combining these with the calculation of the probability of resulting collapse. An overview of the background to probabilistic structural analysis and risk-based inspection planning is given in DNVGL (2019).

Typical inputs to these analyses include a wide range of statistical representations of the following:

- long-term distribution of cyclic loading (fatigue);
- material parameters (yield strength, SN-curve and Paris-Erdogan curve parameters, etc.);
- material and environment related parameters (corrosion rate, sea water temperature, etc.);
- system strength, damage capacity, member geometry, fatigue design factor;
- extreme loading distribution; and
- probability of detection (PoD) linked to the inspection methods used.

As previously mentioned, the use of drones, robots, artificial intelligence and digital twins are emerging and this in the near future will most likely change the approach to inspection planning. The future possibility of autonomous robots that are inspecting and repairing a structure on a permanent basis could lead to minimal need for inspection planning. However, there are several developments in damage detection and location required as well as the ability to link an autonomous robot to computerised integrity-management systems. Some progress in this direction has already been made for the integrity management of bridges, using autonomous robots. Further discussion on such developments is given in Chapter 9.

6.1.4   Inspection Programme

The inspection programme is the detailed work scope for the offshore execution of a series of inspections and investigations to be undertaken in a seasonal campaign in accordance with the long-term inspection plan. This programme should define:

- selection criteria for appointment of sub-contractor (if necessary);
- schedules and the sequence of the areas to be inspected;
- detailed description of inspection method and related procedures (e.g. preparation, calibration, operational requirements);
- level of cleaning required;
- required personnel profiles;
- reporting requirements and format;
- approval process for the completion of tasks (e.g. by supervisor and operators site representative);
- requirements for when anomalies need to be re-inspected with more detailed methods; and
- safety and other procedures to be followed.

An inspection programme should also have some indication of actions to be taken depending on the severity of the anomalies found. For example, significant anomalies would usually be reported directly to the structural integrity management team for their evaluation and possible instruction for immediate mitigation. All anomalies found and any follow-up actions taken during inspection should be documented with all records and reports retained.

A typical inspection programme should detail the specification of <u>how</u> to inspect, <u>where</u> to inspect and <u>what</u> is to be performed during the inspection. The programme would typically include detailed instructions regarding the inspection method, the degree of cleaning required prior to inspection and how the inspection is to be deployed. The exact location required for the inspection needs to be specified, including pictures and sketches.

Inspection results are often described in a report including finding types, finding data, description and comments and documentation of the execution (signature and data). A possible example of a work sheet for inspection of structures is given in Table 37. In addition, requirements for the calibration of the inspection tools could be included.

Table 37 Example of a Field Work Sheet for Inspection of a Structural Part.

Field work sheet for inspection of a structural part

Inspection specification	Reference	Campaign ID	Campaign identification number	Year	2020
		Work package ID	Work package identification number		
		Inspection ID	Inspection identification number	Zone	BW
	How	Planned inspection type and deployment:		For example, GVI, DVI, CVI or a required NDT method. Deployment will mostly be ROV but could be in special cases divers.	
	Where Location	Facility	Facility ID	Picture or sketch of location for inspection	
		Area and group	e.g. hull, jacket, GBS	e.g. bracing, leg, pile sleeve, etc.	*Eddy current*
		Depth	Elevation		
		Detail ID	Brace ID, chord ID, etc.	Weld ID	
		Position	e.g. 1 o'clock		
		Drawing	Drawing ID		
	What	Description:		Passing along member looking for anomalies, perform CP measurement, report anode consumption if greater than 50%, report marine growth (max and average).	
		Additional information		In case of an indication of an anomaly removal of marine growth and performance of a CVI to determine cause and detail is required.	

Inspection report

	Execution summary	Problems (y/n)?	Finding (y/n)?	Finding no.____
	Finding location	**Drawing ID**	**Location ID**	**Addition information**
Execution	Inspection completion date:	*Date*	Inspector name:	*Inspector name*
	Company responsible:	*Company name*	Inspection method used:	*GVI, CVI, EC, FMD etc.*
	Reporter name:	*Reporter name*	Deployment method:	*Diver, ROV, inspector*
Anomaly detail	Anomaly type:	*Crack, flooding, corrosion, dent, bow, thickness & dimension, loose or missing, anode condition etc.*	Anomaly data:	*Position, clock from/to, penetration, length, width, depth, dimension, height, thickness, displacement, amplitude fouling thickness, etc.*
	Immediate actions taken during inspection:	*Grinding, stop holes drilled, cleaning of marine fouling*	Further description and comments:	

6.1.5 Integrity Data Management

The optimal management of inspections and eventually (if needed) repair relies on large amounts of information being collected and stored in order to keep a record of all relevant data about the structure. These data should include the inspections and repairs performed and relevant historical and operational documents. The performance of the repairs can be documented by records of subsequent inspections.

A management system for keeping data about a structure is the backbone of structural integrity management. This should be a system for efficient archival and retrieval of characteristic structural data, weight and load data, any changes to the configuration of the structure and especially inspection data, both when anomalies are found and when inspections have revealed that the structural detail is free of defects. In addition, data about any repair performed is important for further evaluations and assessments of the structure.

The first and often most important data to be registered in the database is the structure's characteristic data from design, fabrication and installation. These characteristic data typically for a steel structure include (API 2014, Standard Norge 2017):

- original and current operator, use and function;
- location, water depth, orientation, platform type, number of wells and risers, manning level;
- design drawings, material specifications, design codes, metocean criteria, deck clearance, weight, loading and equipment arrangement, foundation and soil data and other relevant design data;
- fabrication (as-built) drawings, fabrication and welding specifications, material documents (mill certificates and material traceability) and other fabrication data; and
- transportation records, installation specification and records, soil data, pile and conductor driving records, as-installed configuration summary, damage and anomalies, mooring lines and anchor arrangement, inclining test results, grouting records and other relevant installation data.

The data collected during inspections and investigations in operation is often seen as changes to these characteristic data. Such data typically includes (API 2014, Standard Norge 2017):

- platform history (experienced loading, incidents and events, operation, performance, surveys, repairs and modifications);
- present condition and configuration
  - in-service inspection data (images, videos, reports)
  - present physical survey reports including degree of tilt (level), configuration measurements (distances), as is deck clearance, layout, draught, subsidence, bathymetry, as-is weight, etc.;
- present metocean, load and design specifications,
- anomaly register and anomaly mitigation register (severity, confirmation of implemented measures);
- evaluation data;
- strengthening, modification and repair (SMR) data;
- corrosion protection and monitoring data; and
- structural monitoring data.

A database with data on changes and anomalies from all past inspections will provide insight into the evolving history of these changes and anomalies, which are very relevant to the development of regular inspection plans. Hence, data should be stored in such a way as to aid trend spotting where possible. This is of particular value for inspection data which can be used to establish trends between consecutive inspections and between assets.

In addition, data about the structural integrity management process is normally kept as part of the integrity data and includes the basis for surveillance programmes (such as RBI data) and the as-is analysis models.

These data are in more recent years normally kept in a computerised data management system. However, whatever archive system is used, the aim is that data can be retrieved in a timely manner for as long as the structure is in use.

As mentioned above, structural data can be arranged in many possible ways as long as the aim of easy storage, retrieval and trending is met. One organised option is shown in Figure 57 where the flow of information from the as-built state to the as-is condition is illustrated with respect to the main structural integrity topics such as the design regime, the condition and configuration of the structure and loads on the structure.

The generic information as shown in Figure 57 should include what is needed on a daily basis and in emergency response situations. What a structural integrity engineer will need in an emergency situation will depend on how responsibilities are divided in the particular organisation in such situations. Often organisations seek to organise themselves with the same responsibilities in emergency situations as in daily operations. An example list of generic information is given in the left-hand side of Table 38.

Independently on how the database is organized, it should include a clear indication of where the inspections are to be performed, the finding location and where for example the weight has changed. It is beneficial if all these databases are based on the same location hierarchy (see Table 38 right-hand side for an example).

The purpose of a change register as shown in Figure 57 is to communicate significant anomalies to evaluation and assessment engineers, provide anomalies with a severity status and check off findings when compensating measures have been implemented. The as-is integrity of the structure is in most regulations required to be demonstrated for design situations such as the 100-year wave based on a structural analysis model reflecting the as-is situation. However, many changes may be evaluated as insignificant on their own but may be relevant when seen in combination with other

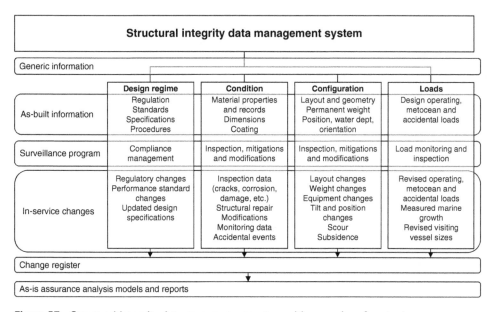

Figure 57 Structural integrity data management system with examples of content.

Table 38 Important Structural Integrity Management Data.

Generic information for daily use and for use in emergency response—examples	Database structure for inspection and surveillance programme and results, as-is weight control, change register and analysis models and reports
Facility data including use and function, manning level, location, water depth, orientation, coordinates, marine system function, structure type and main dimensions, design life, number of wells and risers, other site-specific information	Typical heading in a database for major structures may include: • sub-structure (jacket, semi-submersible, GBS etc.); • topsides support frames; • modules; • bridges; • flares; • helidecks; • crane pedestals; and • templates.
Platform history including operational history, accidental events and damage, anomaly reports, repairs and modifications	
Present condition and configuration	Major structures are often further broken down to components (e.g. members and joints) so that locations can be specified in the surveillance tasks and findings can be registered at the same location
Soil data including geotechnical information, soil type and characteristic values, laboratory tests of samples, pile driving data, pile size anchor configuration and type, anchor holding data	
	In addition, the inventory includes access structures, crane utility, bolts and bolted connections, clamps, caissons, conductors
	Load points and anodes are often included in the database as separate categories for creating statistics and trends in inspection findings and repair needs

changes, especially when the individual effects combine synergistically. The change register can be used to give an overview of these changes not yet included in the analysis model but to be included in the model at a later date.

An as-is assurance analysis model and report overview as shown in Figure 57, containing the included changes, should be established and kept up-to date for all high-consequence structures.

An integrated integrity management database containing all of the above-mentioned databases has the advantage of providing a hierarchy of information relating to both process execution and technical integrity. Such a database should also include key performance indicator (KPI) functionality.

Several computerised systems for structural integrity data management exist which include many of the above-mentioned requirements. One example is the DNV GL ShipManager Hull. This is claimed to be an advanced hull integrity management software available for use in maintenance strategy, planning, inspections, hull surveys, assessments and documentation. The software provides a planned maintenance system for hull structures and improving inspection strategy with risk-based ratings of findings related to the hull integrity. The ShipManager Hull software provides control of the maintenance level of vessels, while also providing documentation to customers, stakeholders and authorities. The software also includes equipment such as anodes, pipes, valves and ladders. Hull integrity findings and monitoring covered in the software include:

• coating condition;
• cracks;
• indents/deformation; and
• buckling and corrosion.

6.1.6   Key Performance Indicators

Key performance indicators (KPIs) are widely used to measure performance against targets, whether these be technical, environmental or financial. Sharp et al. (2008 and 2015) described the background to developing KPIs for offshore structural integrity, relating to aspects which are important for both safety and asset condition. This was achieved based on a hazard approach, which included extreme weather, fatigue, corrosion and accidental damage. It is important that KPIs are measurable and this aspect was incorporated in their development. KPIs can be used to provide a basis for measuring structural performance, particularly for ageing installations where a case for life extension needs to be made. This basis can also be incorporated into structural integrity management programmes enabling the performance of this management to be monitored. This applies particularly to inspections, but repairs are not generally addressed.

It is important to note that eight of the 24 indicators developed for steel structures in Sharp et al. (2008) depend on data from recent in-service inspections (in many cases from more detailed inspections than are normally carried out). A further two are based on data from the inspection data management programme or from performance trials of inspection methods. It is important to note that three of the five KPIs associated with fatigue and all four of those related to corrosion are dependent on inspection. The remaining ten KPIs were developed from assessment of the design requirements but using up-to-date design standards.

A typical example of such a KPI is the number of cracks observed during an inspection. A further two are based on data from the inspection data management programme or from performance trials of inspection methods. An example of this type of KPI is a measure of the quality of inspection method used (e.g. by its probability of detecting cracks).

Key performance indicators have also been developed for offshore concrete structures. Many of these depend on original design requirements, but several others rely on inspection for detection and repair of any damage if severe.

The KPI methodology was later extended by Sharp et al. (2015) to floating installations. A further 15 KPIs were seen to be relevant to mobile installations. Ten of these were dependent on data from inspections for their basis.

6.2   Previous Studies on Long-Term Planning of Inspections

6.2.1   PMB AIM Project for MMS

As mentioned earlier in Section 3.3.3 the US Mineral Management Services (MMS) undertook the AIM (assessment, inspection and maintenance) project to provide guidance on managing the integrity of existing fixed steel platforms in the Gulf of Mexico. The project was carried out in four phases by PMB Engineering Group as a Joint Industry Project (PMB 1987, 1988 and 1990).

Phase IV of the programme (PMB 1990) had two components. The first considered the accuracy of the results from the ultimate strength analyses compared with the actual response from a structure. The second task addressed particularly inspection as part of the processes within the AIM programme which is the most important aspect for this book. This task involved outlining three levels of inspection for both old and new structures, as well as collecting and categorising inspection data from the participants in the JIP. The intent of this part was to provide operators with a usable approach for defining specific inspection programmes. The database for this part of the project would be the failure database, current inspection procedures and survey data to develop rational inspection programmes.

Initial conclusions regarding the current practice at that time were:

- routine inspection plans generally met the API RP2A 18th edition requirements;
- node cleaning for inspection purposes generally exceeded that specified; and
- several operators considered that addition of a baseline survey after platform installation was beneficial even though this was not specified by the API RP2A 18th edition.

Regarding the inspection methods used, visual inspection dominated, with some FMD and CP tests. The analysis showed that visual inspection was the most effective in determining the existence of significant damage. The conclusion was that priority should be given to visual inspection rather than the more expensive methods of NDE such as UT and MPI.

Damage locations were reviewed with damage to braces as 65% of the instances reported, 20% for joints and a small number of cases where damage was at legs or conductors. Repairs and modifications were undertaken in 23 cases of those analysed, with no repair for 13 cases in the study. There is minimal information in the report on the types of repair used.

Most operators found damage through routine inspections rather than by random processes. It was also concluded that the level III inspection (see Section 2.3.2) was not a substitute for the need to carry out a level II inspection even if they were scheduled at the same time. Inspection requirements for the level III inspection relied on the input of engineering analyses. Cost-benefit analyses were also undertaken which showed a wide range of costs depending on the assumptions made.

6.2.2 MSL Rationalization and Optimisation of Underwater Inspection Planning Report

An important deliverable from the MSL JIP (MSL 2000), which was undertaken 10 years after the AIM projects described in 6.2.1, was a set of updated in-service inspection guidelines. These were at the time only available for use by JIP participants, as a key part of their structural integrity management system. This would enable them to plan inspections and define the intervals for their Gulf of Mexico platforms. These guidelines later provided useful input to the development of ISO 19902 clause 24.

The long-term inspection programme (called inspection strategy in this JIP report) was developed through an assessment and evaluation of all available inspection, damage and repair data. The guidelines also addressed other issues such as regulatory and operator requirements, platform decommissioning plans and planning of managing incidents. In addition, the availability of inspection techniques (application and reliability) and the flexibility of scheduling was also considered. The collation of data on inspection from the participants' platforms in the GoM was a major contributor to the development of the long-term inspection programme. For the development of inspection guidelines for a particular structure, an important input was taking into account the likelihood of platform failure. This included such aspects as robustness and damage tolerance, original design criteria and damage susceptibility.

The guidelines were developed to be consistent with the principles within the existing API RP-2A recommendations and were used in the development in the emerging ISO 19902 (ISO 2007). The findings of the JIP supported the existing industry position stated in API RP-2A (API 2014a) that the appropriate use of FMD provides an acceptable alternative to CVI examination and could sometimes be preferable.

The guidelines included the recognised types of surveys such as baseline, periodic and special with similar definitions to those in API RP-2A (API 2014a) and ISO 19902 (ISO 2007). The requirements for underwater surveys included GVI for general platform examination, FMD to identify

Table 39 Inspection Requirements for Structures of Different Risk Level.

Defect and anomaly surveys	Inspection requirement		
	II	III	IV
General visual	X	X	X
Anode	X	X	X
FMD		Part survey	Part survey
Visual corrosion	If CP readings are high	X	
Weld and joint			X
CP		Part survey	X
Debris	X	X	X
Scour	If seabed is loose sand or instability is known or suspected		
Marine growth			X

flooded members, CVI of welded joints, CP and anode surveys, visual checks of corrosion, debris, scour and marine growth.

For the periodic inspections, the guidelines introduced a risk-type matrix showing three categories (II, III and IV in increasing risk levels) of platforms based on the following criteria:

- the susceptibility to mechanical damage by an importance categorisation (high, medium and low); and
- platform vintage in terms of API practice used for design.

For the early pre-RP-2A platforms corrosion and CP history and deck elevations were further used to set inspection levels. The matrix showed three levels of inspections as shown in Table 39 with details of surveys for defects and anomalies. The inspection intervals recommended were 3–5 years for high-risk level structures designed pre-RP-2A, 5–10 years for early RP-2A designed platforms and 10 years for more modern structures. Longer intervals were indicated for structures with a lower risk level.

The guidelines developed as part of the JIP identified where to target inspections on different categories of platforms with the aim of optimising inspection resources without compromising safety. For inspection intervals the JIP demonstrated that the default 5-year inspection intervals were unnecessary for certain categories of structures, provided an appropriate structural integrity management process was maintained. The JIP indicated, as noted above, that these intervals could be extended to 10 years and longer without compromising safety. The JIP also identified differences in the susceptibility to damage of different platforms which led to a replacement of the previous categorisation of structures to the new levels II, III and IV described above.

6.2.3 HSE Study on the Effects of Local Joint Flexibility

HSE in 2001 initiated a study by MSL Engineering Ltd on the effects of local joint flexibility on the reliability of fatigue life estimates and inspection planning (HSE 2001) on fixed offshore structures. The review indicated that the fatigue-life predictions taking into account joint flexibility increased compared to at the time conventional rigid-joint analysis. The corrections to the calculated life depended on the location of the joint within the structure (the ratio of the life calculated using flexible joint modelling to the life calculated using a rigid-joint model). The average ratio on life for each of the framing components considered was:

- transverse frames (A to F): 19.3 factor on fatigue life;
- longitudinal frames (1 &2): 9.2 factor on fatigue life; and
- horizontal framing (–24' elev.): 8.0 factor on fatigue life.

The impact on platform underwater inspection planning was that the implementation of local joint flexibility in the fatigue analysis reduces the requirement for underwater inspection by approximately 75%, depending on the member importance. It should be noted that at present joint flexibility is regularly included in fatigue analysis and inspection planning.

6.2.4   HSE Ageing Plant Report

The HSE report (Wintle et al. 2006) is a landmark report in the field of ageing of different types of plants and indicated the relevance of the ageing and condition of the plant for the inspection plan.

It includes a "bathtub curve" which identifies four stages of equipment life; see Figure 3. The four stages relate to the amount of accumulated damage, the rate of expected degradation and the margins before fitness-for-service is compromised. These may correlate to the age of the equipment and it would be expected that equipment moves progressively from stage 1 to stage 4 as it gets older. The four stages are defined together with the relevant inspection needs (Wintle et al. 2006) are described in Table 40.

6.2.5   Studies on Risk-Based and Probabilistic Inspection Planning

The development of probabilistic analysis of structures (often called structural reliability theory) and its application for offshore structures took place in several JIPs from the 1970s. In the US, JIPs were initiated in 1974 (Moses 1975, Stahl 1975). This was taken further in the Stanford University RMS project that was completed in 1988 (Bjerager and Cornell 1988).

One of the earliest European projects were undertaken by CIRIA (1976) in the UK. Later in 1984 DNV in Norway started the Reliability of Marine structures project which developed significant theoretical background in probabilistic analysis but also developed computer programs for probabilistic analysis (PROBAN) and probabilistic inspection planning (PROFAST). In this project the basis for probabilistic crack growth analysis based on fracture mechanics was established in addition to the concept of probability of detection curves. The basis for probabilistic inspection planning based on structural reliability theory was established early in the 1980s in a JIP led by DNV (Sletten et al. 1982). Updating of fatigue failure probabilities based on events was presented by Madsen (1985), Madsen et al. (1986) and Madsen et al. (1987). Probabilistic inspection planning has since to a large extent been based on this methodology.

Several engineering houses developed the necessary software tools for performing probabilistic inspection planning. Aker Engineering is one notable example of this with their PIA (Probabilistic Inspection Analysis) program (Aker Engineering 1990).

In the 1990s DNV undertook a new JIP to develop guidelines for offshore structural reliability analysis (DNV 1996). This project reviewed the use of structural reliability theory for general use including probabilistic design of structures but also included elements of probabilistic inspection planning.

Table 40   The Phases of the Bathtub Curve and Inspections Needs.

Stage	Description	Inspection needs
1 Initial	After installation there may be a relatively higher rate of damage and issues requiring attention as inherent weakness or faults are revealed.	Baseline inspection is normally required within the first years of operation to reveal faults arising from shortcomings in the material selection, design, fabrication and installation or from early operational errors. The nature of this inspection is confirmatory.
2 Maturity	In the 'Maturity' stage the structures should be predictable, reliable and assumed to have a low and relatively stable damage rate with few issues requiring attention. In this stage most of the initial (unexpected) integrity issues should have been dealt with and the equipment should be in its most stable operating regime.	Inspections in this phase are primarily to confirm the continuing integrity of the structure. Inspection intervals may be extended based on performance or a risk-informed analysis of degradation, failure likelihood and consequences. The nature of this inspection is assurance.
3 Ageing	By this stage the structure is likely to have accumulated some damage, the rate of degradation is increasing and signs of damage are starting to appear. It becomes more important to determine quantitatively the extent and rate of damage and to make an estimate of remaining life. Design margins may be eroded and the emphasis shifts towards fitness-for-service and remnant life assessment of specific damaged areas.	Inspection in this phase should be proactive and is likely to reveal an increasing number of defects; the prime objective is to detect the onset and rate of damage, determine its extent and determine the current condition quantitatively. A more deterministic approach to inspection is appropriate, possibly leading to increased levels of inspection. The nature of this inspection is quantitative.
4 Terminal	The rate of degradation in this phase is likely to become increasingly rapid and is not easy to predict. As the accumulated damage becomes increasingly severe, structures will ultimately need to be either repaired or decommissioned.	For inspection in this final stage, the main emphasis should be guaranteeing adequate safety between inspections while keeping the structure in service. Structural integrity can be managed through making more use of on-line monitoring of the damaged areas, by frequent NDT to monitor the damage and anomalies and repairing these if they become unacceptable. Reliance, therefore, should be placed on inspection data to justify continued operation. Inspection and monitoring are likely to be much more frequent and the accuracy of measurement is even more important. The objective of this inspection is to provide the necessary warning signs of potential structural failure.

HSE undertook a significant review of the use of probabilistic methods in a project undertaken by the Norwegian Technical University (NTNU) and Aker Engineering. This project resulted in the following reports:

- HSE OTO 99 061 (HSE 2002a) gave a review of the theory behind probabilistic inspection planning;
- HSE OTO 99 060 (HSE 2002b) gave a review of the acceptable target probability levels to be used in inspection planning; and

- HSE OTO 99 059 (HSE 2002c) provided a validation of inspection planning methods comparing in-service fatigue performance and the corresponding theoretical predictions, which resulted in recommendations regarding enhancements to existing analytical methods. The theoretical predictions were found to be conservative, predicting 4 to 10 times as many fatigue cracks as observed. However, it was also found that approximately 40% of the observed fatigue cracks were not identified by the fatigue analysis. Fatigue cracks were detected for 15% of the scheduled inspections and in the order of 2% for component normally not scheduled for inspection by the probabilistic inspection planning methods. As a result, the review recommended the need to include some random locations beyond those identified by the fatigue analysis to be inspected. The review also gave some recommendations on improvement of the general probabilistic inspection planning method, primarily related to calibration to experienced cracking on real structures.

It is possible to establish generic pre-prepared inspection plans based on this type of probabilistic methods as described in Faber et al. (2000), Straub (2004), Faber et al. (2005) and Sørensen et al. (2005). These generic inspection plans for different joint types are based on the following parameters:

- calculated fatigue life of detail;
- long-term stress range distribution;
- type of detail (SN-curve);
- fatigue strength measured by the design fatigue factor (DFF);
- strength of a structure, for example, measured by the residual strength ratio (RSR) typically obtained by a non-linear structural analysis;
- importance of the considered detail or member measured by, for example, the residual influence factor (RIF), which is a measure of the importance of a member as $RIF = \dfrac{RSR}{DSR}$ where RSR is the collapse capacity divided by design capacity for an intact structure and DSR is the collapse capacity divided by the design capacity for a structure damaged by a fatigue failure;
- member geometry; and
- inspection, repair and failure cost.

The inspection intervals, the necessary inspection quality and the criteria for when to repair can be determined and optimised with respect to inspection, repair and failure cost (Faber et al. 2005, Straub 2004). The inspection plan is generic as it is a function of the above-mentioned parameters (such as SN-curve, FDF, RSR).

Based on the determined DFF it is possible to establish the corresponding annual probabilities of failure for a specific year. For the joints to be considered in an inspection plan, the acceptance criteria for the annual probability of fatigue failure may be assessed through the damaged strength ratio (DSR) together with the annual probability of fatigue failure. The importance of a joint or member can, as mentioned above, be described by the residual influence factor (RIF) and the relation between the RIF value for the specific member on a specific platform and the annual collapse probability.

The intervals between planned inspections based on probabilistic methods have a tendency to increase as structures age if no cracks are found (see Figure 58 and Figure 59). This is seen as contradictory to the common understanding that fatigue cracks will become increasingly likely with age and older structures will require more frequent inspections.

In order to establish an improved understanding of probabilistic inspection planning of older structures, the Petroleum Safety Authority (PSA) in Norway initiated a review of possible improvements to the method. The work was reported in PSA (2007), Sørensen and Ersdal (2008a) and

Figure 58   Inspection intervals in years for a sample structure (PSA 2007).

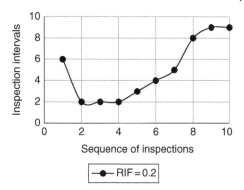

Figure 59   Inspection intervals in years up to 50 years for varying RIF values (PSA 2007).

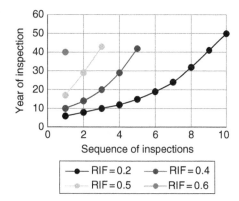

Sørensen and Ersdal (2008b). The following observations were considered for a modified method for probabilistic inspection planning for older installations:

- For an ageing platform, several small cracks are expected to be observed implying an increased likelihood of crack initiation and growth. This was included in the review by increasing the expected value of initial crack size with time due to coalescence of smaller cracks.
- For older platforms, there may be a large number of critical details (possibly with correlated loading, fabrication and fatigue life) that have reached their design fatigue life. The simplified methods normally used in probabilistic inspection planning evaluate the failure of only one member or joint. System effects where several members or joints are failing more or less simultaneously were included in the review.

The most promising method as reported in this review consisted in increasing the rate of defects (crack initiation) at the end of the expected life corresponding to the final stages of the bathtub (ageing and terminal). This approach was shown to imply that inspection intervals decreased at the end of the platform's life, which were more in line with expectations.

The review further discussed methods for establishing an acceptable annual probability of fatigue failure for a particular component depending on the number of fatigue critical components. The acceptable annual probability of fatigue failure of a component was obtained by considering the importance of the component through the conditional probability of its failure. Given this acceptable failure probability, the RIF value and the number of fatigue critical components, the maximum acceptable component annual probability of fatigue failure was calculated using a simple upper bound of the probability of failure.

In 2011 DNV (later to become DNVGL) initiated a JIP to standardise the methodology for risk-based inspection planning which resulted in the recommended practice DNVGL RP-C210 (DNVGL 2019) for inspection planning of fatigue cracks in jacket structures, semi-submersibles and floating production vessels. The recommended practice also provides guidelines for the fatigue analysis needed and gives recommendations for probability of detection curves. This recommended practice is at this time considered to be the industry standard for risk-based inspection planning for offshore structures.

6.2.6 EI Guide to Risk-Based Inspection Planning

The Energy Institute (EI 2020) is at the time of writing this book in the process of publishing a guidance document on risk-based inspection planning for fixed offshore steel structures. This guide provides an alternative and more qualitative description of the RBI process to that of DNVGL RP-C210 (DNVGL 2019) and links RBI more strongly within the context of SIM as it is rooted in industrial practice.

This review provides some useful material and defines inspection as "to determine the physical condition of a structure and the existence and extent of any damage, degradation or defects". Further, the review defines the objective of inspection as "to provide condition data that can be used with other data to confirm that structural integrity is maintained throughout the service life such that damage and degradation mechanisms do not reach an extent that would impair the resistance to defined loading events". This objective is stated to be achieved by providing information about the current extent of damage (known, anticipated and rate) and safety critical areas of the structure whose condition is unknown or uncertain (need for more knowledge and reducing uncertainty). A risk-based inspection plan should result in an overview of what, when, where and how to inspect based on the likelihood and the consequences of damage in order to provide a level of surveillance that ensures structural integrity against defined loading events.

The review provides an extensive overview of the risk-based inspection planning approach and provides guidance on determining likelihood and consequence of failure, which are often an area of debate and uncertainty. The likelihood of failure is described by the utilisation ratio (strength and buckling), the fatigue damage ratio (FDR) and the likelihood or extent of corrosion. The consequences of failure are to some extent linked to the redundancy of the structure but are also linked to the safety of personnel, environment, financial impact and operator's reputation. The guide also contains an example of a risk-based inspection plan for a typical fixed structure together with case studies. The core processes outlined in this guide for structural integrity management are illustrated in Figure 60.

6.3 Summary on Inspection Planning and Programme

6.3.1 Introduction

Inspection is an activity that is important in maintaining the safety of structures in operation, both in reducing uncertainty about their current state and detecting any defects. If defects are found, further inspections, evaluations and analysis are usually necessary to determine the severity of a defect and its effect on structural integrity. In order to achieve the objectives of inspection at a reasonable cost, a detailed inspection plan is needed.

Risk- and probability-informed inspection plans are at present the preferred approach in the industry and are often claimed to be economic. However, calendar-based inspection plans are also still in use and are often found in both standards for fixed and floating platforms and for classed units.

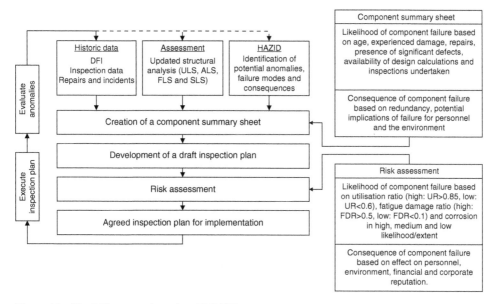

Figure 60 The RBI process, based on EI (2020).

6.3.2 Fixed Steel Platforms

For fixed steel platforms the primary inspection methods are FMD, GVI and CVI. If anomalies are seen in these types of inspection, NDT inspections to verify the extent of the damage may be necessary. An increasing amount of risk or probability-informed inspection planning is used. However, as previously mentioned, calendar-based programs are still used by operators and found in standards.

In the absence of a platform specific in-service long-term inspection plan, standards can often provide a default inspection plan. Example of such plans were included in the previous version of ISO 19902 (ISO 2007) and to some extent transferred to the newer ISO 19901-9 (ISO 2019), as shown in Chapter 2. Reduced inspection than that provided in such default plans is often required to be justified by proper analysis. In ISO 19902 and ISO 19901-9 it is noted that additional inspection may be needed to meet statutory requirements or owner's policy and that the standard does not guarantee fitness-for-purpose by this default inspection plan.

In these standards the annual level I inspection typically contains below water verification of the performance of the cathodic protection system (at least one leg) and a general examination without marine growth cleaning of all structural members in the splash zone and above water. A general examination may include a visual survey to determine the effectiveness of the corrosion protection system, deterioration of coating systems, excessive corrosion, bent, missing, or damaged members, indications of obvious overloading, design deficiencies and any use which is inconsistent with the structure's original purpose. In addition, it is recommended to perform marine growth measurements, estimate the approximate percent depletion of all the anodes and a visual survey of the state of the anodes and reference electrodes. If the level I inspection indicates that underwater damage is possible, a level II inspection should be conducted as soon as conditions permit.

A level II inspection (3–5 years) typically includes a general underwater visual inspection to detect the presence of excessive corrosion, effects of accidental or environmental overloading, scour, sea floor instability, damage detectable in a visual swim-round survey, design or construction deficiencies, presence of debris and excessive marine growth. In addition, the cathodic potentials of pre-selected critical areas should be measured. Detection of significant structural damage during an inspection will initiate a level III inspection to be conducted as soon as conditions permit.

Further, a level III inspection (6–10 years) typically includes underwater cleaning and close visual inspection (CVI) of pre-selected areas based on an engineering evaluation and areas of known or suspected damage. This should include 20 primary members or 5% of the full structure. Flooded member detection (FMD) of at least 50% of primary structural members can provide an acceptable alternative to CVI of pre-selected areas where the structural form and sites of potential damage would lead to flooding. In addition, close visual inspection of pre-selected areas for corrosion monitoring is normally required.

When required, based on results from a level III inspection, a level IV inspection is normally required including underwater NDT of areas and measurement of damage with its extent as determined by the level III inspection or from known or suspected damage. The required marine growth should be cleaned to allow for detailed inspection. The inspection should include 100% of each weld length.

A special inspection may be required involving visual inspection with marine growth cleaning as necessary to assess the performance of prior repairs approximately 1 year after completion of the repair. A special inspection also includes the monitoring of known defects, damage or other anomaly that could potentially affect the fitness-for-purpose.

Unscheduled inspections are performed after the occurrence of an environmental or accidental event. These should include a visual inspection of the whole structure from sea floor to the top of the structure without marine growth cleaning and visual confirmation of the existence of the corrosion protection system with the intent to detect damage and indirect signs of damage. Inspections after accidental events should determine the total extent of any damage and provide information for implementation of mitigation measures such as repairs and down-manning, if required.

6.3.3 Floating Steel Structures

Floating structures include a number of structural and marine systems that are different from fixed structures. The safety-critical elements are the systems which control bilge and ballast, the watertight integrity, the weight control, stability monitoring and station keeping. The watertight integrity includes watertight doors and hatches, seals, valves, pumps, dampers. Failure in any of these components can lead to loss of watertight integrity and possible loss of stability and buoyancy. Hence, this introduces a need for very different types of inspection, as well as a different way of thinking about integrity management in general. In addition, the structural elements of a floating structure are quite different in geometry and form (as they are primarily shell structures, whereas fixed structures are often made of beams and tubular joints).

A significant amount of the inspection tasks for a floating structure can be performed from inside the hull and are as such not part of this book. The underwater inspection tasks on floating structures typically include:

- inspection of the hull outer skin;
- inspection of seawater intakes and outlets; and
- inspection of mooring chains.

Many floating offshore structures in oil and gas production have a class certificate which is provided by one of the leading Classification Societies, such as DNVGL and Lloyds Register. The inspection plans for floating structures have been a key part of the class system and regulation, which has been developed over many years of experience by owners and class societies. These have traditionally been based on a calendar-based system of inspection and typically require a dockside inspection every 5th year. Dockside inspection provides the opportunity for improved access and repair facilities. However, lately risk- and probability-informed methods for developing inspection plans for classed units have been increasingly used and accepted by class societies.

Floating offshore structures used in oil and gas production often stay on location for longer periods and the opportunity for regular dockside inspection is hence not available. The fact that they are permanently stationed offshore places a more onerous requirement on the inspection of these floating structures, for which there is less experience. This lack of experience could be significant for older floating offshore structures and could require a more stringent inspection approach as there is an increased likelihood of deterioration being accumulated so that widespread deterioration becomes more likely.

Several standards have recently been developed to guide the operators of floating facilities in their integrity management including inspection, such as ISO 19904-1 section 18 (ISO 2006), NORSOK N-005 appendices F, G, H and I (Standard Norge 2017) and API-RP-2FSIM (API 2019a), which all follow inspection procedures similar to that of fixed offshore structures. In addition, there is an API-RP-2MIM (API 2019b) which is also highly relevant for the inspection of mooring lines. With regard to the increased likelihood of deterioration for older structures it should be mentioned that the draft API-RP-2FSIM (API 2019a) includes a separate annex specific to life extension which gives guidance specifically on the life extension assessment process.

API-RP-2I (API 2015) and NORSOK N-005 (Standard Norge 2017) recommends inspection intervals for chain mooring systems for floating structures. As mentioned earlier, the maximum interval between major inspections in API-RP-2I is linked to the age of the chain in years. For relatively new chains (i.e. 0–3 years) the recommended interval is 3 years; for slightly older chains (4–10 years) it is 2 years and for chains older than 10 years, the interval is reduced to only 8 months. This short interval is very demanding and costly and hence chains are normally replaced before they reach the 10-year criterion. Inspection of mooring systems can be undertaken visually by an ROV, but this has significant limitations. A more detailed inspection, e.g. by MPI, requires removal of the mooring and dockside inspection with significant cost and operational implications.

6.3.4 Concrete Platforms

The strategies for long term inspection planning (section 6.1.3) also apply in general to concrete structures. Inspection planning should focus on verification of the condition of the structural components so that structural integrity can be verified. However, concrete structures are more damage tolerant than steel structures, which affects the planning of inspections.

Failure of concrete structures is more often related to corrosion of the steel reinforcement and less to fatigue damage. Especially the splash zone is vulnerable to corrosion. However, failure due to corrosion below the waterline is minimised by the presence of anodes. Hence, large areas of a concrete structure such as cell walls and shaft walls below water have a relatively low probability of corrosion damage.

NORSOK N-005 section 2.5.4 (Standard Norge 2017) states that special attention should be given to areas of major importance for the structural integrity. Particularly the atmospheric and splash zones have the highest likelihood of deterioration. Areas of major importance are listed as:

- splash zone;
- construction joints, penetrations, embedments;

- complex connections such as wall-dome connections, cell-wall joints, columns-to-caisson connections; and
- load-bearing areas in the top of shafts.

NORSOK N-005 (Standard Norge 2017) outlines the general inspection requirements and notes that for inspection planning the potential failure mechanisms should be taken into consideration. These can lead to typical types of failure which include reinforcement corrosion, subsequent cracking, spalling and delamination, as well as damaged coatings. The standard also states that appropriate inspection methods should be developed taking into account the likely types of defects and damage which may develop. Visual inspection (GVI and CVI) is recommended as the dominant inspection method to be employed. However, it is also required that the cathodic protection potentials are checked at regular intervals (not specified).

Common practice indicates that the first inspection of concrete structures should be completed within a period of 3–5 years. Based on information gained during the first inspection, the interval can be altered.

Similar guidance for inspection planning is provided in ISO 19903 (ISO 2019b) as described in Section 2.5.2. In general, it seems that standards for integrity management of concrete structures are less specific on inspection needs and intervals compared to those for steel structures, possibly due to the expectance that these are more damage tolerant. However, it is unclear why the standards do not recommend annual checking of cathodic potential similar to that for steel structures.

6.4   Bibliographic Notes

Section 6.1.3 is partly based on Ersdal et al. (2019). Section 6.1.5 is to a large extent based on private communication with Michael Hall.

References

Aker Engineering (1990), *PIA Theory Manual*, Aker Engineering, Oslo, Norway.

API (2008), *API RP-2I In-service Inspection of Mooring Hardware for Floating Structures*, Third Edition, American Petroleum Institute, 2008.

API (2014a), *API RP-2A Recommended Practice for Planning, Design and Constructing Fixed Offshore Platforms*, API Recommended Practice 2A, 22nd Edition, American Petroleum Institute, 2014.

API (2014b), *API RP-2SIM Recommended Practice for Structural Integrity Management of Fixed Offshore Structures*, American Petroleum Institute, 2014.

API (2019a), *API RP-2FSIM Floating System Integrity Management*, American Petroleum Institute, 2019.

API (2019b), *API RP-2MIM Mooring Integrity Management*, American Petroleum Institute, 2019.

Bjerager, P. and Cornell, C.A. (1988), "Specification for a Failure-Path Base Structural Reliability Program", C.A. Cornell Inc, August 1988.

CIRIA (1976), "Rationalization of Safety and Serviceability Factors in Structural Codes", Report No. 63, Construction Industry Research and Information Association, London, UK.

DNV (1996), *Guideline for Offshore Structural Reliability Analysis—General*, DNV, Høvik, Norway.

DNVGL (2019), *DNVGL RP-C210 Probabilistic Methods for Planning of Inspection for Fatigue Cracks in Offshore Structures*, DNVGL, Høvik, Norway.

EI (2020), *Guidance on Risk-Based Inspection for Fixed Steel Offshore Structures*, Energy Institute, London, UK.

Ersdal, G., Sharp, J.V., Stacey, A. (2019), *Ageing and Life Extension of Offshore Structures*, John Wiley and Sons Ltd., Chichester, West Sussex, UK.

Faber, M.H., Engelund, S., Sorensen, J.D. and Bloch, A. (2000), "Simplified and Generic Risk Based Inspection Planning", Proc. 19th OMAE, Vol. 2.

Faber, M.H., Sørensen, J.D., Tychsen, J. & Straub, D. (2005), "Field Implementation of RBI for Jacket Structures", *Journal of Offshore Mechanics and Arctic Engineering*, Vol. 127, Aug. 2005, pp. 220–226.

HSE (2001), *Offshore Technology Report 2001/056: The effects of Local Joint Flexibility on the Reliability of Fatigue Life Estimates and Inspection Planning*, Prepared by MSL Engineering Ltd for the Health and Safety Executive (HSE), HMSO, London, UK.

HSE (2002a), *Offshore Technology Report 1999/061: Review of Probabilistic Inspection Analysis Methods*, Prepared by Aker Offshore Partner A.S for the Health and Safety Executive (HSE), HSE Books, London, UK.

HSE (2002b), Offshore Technology Report 1999/060: Target Levels for Reliability-Based Assessment of Offshore Structures During Design and Operation, Prepared by Aker Offshore Partner A.S for the Health and Safety Executive (HSE), HSE Books, London, UK.

HSE (2002c), *Offshore Technology Report 1999/059: Validation of Inspection Planning Methods Summary Report*, Prepared by Aker Offshore Partner A.S for the Health and Safety Executive (HSE), HSE Books, London, UK.

HSE (2006), *Plant Ageing—Management of Equipment Containing Hazardous Fluids or Pressure*, HSE RR 509, London, UK.

ISO (1998), *ISO 2394:1998 General Principles on the Reliability for Structures*, International Organization for Standardization.

ISO (2006), *ISO 19904:2006 Petroleum and Natural Gas Industries—Floating Offshore Structures*, International Organization for Standardization.

ISO (2007), *ISO 19902:2007 Petroleum and Natural Gas Industries—Fixed Steel Offshore Structures*, International Organization for Standardization.

ISO (2019), *ISO 19901-9 Petroleum and Natural Gas Industries—Specific Requirements for Offshore Structures—Part 9: Structural Integrity Management*, International Organization for Standardization.

ISO (2019b), *ISO 19903 Petroleum and Natural Gas Industries—Fixed Concrete Offshore Structures*, International Organization for Standardization.

Madsen H.O. (1985), "Model Updating in First-Order Reliability Theory with Application to Fatigue Crack Growth", In: Proc. of 2nd Int. Workshop on Stochastic Methods in Structural Mechanics, 1985, University of Paris, Paris, France.

Madsen, H.O., Krenk, S., Lind, N.C. (1986), *Methods of Structural Safety*, Prentice-Hall Inc., Englewood Cliffs, New Jersey, US.

Madsen HO, Skjong R, Kirkemo F. (1987), "Probabilistic Fatigue Analysis of Offshore Structures—Reliability Updating through Inspection Results", International Symposium on Integrity of Offshore Structures (IOS '87), 3rd, Glasgow, UK.

Moses, F. (1975), "Cooperative Study Project on Probabilistic Methods for Offshore Platforms", Technical Report, Amoco Production Company, Tulsa, Oklahoma, US.

MSL (2000), "Rationalization and Optimisation of Underwater Inspection Planning Consistent with API RP2A Section 14", Published by MMS, US, as Project Number 345, November 2000.

PMB (1987), "AIM II Assessment Inspection Maintenance", Published as MMS TAP report 106ak.

PMB (1988), "AIM III Assessment Inspection Maintenance", Published as MMS TAP report 106ap.

PMB (1990), "AIM IV Assessment Inspection Maintenance", Published as MMS TAP report 106as.

PSA (2007), "Safety and Inspection Planning of Older Installations", Prepared by Professor John Dalsgaard Sørensen for the Petroleum Safety Authority Norway (PSA), Stavanger, Norway.

Sharp, J.V., Ersdal, G. and Galbraith, D. (2008), "Development of Key Performance Indicators for Offshore Structural Integrity", Proceedings of the ASME 27th International Conference on Offshore Mechanics and Arctic Engineering, OMAE2008-57203, June 15–20, 2008, Estoril, Portugal.

Sharp, J.V., Ersdal, G. and Galbraith, D. (2015), "Meaningful and Leading Structural Integrity KPIs", Society of Petroleum Engineers. SPE-175519-MS, SPE Offshore Europe Conference and Exhibition, Aberdeen, Scotland, UK.

Sletten R., Mjelde K., Fjeld S., Lotsberg I. (1982), "Optimization of Criteria for Design Construction and In-Service Inspection of Offshore Structures Based on Resource Allocation Techniques", EW322, In: EUROPEC'82. European Petroleum Conference, London, UK.

Stahl, B. (1975), "Probabilistic Methods for Offshore Platforms", In Annual Meeting Papers pp. J1–30, American Petroleum Institute, Texas, US.

Standard Norge (2015), *NORSOK N-006 Assessment of Structural Integrity for Existing Offshore Load-Bearing Structures*, Edition1, March 2009, Standard Norge, Lysaker, Norway.

Standard Norge (2017), *NORSOK N-005 In-Service Integrity Management of Structures and Maritime Systems*, Edition 2, 2017, Standard Norge, Lysaker, Norway.

Straub D. (2004), "Generic Approaches to Risk Based Inspection Planning of Steel Structures", PhD thesis, Swiss Federal Institute of Technology, ETH Zurich, Switzerland.

Sørensen, J.D., D. Straub and M.H. Faber (2005), "Generic Reliability-Based Inspection Planning for Fatigue Sensitive Details—with Modifications of Fatigue Load", Proc. ICOSSAR'2005, Rome, June 2005, Rome, Italy.

Sørensen, J.D. and Ersdal, G. (2008a), "Risk-Based Inspection of Ageing Structures", OMAE 2008. Proceedings of OMAE 2008, the 27th International Conference on Offshore Mechanics and Artic Engineering, Estoril, Portugal.

Sørensen, J.D. and Ersdal, G. (2008b), "Safety and Inspection Planning of Older Installations," *Journal of Risk and Reliability*, Vol 222 No O3, September 2008, pp. 403–418.

Wintle, J., Moore, P., Henry, N., Smalley, S. and Amphlett, G. (2006), *Plant Ageing—Management of Equipment Containing Hazardous Fluids or Pressure*, Prepared by TWI Ltd, ABB Engineering Services, SCS (INTL) Ltd and Allianz Cornhill Engineering for the Health and Safety Executive, Published by HSE books, Sudbury, UK.

7

Evaluation of Damage and Assessment of Structures

If it ain't broke, don't fix it.[1]

—Bert Lance

Without a standard, there is no logical basis for decision making or taking actions.[2]

—Joseph M. Juran

7.1 Introduction

When inspection or monitoring has detected damage or anomalies, the structural engineer needs to make a decision on whether repair or mitigation is needed, or if the structure can be left in its as-is state for a period of time. Repair and mitigation are costly and repair in itself may introduce weaknesses and details that are susceptible to new damage and deterioration. As a result, a proper evaluation of the damage and anomalies and their effect on the component's strength is needed. In some cases, also an assessment of the structural system strength including the damaged component may be required[3].

Evaluation should be performed regularly throughout the life of a platform, often on a 1–5-year basis, by a qualified structural engineer. The process for evaluation may be illustrated as shown in Figure 61. The evaluation should consider all relevant recently detected damage and anomalies in addition to the changes to the platform from the 'register of findings' and from similar platforms where appropriate.

---

1 *Source:* Bert Lance, May 1977 issue of the magazine *Nation's Business.*
2 *Source:* Joseph M. Juran, *Managerial Breakthrough: The Classic Book on Improving Management Performance,* 1955, McGraw-Hill.
3 The difference between what's called evaluation of a damage and what's called assessment is in most literature rather vague. In this book the two are defined as follows:
- Evaluation is focussing on the damage or anomaly and its significance on the structural component. It involves primarily engineering judgement and simplified acceptance criteria and mainly without analysis. If analysis is involved, it is mostly local with respect to the damage, anomaly and the related component rather than an analysis of the complete
- Assessment includes a global structural analysis including the anomaly or change and its effect on the total strength of the structure.

Underwater Inspection and Repair for Offshore Structures, First Edition.
John V. Sharp and Gerhard Ersdal.

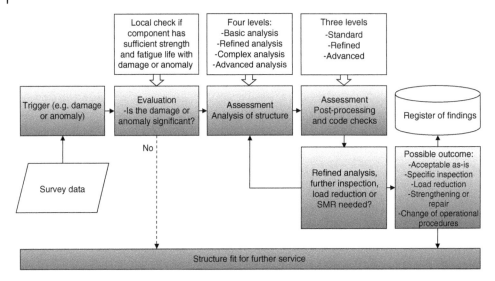

Figure 61 Overall evaluation and assessment approach. *Source:* Based on MSL (2004). Assessment of repair techniques for ageing or damaged structures. Project #502. MSL Services Corporation, Egham, Surrey, UK.

As shown in Figure 61, evaluation involves reviewing the impact to the structural component of any detected damage and anomalies in addition to any changes registered. This typically includes:

- determining the cause of the damage and anomaly;
- comparing identified anomalies with predefined acceptance criteria (e.g. compare corrosion with corrosion allowances, predefined crack sizes, acceptable loading on decks, member damage versus degree of structural redundancy);
- evaluating trends in data to identify any historic developments that need to be assessed;
- deciding whether:
  - further inspections are needed to determine the extent or cause of the anomaly;
  - further inspections on a broader area and on similar components are needed to understand the extent of the anomalies;
  - further analysis and assessments are needed to determine the need for repair or mitigation; or
  - to repair or mitigate without further assessment (e.g. simple repairs that can be performed by the diver or ROV on site);
- demonstrating that structure is fit-for-purpose until the next scheduled inspection and evaluation; and
- updating inspection plans and programs if needed, including monitoring any damage, anomalies and repairs.

Evaluation needs to include preparation of the necessary documentation for the execution of corrective actions and mitigations if needed and ensure that information about the damage or anomaly and possible mitigations are included in the long-term inspection and surveillance plan.

In terms of corrosion control, ISO 19902 (ISO 2007) indicates the following aspects to be considered in the evaluation:

- the assumptions and criteria used in design;
- the details of the system (impressed current or sacrificial anodes) and its past performance;
- cathodic protection readings from monitoring, compared to design criteria; and
- state of the anodes from visual inspection (if a sacrificial system is used).

For any fatigue cracks found, the evaluation could include the use of:

- crack growth rates and remaining life analysis; and
- failure assessment diagrams (FAD) to determine crack stability.

Evaluation typically includes engineering judgement based on specialist knowledge or operational experience, simplified (screening) analysis, or reference to research data, detailed analysis of similar platforms, and so forth. If the evaluation results are unacceptable in terms of not meeting the criteria, then further analysis (assessment) may be required. Alternatively, mitigation measures which reduce the likelihood of structural failure need to be implemented as described in the next section.

Standards such as ISO 19902 (ISO 2007) require that "*all available data on the structure or group of structures shall be evaluated using engineering judgment, operational experience, analysis and predictive techniques, as appropriate*". Further, such standards will require that an assessment is performed if the evaluation of the anomalies and changes trigger any of the assessment initiators provided in the standards.

Risk matrices including appraisal of the consequence of failure of the structural part and the complete platform may be used to differentiate between the importance of different anomalies and changes. Probability-informed inspection planning methods (see Section 6.2.5) can be used to give the right priorities for further inspection, taken into account the anomaly or change. However, according to ISO 19902 (ISO 2007), such analysis "shall be used with care, due to uncertainties in the input data and the degree of uncertainty of fatigue predictions". For further details on assessment, see Ersdal, Sharp and Stacey (2019).

7.2 Previous Studies on Evaluation of Damaged Tubulars

7.2.1 Remaining Fatigue Life of Cracked Tubular Structures

The UK Department of Energy (D.En) and later HSE initiated several significant projects on the evaluation of damaged tubular structures starting in the 1980s.

OTH 87 259 and OTH 87 278 on Remaining Life of Defective Tubular Joints

Two reports, OTH 87 259 and OTH 87 278, were prepared by the Marine Technology Support Unit (MaTSU) for the UK D.En. to review the remaining life of tubular joints with fatigue cracks (Department of Energy 1987a and 1987b). These show that the remaining life of such components may be assessed from the knowledge of the crack growth so far and the maximum tolerable crack size. This remaining life can be established either by fracture mechanics analysis or from tests of similar components under similar simulated operational conditions. At the time these reports were written, it was considered that the fracture mechanics analyses of complex nodes were limited and as a result, reliance was given to the use of laboratory fatigue crack growth data.

As a result, the OTH 87 259 (Department of Energy 1987a) report proposed a method to assess a significant crack identified by inspection based on the data on crack growth from over 100 tests of fatigue cracks. It is in the review noted that fatigue cracks go through a number of stages in terms of number of stress cycles:

- N_1: first detectable crack;
- N_2: visually detectable crack;
- N_3: through-thickness crack; and
- N_4: loss of stiffness of the joint or final separation of the member (end of test).

Offshore structures are designed against fatigue failure using S-N curves derived from tests on tubular joints where failure is normally defined as penetration of wall thickness (N3).

Data on the remaining number of cycles after a detectable crack to a through-thickness crack will provide insight into the remaining fatigue life of a significantly cracked joint and was the focus of this review as shown in Figure 62. Data on remaining fatigue cracks of length 0.5 times the thickness, 2 times the thickness and 4 times the thickness were studied. Cracks of the length of 0.5 times the thickness were assumed to be the shortest crack lengths that could be identified (with typical thicknesses at the time of 50 mm this represent a 25 mm crack length). The lower bound of the remaining life from a $0.5 \cdot t$ crack was reported as 48%, from a $2 \cdot t$ crack 27% and from a $4 \cdot t$ crack only 13% (t represents the thickness).

In OTH 87 278 (Department of Energy 1987b) a power-law formula for the crack depth–to-thickness ratio was proposed to predict the remaining fatigue life (N), given by $a/t = (N/N_3)^p$ (a represents the crack depth). The choice of the exponent p was discussed in detail (typically it is in the range of 3–5).

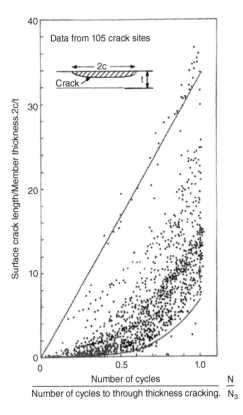

Figure 62 Database of surface crack development in tubular joint fatigue tests. *Source:* Department of Energy (1987a). Offshore Technology Report OTH 87 259 Remaining life of Defective Tubular Joints – and Assessment based on Data for Surface Crack Growth in Tubular Joints Tests. Prepared by Tweed, J.H. and Freeman, J.H., 1987 for the Department of Energy, HMSO, London, UK.

The review OTH 87 259 (Department of Energy 1987a) further indicated that the method may be used to make predictions of the number of surface cracks of particular lengths to be expected when performing inspections, as shown in Figure 63. As can be seen in this figure, for a structure with all welds designed with 20 years fatigue life, after 20 years it is expected that there would be no through-thickness cracks, approximately 3 cracks of lengths 4 times the thickness (t), approximately 5 cracks of lengths $2 \cdot t$ and 20 surface cracks of $0.5 \cdot t$.

Structural Integrity Management of Offshore Installation Based on Inspection for Through-Thickness Cracking

A study by HSE and MaTSU based on the reviews OTH 87 259 and OTH 87 278 was reported by Sharp et al. (1998). The limited life between stages N_3 and N_4 is shown in Figure 64 together with the rapid loss of stiffness after N_3 and approaching N_4. This limited gap between N_3 and N_4 is not necessarily the case for some other geometries, such as brace over chord diameter (normally referred to as β) of one. The review pointed out that although N_3 is a well-defined stage, this is not necessarily true of N_4 depending at which point the test was terminated.

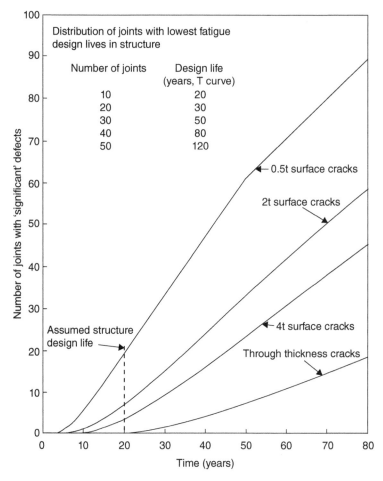

Figure 63 Prediction of number of joints with significant cracks. *Source:* Department of Energy (1987a). Offshore Technology Report OTH 87 259 Remaining life of Defective Tubular Joints – and Assessment based on Data for Surface Crack Growth in Tubular Joints Tests. Prepared by Tweed, J.H. and Freeman, J.H., 1987 for the Department of Energy, HMSO, London, UK.

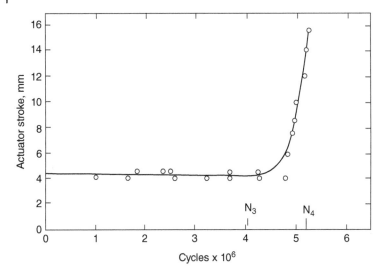

Figure 64   Limited life between N_3 and N_4 for a typical joint test (Sharp et al. 1998). Actuator stroke is an indication of the stiffness. *Source:* Sharp, J.V., Stacey A., Wignall C.M. (1998). Structural integrity management of offshore installations based on inspection for though thickness cracking, OMAE conference 1998.

Figure 65 shows the distribution of N_4/N_3 ratios for a number of different tubular joints tested in fatigue. This shows some variability with the modal range lying in the band 1.2 to 1.3. The cumulative frequency indicates that there is a 40% probability of N_4 occurring at a value of N_4/N_3 of 1.2 (i.e. only 20% in life beyond N3). This again demonstrates the limited life after N3 which is of particular relevance to inspection using FMD and the importance of the reliability of FMD detecting a through-thickness crack when in use. Later analysis of this and more recent data (Zhang and Wintle 2004) indicated a mean value of N_4/N_3 of 1.4, although this varied with geometrical parameters.

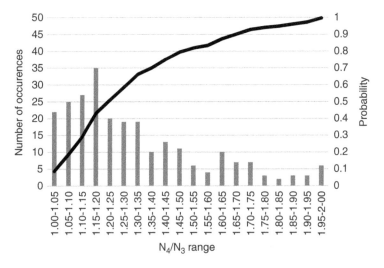

Figure 65   Distribution of N_4/N_3 ratios for 277 different tubular joints. *Source:* Based on Sharp, J.V., Stacey A., Wignall C.M. (1998). Structural integrity management of offshore installations based on inspection for though thickness cracking, OMAE conference 1998.

Flooded member detection (FMD), as further described in Section 4.3.6, is a well-established inspection technique relying on the presence of a through-thickness crack allowing water to penetrate into the previous air-filled space inside the member. Hence the growth of a crack in the region N_3 to N_4 is of particular interest. At penetration of wall thickness (N_3) the crack is usually difficult to detect using global visual inspection (GVI), unless special measures are taken (e.g. cleaning and more close-up visual inspection). The crack growth after through-thickness (N_3) often involves crack branching and deviation of the crack from the weld toe (resulting in complex crack shapes). A typical aspect ratio for crack depth (a) over crack length ($2c$) is ($a/2c$)~0.06 at N_3 (Sharp et al. 1998). Many secondary cracks also develop at lives beyond N_3 (Figure 66) until the N_4 stage is reached; at this stage the joint loses stiffness and load is transferred to other members. The crack, however, can continue to grow until member severance occurs.

RR 224 Growth of Through-Wall Fatigue Cracks in Brace Members

TWI Ltd performed further reviews for HSE (HSE 2004) of the remaining life of tubular brace members based on the same model as used in OTH 87 259 (Department of Energy 1987a), OTH 87 278 (Department of Energy 1987b) and Sharp et al. (1998). This review indicated that the remaining number of cycles at the through-thickness cracking stage was significant and the number of cycles at N_4 was approximately 1.1 N_3, see Figure 67, which is smaller than the reviews mentioned above indicate.

TWI Review for HSE on Assessment of Fatigue Data for Offshore Structural Components Containing Through-Thickness Cracks

The most recent study on this topic was performed by Zhang and Wintle (2004) as reported in Zhang and Stacey (2008). This review includes a comprehensive examination of published work

Figure 66 Tubular joint with large degree of cracking. *Source:* John V. Sharp.

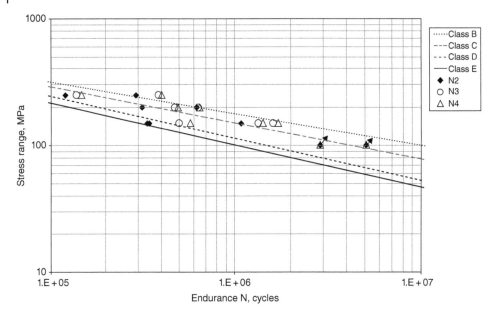

Figure 67 Fatigue test results including N_2, N_3 and N_4 measured cycles. *Source:* HSE (2004). Growth of through-wall fatigue cracks in brace members, Prepared by TWI Ltd (Dr Marcos Pereira) for the Health and Safety Executive (HSE) as Research Report 224. HSE Books. Sudbury. UK. Licensed under Open Government License.

containing data on fatigue lives beyond through-thickness cracking. In total, a database of 281 relevant tests were developed of which fifteen were tested under free water corrosion, eight under cathodic protection and the rest in air. The database was used to perform a statistical assessment of the effects of different testing conditions and geometrical parameters on the remaining fatigue life beyond through-thickness cracking.

Statistical analysis was carried out on N_3 and N_4 endurance of all data collected and fitted to a SN-diagram with a slope of the curve forced to 3.0, similar to the T' design curve for tubular joints (HSE 1995). The N_4/N_3 ratio for the resulting SN-curves were found to be 1.38 and 1.25 for the mean and mean-2SD curves, respectively. However, the review indicates that the scatter of the data is significant to such a degree that the mean endurance is about a factor of 10 longer than that for the mean-2SD curve and even the mean-2SD for the N_4 endurance falls below the T' design curve.

The review indicated that the ratio $Re=(N_4-N_3)/N_3$ illustrating the remaining life after through-thickness cracking depends strongly on chord thickness, loading mode, type of joint and testing environment, although the data showed a large amount of scatter. As a result of this complexity of interactions of many parameters on remaining life of through-thickness cracks, as well as the limited number of tests in some conditions, the review admitted that it was difficult to draw sound conclusions. However, the following conclusions and trends were indicated based on the assessments:

- taking all the results together, the mean value of Re was 0.4 (corresponding to a value of N_4/N_3 of 1.4);
- under axial loading, the stress ratio did not show a significant influence on Re;
- the chord-to–brace diameter ratio (β) did not exhibit a significant effect on Re under axial and OPB loading;

- Re was found to decrease with increasing values of the chord-to-brace thickness ratio (τ), especially under OPB loading;
- Re was found to be dependent on chord wall thickness, decreasing under OPB loading;
- compared to axial loading, higher Re values were achieved under OPB and IPB loadings, especially under OPB loading for joints with a small chord size;
- under OPB loading, T joints achieved a higher Re value than K/KT joints when compared at a chord wall thickness of 6 mm. However, the difference in Re between the two types of joints became small when compared to a chord wall thickness of 16 mm;
- specimens tested in free seawater corrosion and cathodic protection had slightly higher Re values. However, it should be noted that very few specimens were tested in this environment; and
- girth welded joints in plain tubes gave a low Re value, with a mean of only 0.086.

The main purpose of the analysis of these data has been to check the effectiveness of using flooded member detection (FMD) in terms of remaining life after through-thickness cracking.

Although these data provide some indicative information for evaluation, a proper fracture mechanics crack growth analysis in accordance with BS 7910 (BSI 2013) or similar will in most cases be required for an improved evaluation of remaining fatigue life of a cracked joint.

7.2.2 Static Strength of Cracked Tubular Structures

Several projects have looked into the ultimate (static) strength of cracked tubular joints from the 1980s and onwards. Several of these were initiated by UK Department of Energy and later UK HSE. A review of these projects was published by HSE (HSE 2002) named "The Static Strength of Cracked Joint in Tubular Members" (see more on this project later in this section).

A series of research programmes combining finite element analysis, fracture mechanics procedures and model scale testing was undertaken by several universities and TWI. Work by Burdekin and Frodin (1987) involved a combination of finite element analysis and model scale experimental tests on double T joints with through-thickness cracks in the chord (up to 33% of the joint perimeter). Failure by ductile tearing was demonstrated in experimental tests on grade 275 steel. This provided the first demonstration of the significant contribution of mode III loading in T joints with through-thickness cracks under tension loading. Later work by Cheaitani and Burdekin (1994) involved a similar combination of finite element analysis and model scale testing applied to 450 K joints under balanced axial loading, with through–wall thickness cracks at the crown of the tension brace (lengths up to 38% of the weld toe perimeter length). The experimental tests demonstrated that for this configuration, the effect of the presence of the crack was to undermine the support provided by the tension brace against buckling failure of the chord under the compression brace. When this buckling failure started to occur, the severity of conditions at the crack tip increased substantially and tearing started to develop.

The above investigations demonstrated a progressive reduction in the ultimate strength of the cracked joints with crack size, compared to the baseline of an uncracked joint with the same geometry and loading configuration. It was recommended that a safe lower bound assessment of the plastic collapse strength for a given value of crack length could be made by applying an area reduction factor (ARF) to the HSE characteristic strength formulae for the uncracked joint geometry. However, it was further recommended that fracture should be considered separately. This area reduction factor for plastic collapse was recommended as follows:

$$ARF = \left[1 - \frac{cracked\ area}{intersection\ area}\right] \cdot \left(\frac{1}{Q_\beta}\right), \tag{1}$$

where Q_β is a geometric factor. A recommended procedure for defect assessment in offshore structures was published by Burdekin and Cowling (1998).

Two reviews of the work in this field completed up to 1996 were included in Stacey, Sharp and Nichols (1996a and 1996b). The static strength of tubular joints is a design criterion using formulae in codes and standards. However, these codes and standards are based on tests of uncracked joints and normal procedures for the assessment of static strength do not make allowance for the presence of cracks.

Further work was undertaken to extend and validate these initial findings. A major Joint Industry Project organised by TWI had large-scale experimental tests on grade 355 steel together with joint finite element analysis by TWI and UMIST (Hadley et al. 1998). The main part of this test programme was concerned with double T joints of a larger size than the earlier UMIST model scale tests. This work covered through-thickness cracks in the chord (15% and 30% of the weld toe perimeter). The results of this study and other studies were reviewed in a project initiated by HSE (2002) described later in this section.

A paper by Sharp et al. (1998) indicated that data has shown that the ultimate capacity of joints is dependent on the load-bearing cross-sectional area and hence on the crack-aspect ratio when cracks are present. This data indicated that at N_3 the loss in static strength is likely to be in the range of 15–30% depending on geometry. At N_4 the loss is much greater—up to 90% (see Table 41).

The implications from these results for managing structural integrity are of considerable interest, particularly when either GVI or FMD is used as the main inspection technique with detection only at the through-thickness stage or beyond. The interval between inspections is also relevant, since on a five-year cycle with a 25-year design life, a fatigue crack just missed on the first inspection would have reached the $1.2 \cdot N_3$ stage by the next inspection with more significant potential loss in static strength capacity as seen in Table 41 and hence less ability to resist severe wave loading without failing. In planning inspections using FMD, the potential loss in static strength capacity should be taken into account.

It should be recognised that the presence of a large crack in one tubular joint is generally unlikely to lead to structural collapse due to redundancy in the structure. However, two or more cracks in joints within the same failure path could have more serious effects on the platform integrity. In both K and X braced framing there are similar joint types with similar stress concentration factors (SCFs). Most analyses of the damaged state do not allow for more than one component suffering damage at the same time.

Studies were undertaken by MMI reported in HSE RR344 (HSE 2005) on the ultimate strength of multi component damage to structures, which is further described in Section 7.2.3.

Table 41 Variation in Static Strength Capacity with Crack-Aspect Ratio. *Source:* Sharp, J.V., Stacey A., Wignall C.M. (1998). Structural integrity management of offshore installations based on inspection for though thickness cracking, OMAE conference 1998.

Stage in fatigue life	Crack-aspect ratio ($a/2c$)	Reduction in static strength capacity
N_3	0.04–0.10	15–30%
$1.2 \cdot N_3$	0.03–0.06	20–50%
N_4	0.015–0.025	40–90%

The repair options for joints having reached the N_3 stage in fatigue is generally limited to clamps, welding and possibly hole drilling (see Chapter 8). Clamps are often chosen as these provide an alternative load path and hence no further fatigue cracking should take place in the damaged detail. It is important, however, to recognise that these types of repair require significant time for design, fabrication of hardware, installation planning and implementation, which can take up to several months, particularly if it is necessary to wait for a suitable weather window. During this period through-thickness cracks will continue to grow and hence there is an enhanced risk of further structural damage which may not be acceptable.

Most of the experimental work described above has been undertaken on conventional structural steels of grade 275 or 355. A series of tests on high-strength steel (yield strength 700 MPa) including six T and three Y joints has been completed at University College London (Talei-Faz et al. 1999). These joints with a chord diameter of 457 mm had been subjected to prior fatigue testing producing cracks representative of what might broadly occur in service. The results from these tests are given in HSE (2002b) as described below.

A major programme was completed by UMIST and EQE which included finite element analysis studies on the behaviour of circumferential cracks in tubular members subject to axial tension, typical of through-thickness cracks in tubular members which would be detected by FMD (Burdekin et al. 1999). This work found that the widely adopted solutions at the time on the effects of cracks on the ultimate strength of tubular members were not appropriate for use in conjunction with a fracture mechanics treatment.

OTO 1999 054 Detection of Damage to Underwater Tubulars and its Effect on Strength

A review of static strength of cracked tubular joints was published in the report OTO 1999 054 (HSE 1999). Cracks at the chord saddle of a tension-loaded DT joint were studied and the reduction in strength was found to be proportional to the crack area for low β values. For joints with higher β values, the reduction in strength was found to be greater and dependent on the geometric modifier Q_β. The area reduction factor was found to be as given in Equation 1. In case of a DT joint, the intersection area would be the product of the length of the intersection and the chord thickness.

OTO 00 077 Fracture Mechanics Assessment of Fatigue Cracks in Offshore Tubular Structures

HSE report OTO 00 077 (HSE 2002a) described the modelling procedures for plain plates and T-butt joints containing semi-elliptical cracks. Two methods for stress intensity factor (SIF) evaluation were developed, including displacement extrapolation and the J-integral. A very extensive parametric study of plain plates and T-butt joints using this modelling and post-processing procedures was included. This encompassed a variety of crack sizes and attachment geometries.

Fatigue crack growth calculations carried out using the solutions developed in this study highlighted the effects of the weld angle, weld toe grinding, degree of bending (DOB) and chord wall thickness on the predicted fatigue life. The results from these calculations underestimated the fatigue life to be approximately half that of the experimental tests. This discrepancy was shown to be primarily due to ignoring the combined effects of load shedding and the intersection stress distribution.

OTO 00 078 Static Strength of Cracked High-Strength Steel Tubular Joints

The OTO 00 078 review (HSE 2002b) included nine static strength tests on high-strength steel tubular welded joints (T and Y) made from SE702 (yield strength 700 MPa). These consisted of six T joints and three Y joints, which were tested to failure (through-thickness fatigue crack) in either axial or out-of-plane bending loading. The diameter was 457 mm and thickness 16 mm. These specimens were available from two previous fatigue test programmes conducted on a high-strength jack-up steel (SE702). All test specimens had at least one through-thickness fatigue crack at the weld toe. In terms of the normalised static strength capacity, it was found that the mean curve for the high-strength steel specimens was slightly lower (by about 5%) than the yield strength for 355D steel as per BS7191 (BSI 1989). It was concluded that the slight difference between the results for the SE702 specimens and similar tests on 355D material could be related to the crack path as it was noted that for the SE702 tests, the crack remained close to the weld toe at all times. For the 355D tests, the crack often branched away from the weld toe and grew into the parent plate.

OTO 01 080 The Static Strength of Cracked Joints in Tubular Members

The report OTO 01 080 performed by UMIST for HSE (HSE 2002c) included a review of research between 1980s and 2000 into the effect of cracks on the static strength of tubular joints, including both plastic collapse and fracture for a range of different types of joint geometry (T, Y, DT and K joints). Cracks in the weld toe in the chord and as circumferential cracks in brace members were included and brittle fracture and ductile tearing fracture were considered. The review indicated that in general, joints in tension usually fail by tearing and fracture failure, whilst failure of joints in compression often occurs by buckling of the walls of the tubular member. The report concluded that the plastic collapse strength of cracked tubular joints can be estimated with use of an area reduction factor (see Equation 1). The fracture strength of the cracked joint or member needs to be determined by a fracture mechanics analysis, preferably by use of a geometry-specific finite element analysis that includes the crack. Approximate methods based on BS7910 (BSI 2019) defect assessment procedures were also described including recommendations for determining the stress intensity factor ratio K_r and the collapse load ratio L_r.

Both the experimental test results and the finite element analysis predictions included in this review showed significant scatter. However, these also showed a general trend for the strength of cracked tubular joints to follow the uncracked characteristic strength multiplied by the above-mentioned area reduction factor (see Figure 68). The test results in Figure 68 which fall significantly below the HSE characteristic strength with the before mentioned reduction factor, are those of Gibstein (1981), Machida et al. (1987), Skallerud et al. (1994) and some of the UCL high-strength steel tests. The results plotted in Figure 68 represent the initial crack length for the test and the maximum load reached, by which time the crack length would have increased considerably from the initial condition.

The review highlights that it was vitally important to recognise that any prediction based on the strength of uncracked joints modified on an area reduction basis considers only plastic collapse behaviour and takes no account of fracture. The tests by Gibstein (1981) and Machida et al. (1987) were mentioned as examples of failure by brittle fracture which should have been assessed taking this possibility into account. However, the review indicated that virtually all of the other tests showed ductile crack extension and tearing behaviour before final failure and the implications of this should also be taken into account.

The review further indicated that the high-strength steel tests by UCL (Talei-Faz et al. 1999) did follow the same general pattern as those for the general structural steel tests. However, it was noted that the plastic collapse loads for the high-strength steel would be proportional to the yield stress and hence higher than those for the general structural steel. As a result, to achieve maximum failure loads in the experimental tests for high-strength steel, the fracture toughness would also have to be increased in the same proportion. If such an increase in toughness is not possible, the actual failure loads are more likely to be influenced by fracture rather than plastic collapse. As a result, these are likely to fall below the simple area reduction correction factor prediction. This is likely to be the explanation for the results falling below the area reduction line in Figure 68 including two of the high-strength steel UCL tests.

OTO 01 081 Experimental Validation of the Ultimate Strength of Brace Members with Circumferential Cracks

The HSE report OTO 01 081 (HSE 2002d) describes a series of twenty static strength tests which were carried out on both small-scale and large-scale plain tubular members. These had either through-thickness cracks or surface cracks lying in the circumferential direction. The through-thickness cracks had lengths varying from about 10% to 80% of the circumference of the tubes. The

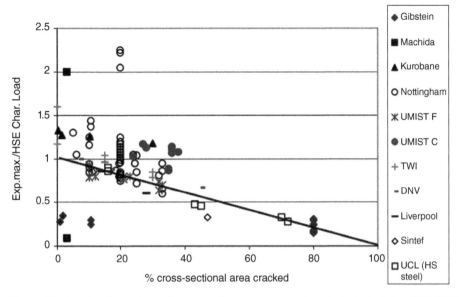

Figure 68 Results of experimental tests on ultimate strength of cracked tubular joints (HSE 2002c) where the uncracked tension strength has been used as the basis for T and DT joints, whereas the uncracked compression strength has been used as the basis for K joints. *Source:* HSE (2002c). Offshore Technology Report OTO 2001/080 The static strength of cracked joints in tubular members, Prepared by UMIST (Professor F M Burdekin) for the Health and Safety Executive. HSE Books, Sudbury. UK. Licensed under Open Government License.

surface crack tests had depths of 80% of the thickness and lengths corresponding to the through-thickness cases.

The results were analysed and compared to the BS7910 (BSI 2013) procedures for assessing ultimate strength or fracture behaviour and also to modified procedures which were based on previous finite element analysis research at UMIST. It was found that improved assessments were given by the UMIST modified procedures. For through-thickness cracks, the results indicated that the BS7910 approach could be non-conservative whilst for part thickness cracks, the results indicate that BS7910 may be over-conservative.

7.2.3 Effect of Multiple Member Failure

MMI undertook a study for HSE RR344 (HSE 2005) on the ultimate strength of multi component damage to structures, including assessment of the integrity over typical inspection intervals. Member failure in jacket structures, leads to stress redistribution which can enhance the potential for failure in other members in the same structure; eventually, this may lead to collapse. This study included analysis of three four-legged platforms (single diagonal, X and inverted K braced framing). These were analysed using both linear-elastic and non-linear pushover analyses to enable the ultimate strength of the structures to be determined in all single-member failure conditions. In addition, many dual failed member combinations for each type of jacket were similarly analysed. The most significant stress redistributions combined with non-linear pushover analyses of the damaged structures were used to assess the platform reliabilities. These were determined for three cases for a structure with:

- single-member failure only;
- dual-member failure but ignoring the effects of stress redistributions; and
- dual-member failure incorporating the effects of stress redistributions.

It was shown that for the cases examined the effect of stress redistributions and dual-member failure reduced the predicted platform reliability. The most dramatic reduction (by two orders of magnitude) was for the single diagonal braced structure where the failure of the second brace had a marked effect on the ultimate strength. This demonstrated the low redundancy of this type of structure. The ultimate strength of the X braced structure was in contrast not significantly affected by the failure of the second member. It should be noted that Ersdal (2005) found that the reduction in the ultimate strength of an X-braced fixed steel structure could as a worst case be estimated by a factor of 0.8^n, where n is the number of braces that had failed.

The HSE RR344 report recommended that for all structures some non-linear pushover analyses including the effects of dual-member failure and stress redistribution should be performed to identify cases which, if ignored, could result in an overly optimistic prediction of platform failure. If such cases are found, then these should be included in reliability assessments used for inspection planning.

The report also analysed the effect of inspection intervals. If these intervals are sufficiently short to detect any damage, the structure can be repaired in a timely fashion, reducing the damaged structure's vulnerability to storm loading. In addition, this approach also acts to limit the acceleration of fatigue cracking. The report noted that even for a 2- to 3-year inspection interval, there would be a high probability of collapse for low redundancy structures (diagonally braced) due to the probability of two members having failed. The report concluded that such structures would be ideal candidates for the fitting of continuous monitoring systems which could detect a failed member between inspections and enable a repair to be undertaken.

The report also advocated further research on this topic as only single-wave direction analyses had been performed in this study. Particularly for acceleration of fatigue the effect would need to be averaged over all directions.

Ersdal (2005) investigated a similar approach to multiple member failure in one fixed steel structure typical at that time on the Norwegian continental shelf, with a 4-year inspection interval. The

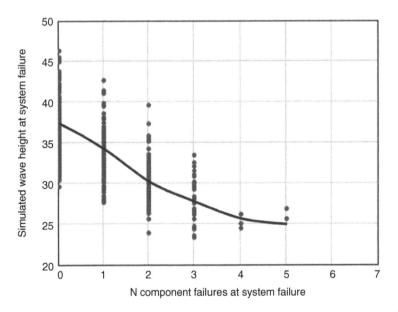

Figure 69 Wave height at system failure with increasing number of component failures. *Source:* Ersdal, G. (2005). Assessment of existing offshore structures for life extension. Doctorate thesis, University of Stavanger, Norway.

wave height experienced by the structure at failure was estimated as shown in Figure 69. As clearly indicated in this figure, when one or more components have failed due to fatigue prior to overload failure, smaller wave heights caused the structure to fail. This clearly demonstrated the weakening of the structure as the number of component failures increased.

The effect of multiple member failure on the inspection intervals is presented in Section 6.2.5, indicating the need for a decreased inspection interval when several components are regarded as critical. It also suggests a method to determine the necessary intervals.

7.2.4 Corroded Tubular Members

Lehigh University in Pennsylvania, US, ran an important project on the remaining strength of damaged structural members under their Advanced Technology for Large Structural Systems (ATLSS) engineering research programme in the 1990s. Several projects, including the project "Residual Strength of Damaged and Deteriorated Offshore Structures" were performed for the offshore industry. Many financed by the US governmental agency MMS and some also by oil companies and European institutions.

Offshore structures are subjected to a corrosive environment and despite the use of protective coatings and cathodic protection systems, corrosion damage does occur. The effect of corrosion is primarily a reduction of the wall thickness. For a tubular section this leads to the following effects:

- reduced gross section area and hence a reduction in load-carrying capacity;
- increasing susceptibility to local buckling in the corroded area;
- introduction of an eccentricity and an internal moment if corrosion is asymmetrical around the circumference, causing additional compressive stresses; and
- variation in wall thickness, which can create a stress concentration in the corroded areas.

Corrosion on tubular members is typically classified either as uniform corrosion or patch corrosion (non-uniform). Uniform corrosion implies a relatively constant reduction of wall thickness around the entire circumference. Evaluation of the capacity of a tubular member exposed to uniform corrosion is relatively straightforward as the standardised capacity formulae can be used adjusted to the thinner wall thickness. However, if the uniform corrosion is present only on parts of the full length of the tubular member, this may be slightly conservative. Patch corrosion is more difficult to evaluate. Patch corrosion is asymmetrical around the circumferential resulting in an eccentricity due to the shift in the centre of gravity of the cross-section as illustrated in Figure 70.

Ostapenko et al. (1993) performed tests of two salvaged tubular members with corrosion damage from decommissioned Gulf of Mexico fixed offshore steel structures after an estimated twenty to thirty years of service. In the tests both specimens failed by local buckling in corroded areas and the report recommended using an equivalent thickness based on the area with the greatest amount of corrosion (average over a rectangular area extending on radius around the circumference and one-third radius in the longitudinal length). The specimens failed at loads of 42.6% and 77.3% of the capacity (area times yield stress).

Hebor (1994) performed more tests on the strength of corroded tubular members. Twelve steel tubular test specimens with various forms of single patch of simulated non-uniform corrosion damage were tested as part of this master thesis project. The ratio of corroded wall thickness–to–original thickness, t_r/t and the angle exposed to corrosion were used to define the severity of the corrosion. Length-to–radius of gyration (L/r) ratios for the specimens ranged from 10 to 19, with corresponding length-to-diameter (L/D) ratios ranging from 3.5 to 6.4. The low L/r ratios of the

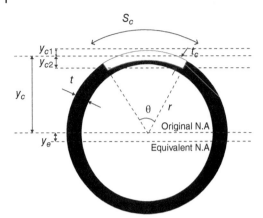

Figure 70 Model of patch corroded tubular member.

specimens ensured that the mode of failure would be local buckling and not overall global buckling in the corroded specimens. The experimental test matrix also included non-corroded stub-columns and repaired "corroded" specimens. The experimental study was accompanied by appropriate analytical analyses and a parametric study using non-linear finite element analysis. The experimental test results showed a capacity reduction of up to 32% for corrosion damaged specimens. The reduction in strength was most severe for specimens with larger patch dimensions and reduced wall thickness. The behaviour and performance of the repaired specimens was found to be influenced by the bond strength of the grouting material. However, the repair was determined to be successfully implemented with relatively low bond strength grout.

A comparison between the damaged and un-damaged columns showed that the capacity was greatly reduced due to the corrosion. The reduction was reinforced by the fact that an eccentricity arose due to the geometric change resulting from asymmetrical metal loss. In addition, high stresses occurred in the corroded area.

The work of Hebor (1994), also reported in Hebor and Ricles (1994), emphasized inelastic local buckling of short tubulars with a single patch of corrosion subjected to monotonic axial loading. The work performed by these researchers resulted in a recommended set of residual strength equations.

$$\frac{P_u}{P_y} = 1.0 - 0.001\frac{D}{t} + 0.052\frac{t_r}{t} - 0.0026 \cdot \theta + 0.0028\frac{t_r}{t} \cdot \theta \leq 1.0$$

These equations were recommended to be used to predict the approximate ultimate strength.

Different ways of simulating the patch corrosion damage are possible: cut-off, rounded and sinusoidal, as shown in Figure 71.

The type of patch corrosion profile has been shown (Atteya et al. 2020) to be important for the calculation of the ultimate strength of the tubular column. This is likely to be a result of higher stresses occurring in the corroded area when the cut-off is sharper. It should be noted that the work of Hebor (1994) and Hebor and Ricles (1994) are based on a rounded patch.

Ostapenko et al. (1996) conducted further experiments on 11 specimens with patch corrosion damage simulated by grinding. The specimens were divided into two groups: (1) P5 to P9 with a slenderness ratio of 14.7, and (2) specimens P10 to P15 with a slenderness ratio of 6.4. Their corroded patch profile was of the sinusoidal type. The primary failure mode of these test specimens was local buckling.

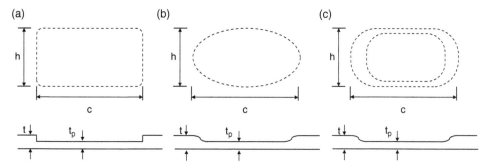

Figure 71 Patch corrosion profiles: (a) cut-off, (b) rounded and (c) sinusoidal.

Hestholm and Vo (2019) performed additional experimental tests of tubular columns with simulated corrosion and compared the results to standardised capacity formulae for tubulars with corrosion patches. Hestholm and Vo (2019) simulated cut-off profiles for the patch corrosion.

An overview of all known test specimens of tubular columns with patch corrosion is illustrated in Figure 72.

Standards such as NORSOK N-004 (Standard Norge 2004) included patch corrosion into the strength calculations by including the corrosion in the formulae for dented tubulars by an equivalent dent. The equivalent dent depth that is intended to illustrate the corrosion damage is given by:

$$\delta' = \frac{1}{2}\left[1 - cos\left(\pi \frac{A_{corr}}{A}\right)\right]D$$

where δ' is the equivalent dent depth, D is the diameter of the tubular column, A is the cross-section area of the tubular column and A_{corr} is the cross-section area of the corroded tubular column.

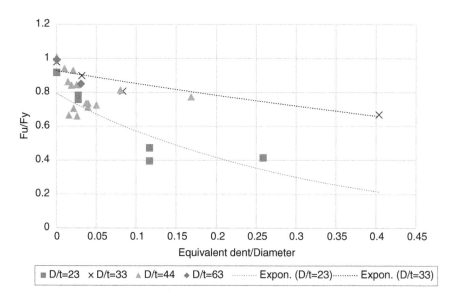

Figure 72 Experimental results of patch corroded tubulars.

Later, studies by Lutes et al. (2001) indicated that the American Petroleum Institute design formulae for inelastic buckling were sufficiently conservative to predict the ultimate capacity of typical tubular bracing members without explicitly considering patterns of corrosion. More recently, Øyasæter et al. (2017) have studied modelling patch corrosion in finite element analysis and comparing these to analytical formulae.

Vo et al. (2019) found that when comparing experiments with the formulae provided in NORSOK N-004, the internal bending moment due to eccentricity needed to be included, which is consistent with the findings of Hebor (1994). If this bending moment was included, the NORSOK N-004 (2004) formula would provide conservative results. The eccentricity will introduce an internal bending moment due to the asymmetric corrosion:

$$y_e = \frac{-A_c}{A - A_c} \cdot \frac{2 \cdot \left(\frac{D}{2} - \frac{t_c}{2}\right) \cdot \sin\left(\frac{\theta}{2}\right)}{\theta}$$

Atteya et al. (2020) has used the tests discussed so far and compared the results to modern non-linear finite element analysis and the ISO 19902 (ISO 2007) and NORSOK N-004 (Standard Norge 2004) formulae as shown in Figure 73 and Figure 74.

In general, the results of Atteya et al. (2020) indicate that ISO 19902 (ISO 2007) and NORSOK N-004 (Standard Norge 2004) still can be used with reasonable accuracy for evaluations of bracing members with corrosion provided that the internal bending due to corrosion is included. However, more accurate results can be obtained by non-linear finite element analysis, provided that a calibrated finite element model is used. The calibration of the finite element model should be performed to match the results of tests such as those mentioned in this section.

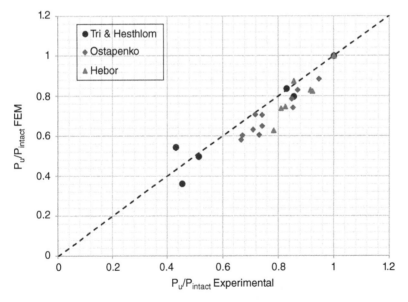

Figure 73 Q-Q plot comparing experimental and FEA capacities of tubulars with simulated patch corrosion damage (bias 1.1, CoV 8.5%). *Source:* Based on Atteya, M., Mikkelsen, O., Oma, N. and Ersdal, G. (2020). Residual strength of tubular columns with localized thickness loss. In Proceedings of the 39th International OMAE2020 Conference. ASME.

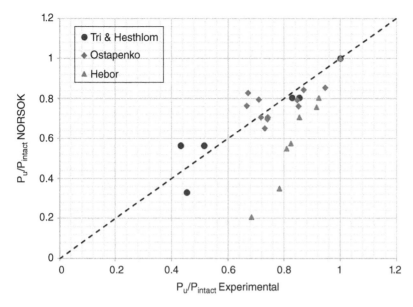

Figure 74 Q-Q plot comparing experimental and NORSOK N-004 capacities of tubulars with different amount of simulated patch corrosion damage (bias 1.13, COV 25%). *Source:* Based on Atteya et al (2020).

7.2.5 Dent and Bow Damage to Underwater Tubulars and Their Effect on Strength

Report OTO 99 084 (HSE 1999) reviewed the effect of dent and bow damage to underwater tubulars and the detection of these. The actual work was performed by MSL. The contents of this report are also discussed in Section 4.2.7.

Report OTO 99 084 (HSE 1999) gave an overview of all earlier experimental and analytical work on dented and bowed tubulars and provided a comparison of these results with the design formulae suggested for implementation in ISO 19902 (ISO 2007), which was the suggested approach at the time. The review states that up to the release of this report, no formal procedures were available for determining the residual strength of damaged offshore components. Several relevant methods and computer software were available in the public domain, but none of these were generally accepted and included in standards.

The potential for damage to tubular members on fixed steel structures due to vessel impact or dropped objects was recognised in the middle 1970s. This resulted in several studies on the effect of dents and bows on strength, which investigated them both through physical tests and by modelling.

The first studies of damaged tubular members (in the form of lateral bending and local denting) were reported by Smith et al. (1979) for small-scale (diameter of 50 mm) cold-drawn tubes. Later further work was performed by Smith et al. (1980), Smith (1983) and Smith et al. (1984). These investigations reported a significant loss in strength in tubulars with relatively minor damage. They also compared small-scale tests to large-scale test, which showed no significant scale effects.

Further studies were carried out by Taby et al. (1981) and Taby and Moan (1985) where more tests were performed for larger-scale models (diameters up to 150 mm). The computer program (DENTA) for calculating the load deflection behaviour for dented tubulars were also developed (Taby and Moan 1987). Taby (1986) also included tests for higher diameter-to-thickness (D/t) ratios suitable for semi-submersible structures and for lower D/t ratios suitable for jack-ups. These specimens had a D/t ratio in the range from 12 to 135 and a reduced slenderness ratio in the range 0.5 to 1.0.

The slenderness and reduced slenderness ratio for a tubular member are defined as:

$$\lambda = \frac{L}{r} \, \bar{\lambda} = \frac{L}{r \cdot \pi} \sqrt{\sigma_y \Big/ E}$$

where L is the length of the member, r is the radius of gyration, σ_y is the yield strength and E is Young's modulus.

During the mid-1980s the first set of tests involving the use of dynamic loading to generate damage was conducted by Cho (1987). These were followed by an impressive series of 160 specimens test by Allan and Marshall (1992). These tests were performed on 100 mm diameter tubulars of various thicknesses. Landet and Lotsberg (1992) added even more tests with combined axial compression and bending.

In addition, several tests were performed as part of the ATLSS projects notably Rickles et al. (1992), Ostapenko et al. (1993), Bruin (1995) and Ricles et al. (1997). Padula and Ostapenko (1987) included tests of the axial load capacity under fixed-fixed end condition of two large diameter tubular steel columns with dents (2.7% and 5.5% of the specimen diameter). One of the specimens showed no reduction in ultimate axial capacity when compared to the undented specimen. The other specimen showed a 6% reduction in axial capacity.

Padula and Ostapenko (1989) included a simple method for computing axial load versus axial shortening for tubular members with dents and out-of-straightness, developed based on a regression analysis of published test results and finite element analyses.

Ostapenko et al. (1993) performed tests of twenty tubular columns to investigate the residual strength of dented and corroded columns. Eighteen of these were large-scale tubulars where 11 were salvaged from actual offshore structures, seven were fabricated and two were small-scale tubulars. All fabricated and small-scale manufactured specimens and the four salvaged specimens were dented at mid-length (5% to 15% of diameter). Various end fixities were applied. The seven remaining salvaged specimens were tested without denting and three of these were corroded, one was badly bent and three were straight and had no sign of corrosion. The report indicates that analytical methods could relatively accurately predict the ultimate strength for dented and undented columns. It is further reported that the analytical formulae were slightly conservative in estimating the post-ultimate behaviour. However, the capacity and behaviour of the two patch corroded specimens failing by local buckling was not well predicted by the analytical methods. A stress amplification factor (k) to account for the lack of symmetry in the patch corroded cross-section and the general out-of-straightness at this location was introduced and a four-step method to achieve axial capacity of local buckling was established (1: determine average thickness over a rectangular area extending one radius around and one-third radius in the longitudinal direction, 2: compute local buckling stress using this average thickness according to recognized formulae, 3: compute the cross-sectional properties necessary for calculating the stress amplification factor k, 4: compute the capacity by the effective area and local buckling stress divided by the stress amplification factor k).

Many of these tests are shown in Figure 75 sorted according to their D/t ratio. A clear scatter can be seen which will partly be due to different slenderness, different level of bows and other factors not clearly indicated in this figure but also randomness in test results. Mean lines for D/t=33 and D/t=120 are indicated. A significant reduction in axial capacity is seen even for small dents and the few tests performed with dent depts up to 0.5 diameter indicate more or less a total loss of capacity.

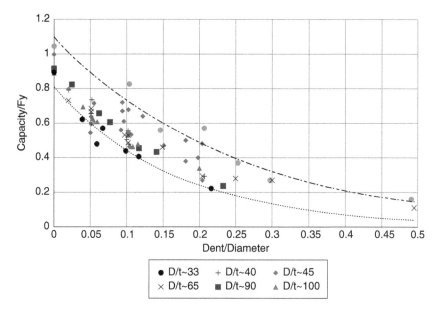

Figure 75 Capacity of unrepaired dented joints.

Further studies have been performed on the strength of repaired dented tubulars, which will be discussed further in Chapter 8.

OTO 99 084 (HSE 1999) further used this database of tests after a certain screening to evaluate the proposed design formulae for ISO 1992 and other possible analytical results. The review concluded that the DENTA procedure is capable of considering the full range of geometrical, material and dent and bow damage parameters likely to apply to North Sea structures. The proposed ISO formulations have a similar range of capabilities, but the approach has a cut-off that limits its application to cases where the dent depth–to-thickness ration is less than 10.

Frieze et al. (1996) studied the prediction accuracy of the ISO formulation and the DENTA program as shown in Table 42. It can be seen that once a significant number of test results is considered, CoV > 0.14 in the case of compressive residual strength. Under combined loading, the CoV is around 0.22. The so-called ISO MTCG (ISO panel TC67/SC7WG3/P3 Members Technical Core Group) data is a combined database used by the ISO panel responsible for this part of the standard developed by Chevron Petroleum Technology in 1996.

A subset of these data has been recalculated according to the ISO formulae (which are identical to the formulae in NORSOK N-004) and Q-Q plotted against the results of the laboratory tests in Figure 76.

Other theoretical investigations were performed by Ellinas (1984) and Ueda and Rashed (1985). Most experiments and theoretical work were performed for tubular elements with geometry usually found in fixed steel structures.

ISO 19902 (ISO 2007) contains a section on the effect of dents on tubular members. Equations are given for the effect on strength and stability for dented members subjected independently to axial tension, axial compression, bending or shear. The studies by Vo et al. (2019) indicates that the exclusion of eccentricity formed by the dent needs to be included in the calculation of strength; otherwise, NORSOK and ISO formulae significantly overestimate the strength. Similarly, the bow needs to be included in the calculations for strength to match the experimental results.

Table 42 Prediction Accuracy.

Procedure	Source of comparison (number, bias and CoV)		
	Taby et al. (1981)	Taby and Moan (1985)	ISO MTCG (22)
DENTA	21	44	240
	1.020	0.992	1.019
	0.091	0.082	0.140
ISO MTCG			191
Axial compression			1.103
			0.146
ISO MTCG			74
Combined loading, dent in compression			1.203
			0.213
ISO MTGC			9
Combined loading, dent in tension			1.545
			0.225

Source: Frieze, P.A., Nichols, N.W., Sharp, J.V. and Stacey, A. (1996). Detection of Damage to the Underwater Tubulars and its Effect on Strength, OMAE Conference, Florence 1996.

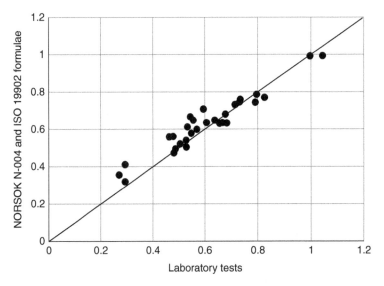

Figure 76 Q-Q plot of results of laboratory tests versus calculations according to NORSOK N-004 and ISO 19902. *Source:* Based on Vo, T., Hestholm, K., Ersdal, G., Oma, N. and Siversvik, M. (2019). Buckling capacity of simulated patch corroded tubular columns – laboratory tests. Second Conference of Computational Methods in Offshore Technology (COTech2019). IOP Publishing.

7.2.6 Studies on Assessment of System Strength

If linear analysis of structures fails to provide code acceptance, non-linear analysis may be an satisfactory approach to determine the strength of structures. Theories for the non-linear collapse (push-over analysis) of jackets have been established by Søreide and Amdahl (1986). The approach was further developed by Hellan (1995) and Skallerud and Amdahl (2002).

Tests to investigate the non-linear collapse (pushover analysis) of jackets were undertaken in a Joint Industry Project supported by the UK Department of Energy and a number of operators of offshore structures (Bolt 1995). This programme of tests (JIP Frames project), established in 1987, was undertaken initially by the Steel Construction Institute and later by BOMEL and involved testing large instrumented two-dimensional frames to collapse. Each of these frames was approximately 15 m high and 6 m wide, involving both X and K joints. A loading assembly including jacks was also constructed to allow the frame to be tested to failure. It was found that the frames exhibited different sequences of member and joint failure and provided valuable insight into jacket response to large loads.

These tests allowed the generation of non-linear response curves for the frames and the results allowed comparison with predictions from numerical models. Individual tests on X and K joints of a similar geometry were also undertaken to compare with the performance of the joints in the frame.

Phase II of the JIP Frames Project included four large-scale collapse tests of K-braced frames in which both gap and overlap K joints were the critical components. The local failure modes differed from typical isolated component tests yet were representative of structural damage observed following Hurricane Andrew. These frame test results therefore provided important insight to the ultimate response of jacket structures.

There was a follow-on project which involved benchmarking of non-linear computer programmes to predict the behaviour of the tests (Nichols et al. 1994). It was important that this benchmarking was undertaken blind and those involved in the analyses were only provided the actual test results at the end of the trials. The range of predicted versus experimental reserve and residual strengths was found to be varied depending on the type of behaviour. For each test at least one programme gave a good prediction of the overall performance. However, many of the analyses over-predicted the actual capacity. In one of the frame tests, joint failure dominated and many packages failed to predict this.

These tests and the benchmarking of non-linear computer programmes provided new information on the performance of frames containing joints and brace members and enabled the programmes to be refined to match actual performance. It was recognised that these tests were large but not at full scale. The USFOS programme (SINTEF 1988) was one such programme that learnt from this data and is recognised as one of the leading programmes in this field to this day. A guideline for performing non-linear collapse analysis as a result of these projects was developed in the Ultiguide project (DNV 1999). Later, recommended practice DNVGLRPC208 for non-linear analysis was developed by DNVGL (2019).

Non-linear pushover analysis may be performed in both a limit state and partial factor format and an allowable stress format. The former is used in DNVGL-RP-C208 (DNVGL 2016) and the latter is used in, for example, API RP-2A (API 2014). The format for these approaches is illustrated in Table 43.

An illustrative Q-δ curve (load versus deflection curve) is shown in Figure 77, for both an intact structure and a damaged structure.

The type of framing selected at the design stage can have a significant influence on the non-linear structural strength. The non-linear strength is an important input in designing inspection plans for structures and, for example, a high RSR value may compensate for limited inspection using a global inspection approach such as GVI. The non-linear strength will also influence the need for repair as further discussed in Section 7.2.8.

7.2.7 PMB AIM Project for MMS

As mentioned earlier in Sections 3.3.3 and 6.2.1, the US Mineral Management Services (MMS) undertook the AIM project to manage the integrity of existing fixed steel platforms in the Gulf of Mexico. The project was carried out in four phases by PMB Engineering Group as a Join Industry Project (PMB 1987, 1988 and 1990).

Table 43 Format for Non-Linear Pushover Analysis.

Limit state and partial factor format	Allowable stress format
Ultimate strength (or ultimate collapse capacity) R_{ult} is calculated based on characteristic values of strength linked to the characteristic values of loads.	The reserve or damaged strength ratio (RSR and DSR) are used as the ratio between the design loads and the collapse capacity of the intact or damaged structure. The RSR is defined as $RSR=Q_u/Q_d$ and the $DSR=Q_r/Q_d$ where Q_u is the load that ultimately results in the collapse of an intact structure, Q_r is the similar load for a damaged structure and Q_d is the design load for that structure (typically at the annual probability of 10^{-2}). Typical values of the required RSR range from 1.6 to 2.4 for critical structures (manned structures).
The limit state check is then $(\sum Q_i \cdot \gamma_i) \cdot \gamma_m \leq R_{ult}$ where Q_i is the load i, γ_t is the partial factor for load i and γ_m is the material factor. An extra safety factor for non-linear analysis is often recommended due to the uncertainties introduced in advanced analysis.	

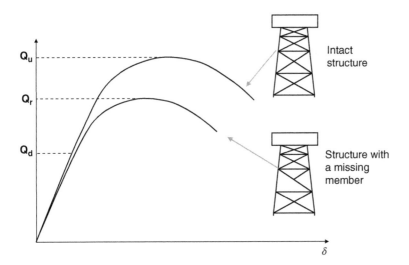

Figure 77 Illustrative Q-δ curve for a jacket structure with an indication of the design load level ($\boldsymbol{Q_d}$) and the collapse load level in intact ($\boldsymbol{Q_u}$) and damaged ($\boldsymbol{Q_r}$) situation, where $\boldsymbol{Q}$ along the vertical axis is load and δ along the horizontal axis is deformation.

The AIM Phase III programme (PMB 1988) concentrated on assessments of damaged conditions, consequences of the platform failure database and AIM evaluation guidelines. For the damaged condition assessment an eight-legged platform (C) was selected, installed in 1970. A detailed structural assessment was carried out comparing several damage states with the reliability of the intact structure. Five cases were evaluated, which were:

- damaged members near waterline (e.g. from boat impact): it was found that the platform capacity dropped by between 5 and 15% depending on how many members were damaged;
- damage near base of jacket (e.g. from dropped objects), in the form of completely severed members: in this case the platform capacity dropped by 10–12%;

- interior horizontal damage (missing interior horizontals and a damaged transverse diagonal): in this case the platform capacity dropped by about 20%, with the damaged diagonal contributing most to the loss in capacity;
- corrosion damage—this case considered overall corrosion damage together with localised corrosion in the splash zone: the overall corrosion lead to a 35% drop in capacity and the splash zone corrosion to a 20% drop in capacity; and
- foundation defects—two piles were considered underdriven by 37 m (out of 82 m): it was found that capacity dropped by between 5 and 17% depending on loading direction.

Given the present-day structural analysis methods, it is slightly unclear how these reductions in capacity were obtained. However, they are interesting as indications of the effects of damage.

In the AIM III report (PMB 1988) it was considered that the cases examined provided the start of a "damaged condition library" for offshore structures. These assessments offered an opportunity to make an initial evaluation of a damaged structure.

The second part of Phase III was intended to provide operators with a starting point in assessing the potential consequences of a platform failure, thereby helping to develop the basis for decisions on structural integrity. Consequences included injuries to personnel, pollution, salvage and structural replacement. The outcome of this part of phase III has limited interest for inspection and repair.

The third part of Phase III was to develop a platform database of the consequences of past failures of platforms in the Gulf of Mexico. The database would summarise known failures due to storms. In some cases, damage from these severe storms was later repaired and the platforms returned to service. However, the database included only damaged structures taken out of service. Repaired damage was excluded from the database.

7.2.8 MSL Significant JIP for MMS

As part of a Joint Industry Project, MSL (2003) prepared guidelines for defining if a sustained damage to existing fixed steel structures (jackets) was significant or not and gave further advice on how to perform more in-depth examination if needed. The study related the effect of damage to the effect it had on the reserve strength ratio (RSR) values.

The review considered denting and bowing damage due to vessel impact and dropped objects, uniform corrosion, local corrosion (holes) and cracking or members and joints to be the most frequent damage on fixed steel structures.

In the review, robustness (redundancy) of fixed offshore steel structures was categorised by their type of framing (see Figure 78), their vintage (pre- or post-1980) and the number of legs.

The platform robustness for these configurations of fixed offshore steel structures was in the review defined as shown in Table 44.

The guideline includes a review of the capacity of damaged members for various damage types based on standards such as ISO 19902 (ISO 2007) and NORSOK N-004 (Standard Norge 2004), including:

- impacted tubular members;
- tubular members with uniform corrosion;
- tubular members with local corrosion; and
- cracked members and joints.

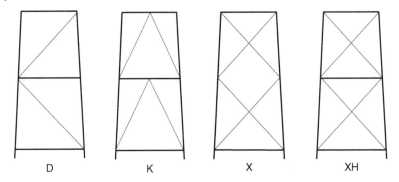

Figure 78 Configurations of fixed offshore steel structures (D: diagonal braced, K: K-braced, X: X-braced and XH: X-braced with a horizontal).

Table 44 Definition of Platform Robustness (MSL 2003),

Platform robustness		Number of legs				
		3	4	6	8	Multi-legged
Brace configuration	K	Non-robust	Non-robust	Non-robust	Moderately robust	Moderately robust
	D	Non-robust	Non-robust	Moderately robust	Moderately robust	Robust
	X	Non-robust	Moderately robust	Moderately robust	Robust	Robust
	XH	Moderately robust	Moderately robust	Robust	Robust	Robust

Eight typical jacket platforms of the GoM were studied with various forms of damage either in the splash zone or in the submerged zone. The conclusions of the analysis of these platforms were summed up as shown in Table 45.

In the review it was found that for corrosion damage the strength loss of the global structure was directly proportional to the remaining thickness, which, for example, implies that a 25% thickness loss would lead to a 25% reduction in the RSR value. The total reduction of the RSR due to member and joint local damage was found to be dependent on the structural robustness (non-robust or moderately robust), location of the damage (submerged zone, atmospheric and splash zone) and the type of member (brace or leg) where the damage had occurred.

In the project, significant damage was defined as damage causing the capacity of the structure to reduce by 10% or more. As a result, the following damage scenarios are likely to be of no significance:

• general corrosion where the reduction in thickness is less than 10% of the original thickness;
• minor to moderate damage to horizontal chords of K-bracings (with some limitations);
• damage to members that does not directly or indirectly participate in the collapse mechanism; and
• damage to structures caused by failure of the soil.

Table 45 Effect on Damage on Non-Robust and Moderately Robust Fixed Offshore Steel Structures (MSL 2003).

	Non-robust platforms		Moderate robust platforms	
	Submerged zone	Atmospheric or splash zone	Submerged zone	Atmospheric or splash zone
50% loss of member	<15% reduction in RSR	<15% reduction in RSR (except legs) <25% reduction in RSR (legs)	<15% reduction in RSR	<10% reduction in RSR (except legs) <15% reduction in RSR (legs)
100% loss of member	<35% reduction in RSR	<25% reduction in RSR (except legs) <45% reduction in RSR (legs)	<25% reduction in RSR	<20% reduction in RSR (except legs) <35% reduction in RSR (legs)
Failure of pile-leg weld at jacket top	<40% reduction in RSR		<35% reduction in RSR	

The evaluation and assessment procedure proposed in the review included the following steps:

- categorisation of the typical damage expected to be found in an offshore survey (dent and bows, uniform or local corrosion, cracked member or joint);
- definition of the damage scenario (location, type and amount) based on the survey information available. The review recognises that, for example, an inspection finding from a flooded member detection may not provide sufficient information for determining the cause and amount of the damage.
- deriving an equivalent wall thickness to represent the experienced damage and the member capacity reduction based on formulae and tables provided in the report;
- derive an expected reduction in the RSR value as a result of this damage depending on the robustness of the structure, the estimated extent of damage and the location of the damage, as shown in Table 46;

Table 46 Expected Reduction in RSR based on Robustness, Extent and Location of Damage (MSL 2003).

		Expected reduction in RSR			
Robustness	Estimated extent of damage to member	Submerged member	Splash zone member	Leg	Weld
Non-robust	50%	15%	15%	15%	40%
	100%	35%	25%	45%	
Moderately robust	50%	15%	10%	15%	35%
	100%	25%	20%	35%	
Robust	50%	15%	10%	15%	35%
	100%	25%	20%	35%	

Source: MSL (2003). Guideline for the definition and reporting of significant damage to fixed steel offshore platforms. MSL Services Corporation, Houston, Texas, USA.

- compare these with tolerable RSR reductions which depends on the vintage of the structure (pre-1980 and post-1980) and the exposure category (manned, manned-evacuated or unmanned). For post-1980s fixed offshore structures, the acceptable reductions in RSR are 20% for manned, 47% for manned-evacuated and 60% for unmanned platforms. Similarly, for pre-1980s structures, the acceptable reductions in RSR are 10%, 40% and 55% respectively.
- decide whether strengthening, mitigation and repair are needed; the review recommends to base these decisions on the expected reduction in the RSR value and the acceptable reductions.

7.2.9 MSL Assessment of Repair Techniques for Ageing or Damaged Structures

Section 4 of the MSL report "Assessment and Repair Techniques for Ageing and Damaged Structures" (MSL 2004) provides good guidance for the assessment of structures with respect to deciding the need for and the type of strengthening, mitigation and repair that should be implemented when structures are found to have damage or anomalies. According to the review, there are two main purposes of conducting assessments:

- to explore the need for and extent of strengthening, mitigation and repair (SMR); and
- to demonstrate the adequacy of the selected SMR scheme (which is covered in Chapter 8 of this book).

The review indicates that a structural assessment is normally required to decide whether SMR is needed and, if so, to what extent. Such an assessment is described to include a structural analysis to establish member loads in design events and checks of intact or damaged component capacity. The review also highlights that these structural analyses and component checks may be performed by refined or advanced analysis techniques if standardised and simplified analysis fails to show that the structure is fit for further service. In many cases, refined and advanced analysis and component checks may demonstrate that the structure can be used further without strengthening, modification or repair.

The review further points out that the quality of the input data for the structural analysis and component checks is significant for the quality of the assessment. It should be based on the as-built structural geometry and should include all subsequent changes, modifications and repairs. Such changes and modifications can sometimes reduce the loading, for example, when conductors, risers or caissons have been removed or similar changes have been implemented. Also, the original deck loading is described to often be conservative and no longer factual. However, examples are also given of cases where deck loading has been significantly increased after the original design. The soil strength is also mentioned as an area where more refined analysis can be performed, for example, based on new soil data or re-analysis of existing soil data. It is also pointed out that having the actual material certificates can be valuable in defining the yield strength with more accuracy and often at a higher value. An increase of the yield strength of 15% or higher is indicated as possible (MSL 2004).

Four possible levels of structural analysis are described in the review:

- basic analysis (similar to standard analysis as performed during design of new structures)—these analyses should take into account the subsequent modifications and should include any damaged members in a simplified way (remove the member from the analysis).
- simple refined analysis where the structural model and load model is refined, for example, by including member eccentricities and offsets in the model. The stiffness of damaged members should be reduced to a realistic value;

- complex refined analysis where more updates to the structural model and load model are included, for example, by including local joint flexibilities in the model; and
- advanced analysis where non-linear analysis is performed.

At the time of writing this book most structural analysis of fixed offshore structures is performed by complex refined analysis and, as such, the choice is n w more between this type of analysis and a non-linear analysis.

Similarly, the review proposes three levels of component checks:

- standard code check according to recognised standards such as API RP-2A (API 2014) or ISO 19902 (ISO 2007);
- refined component checks that includes the use of yield stresses from material certificates (mill certificates), review of member lengths for more accurate buckling analysis, review of SCF for more accurate fatigue analysis, appraisal of the capacity of damaged members by assessing existing test data and fracture mechanics studies. In addition, the review suggests the use of reliability analysis for component checks; and
- advanced component checks where finite element analysis of the components is suggested to be performed or laboratory tests are commissioned to provide data for evaluations of the strength.

These three levels of component checks are rather typical of what is used in analysis of existing structures at the time of writing this book. However, laboratory tests are normally seen as very expensive and only used as a last resort.

7.3 Previous Studies on Evaluation of Damaged Plated Structures

7.3.1 Introduction

Corrosion is one of the primary ageing factors of ship-shaped structures as hull structural members are exposed to corrosive environments and thickness diminution due to corrosion, which may be unavoidable. Hence, as stated by Nakai and Yamamoto (2008), "In order to ensure the structural integrity of ship-shaped structures, it is of crucial importance to understand the corrosion process, estimate the corrosion rate and evaluate the effect of corrosion wastage not only on overall strength but also on local strength accurately".

Corrosion wastage of plate elements can reduce their ultimate strength. A corroded surface on a plate has a random distribution of thickness over the area of the plate. These variations in thickness will lead to local stress concentrations and plastic hinges may be formed that will affect the buckling behaviour. In addition, the loss of thickness due to corrosion may form eccentricities that could influence the capacity of the plate.

Corrosion wastage can be in the form of general uniform corrosion or as localised pitting or patch corrosion. The strength of a plate with general uniform corrosion is normally estimated by simply excluding the thickness that is lost due to corrosion. However, this will not fully take into account the rough surface of a corroded plate. Modelling of the corrosion process is further reviewed in, for example, Paik (2004), Paik et al. (1998, 2003a, 2003b, 2004a), Guedes Soares and Garbatov (1999), Garbatov et al. (2004, 2005) and Qin and Cui (2002, 2003).

Fatigue cracking is also one of the primary ageing factors of ship-shaped structures and semi-submersibles. Guidance on evaluating the strength of cracked stiffened plates can be found in Paik and Melchers (2008).

In addition, in ship-shaped structures, longitudinals in side-shells may be bent or twisted due to a local impact resulting in a local reduced structural capacity. In semi-submersibles, the columns are most vulnerable to local denting. A bent and twisted ring stiffener or girder will reduce the global buckling capacity of the column. Guidance on evaluating the strength of dented stiffened plates can also be found in Paik and Melchers (2008).

7.3.2 SSC Studies on Residual Strength of Damaged Plated Marine Structures

In 1995 SSC undertook a project, SSC 381, to review the residual capacity of ship structures after damage (SSC 1995). The key elements required to undertake an engineering analysis to evaluate the residual strength were defined and methods such as fracture analysis and ultimate strength analysis were described. The types of damage identified in this review are discussed in Section 3.4.2.

The principle of reserve and damaged strength ratio often used for fixed offshore steel structures was in this review used also for ship structures, although reserve strength ratio was in this report named reserve resistance factor (REF) and the strength in damaged condition was illustrated by a residual resistance factor (RIF) defined by the collapse load in the damaged condition divided by the collapse load in the undamaged condition.

SSC 381 (1995) describes the following methods to evaluate residual strength of damaged marine structures:

- an indirect method, which is a three-step method using service life experience and semi-empirical assessment criteria for local damage to determine the ultimate strength; and
- a direct method, which includes identification of potentially critical locations, determination of the type and extent of damage at these locations and the ultimate strength analysis of the global structure by linear and non-linear finite element analysis.

Indirect Method
The first step in the indirect method is the identification of critical locations and applicable failure criteria for damage at these locations. The next step is the local analysis to determine the extent of damage based on the applied load and the mode of failure. The last step is determination of the ultimate strength of the structure taking into consideration the damage components. A large amount of assessment methods and diagrams are provided for various types of damage. These include:

- overview of allowable permanent deformation criteria based on both in-service experience and analytical methods based on calculation of maximum structural strain and the limiting fracture strain;
- linear elastic fracture mechanics (LEFM) assessment of large structures subjected to low stresses (less than half the yield stress) and elastic-plastic fracture mechanics (EPFM) for higher stresses. However, fracture assessment diagrams in accordance with BSI PD 6493[4] are recommended accounting for failure in the fracture and collapse modes based on the stress intensity factor K from a LEFM analysis or the crack tip opening displacement (CTOD) procedure. These evaluations are based on primary stresses, secondary stresses and peak stress, flaw description, material properties such as fracture toughness and tensile properties;

---

4 BSI PD 6493 is later replaced by BS 7910 (BSI 1999).

- two-dimensional methods for ultimate strength assessment evaluating the hull girder strength on the basis of individual strength behaviour of the components (gross panel elements and hard corner elements) making up the hull girder; and
- damage is included as loss of stiffness and strength of the local components, which results in an overall reduction of the ultimate hull girder strength. For severe damage, it is assumed that the whole damaged structure is fully ineffective in carrying any further load.

It is recognised that the indirect method is based on numerous assumptions and that the method is very conservative. The identified limitations of the indirect method included:

- it is based on the assumption that the location and form of local damage will be available from service life experience (truer for merchant ships than for ship-shaped offshore structures);
- damage like cracks are modelled by removing damaged portions without any consideration for the residual stiffness or strength the member still possesses; and
- interaction of failure modes are not allowed for in this method.

Direct method
A 3-D linear finite element analysis of the whole structure is performed and results of this analysis are to be used at the local level for a non-linear fracture and ultimate strength assessment. To identify locations susceptible to damage, the report suggests that ship motions and load programs are used in conjunction with the finite element analysis. When locations prone to cracking are determined, calculations are made to determine the time to failure of the components, typically based on the Paris crack growth equation. Determination of the critical crack length is recommended to be performed according to BSI PD 6493.

To integrate the effect of local failure upward to the global structural level and ensure that correct load transfer paths and internal load distributions are calculated, a full three-dimensional non-linear finite element analysis is recommended to be carried out.

The direct method is recognised as time consuming, requiring a considerable amount of technical expertise but is the most rational and complete procedure for residual strength evaluations of damaged ship structures.

An example of an ultimate strength analysis of a typical tanker is provided. The evaluation includes:

- typical locations, type of damage and criteria for critical crack length and crack growth rate;
- load analysis;
- determination of allowable crack length according to BSI PD 6493; and
- determination of crack growth according to PD 6493 (based on a 0.005 mm assumed initial crack length) providing the crack length as a function of number of voyages.

Determination of critical crack length by K-method:

$$a_m = \frac{1}{2\pi} \left(\frac{K_{mat}}{\sigma_1} \right)^2$$

Determination of critical crack length by CTOD-method:

$$a_m = \frac{\delta_{mat} \cdot E}{2\pi \cdot \left[\frac{\sigma_1}{\sigma_y} - 0.25 \right] \sigma_y} \quad for \ \frac{\sigma_1}{\sigma_y} > 0.5$$

$$a_m = \frac{\delta_{mat} \cdot E}{2\pi \cdot \left[\dfrac{\sigma_1}{\sigma_y}\right]^2 \sigma_y} \quad for \ \frac{\sigma_1}{\sigma_y} \le 0.5$$

In the example provided, the allowable crack lengths were 56 mm for the K-method and 106 mm for the CTOD method.

Later, the SSC in their report SSC 416 (SSC 2000) included advice for the evaluation of hull structures (appendix B):

- the hull girder strength is confirmed on the basis of the actual hull girder section modulus, which may be assessed initially using an allowable area at deck and bottom;
- buckling found during the survey should be taken as an indication of areas which require stiffening or renewal of material;
- any fractures found should normally be repaired by part renewal of material or by welding; structural modifications may also be advisable to avoid repetition of fractures;
- areas of heavy wastage due to general corrosion need to be analysed, for example, by the percentage thickness reduction and a buckling criterion—if wastage is in excess of the allowable limit, steel renewal may be needed;
- local corrosion or pitting of the shell that can lead to possible hull penetration needs to be studied;
- isolated pits are not believed to significantly influence the strength of plates or other structural members but may cause a potential pollution or leakage problem;
- large areas of pitting of a structure will influence the strength and must be considered when assessing the residual mean thickness of material; and
- the bending capacity reduction obtained from testing of plates with uniform machined pits suggests that capacity reduction is roughly proportional to the loss of material (pitting diagrams may be used in combination with measurements of pitting depths).

Here, detailed methods for assessment are not provided, but reference is made to Tanker Structure Cooperative Forum "Guidance Manual for the Inspection and Condition Assessment of Tanker Structures", "Guidance for the Inspection and Maintenance of Double Hull Tanker Structures" and "Guidance Manuals for Tanker Structures" (TSCF 1986, 1995 and 1997).

7.4 Previous Studies on Evaluation of Damaged Concrete Structures

7.4.1 Department of Energy Assessment of Major Damage to the Prestressed Concrete Tower

A review of major damage to concrete offshore structures resulting from ship impact from different types of vessels was undertaken for the Department of Energy (1988). It also devised methods for the repair of different levels of damage following impact. The different types of potential collisions included were:

- sideways collision with a 2500 tonne displacement supply vessel (in the context of modern supply vessels this is a small size vessel);
- bow collision with a cargo ship (875 tonne displacement);

- bow collision with a channel ferry (8900 tonne displacement);
- bow collision with a large cargo ship (21,300 tonne displacement); and
- bow collision with a large oil carrier (127,000 tonne displacement).

Such damage can have effects both on strength and on seawater ingress, which could lead to corrosion or loss of operational capability. For example, accidental damage to a concrete tower can cause seawater leakage. This can lead to loss of pressure control and consequent overstress of the cell-to-tower connection or direct loss of strength in the area of the damage. In many concrete platforms the stresses in the base cells and the adjacent parts of the leg are influenced by the internal pressure in the cells and loss of the pressure regime within the cells (drawdown) can lead to possible overstress of the structure under severe weather. This pressure is controlled by the water level in a ballast water header tank situated within one of the legs. The normal underpressure of the cells is around 3–4 bar in relation to the adjacent sea water, which results in beneficial compressive stresses in the towers; this could be lost with water ingress to the cells (Ocean Structures 2009).

The report also analysed the damage in more depth, reaching the following conclusions:

- Assessment of the sideways collision showed that a punching shear failure was unlikely, although it was noted that high-stress concentrations arose from impact with the half round steel fenders on the supply vessel. It was also noted that supply vessels were getting larger and such vessels were capable of more severe damage. It should be noted that the bows of modern supply vessels are different from those analysed in 1987 (bulbous bows and X-bows), and in recent years many supply vessels are ice strengthened which in both cases will have a very different impact load scenario.
- For the bow collisions the report identified the shape of the bow and its elevation, which might impact a concrete tower. In each case the degree of punching shear failure was estimated for realistic impact speeds. In terms of the structural integrity of a damaged platform, analyses were performed on typical offshore towers to assess the effects of major damage. This included initially an elastic analysis on a damaged tower to identify the distribution of stresses in the region of the damage and to identify areas where failure of the concrete might occur. Secondly, an ultimate limit state analysis was performed to investigate the load-carrying capacity of a tower with various degrees of damage. This was coupled with a wave force analysis to determine the limiting sea sate which the damaged structure could withstand.

The tower selected for analysis was from the Cormorant A platform. The tower size was 8 m in diameter, 650 mm wall thickness. It was decided to restrict the study to holes which might result from a moderately severe impact and square holes of different dimensions. The analysis indicated that areas of yielding in compression would be very localised and confined to the area adjacent to the hole for the scenarios investigated.

The ultimate limit state analysis of the damaged tower was performed on a prestressed tower containing various degrees of damage. In this case the tower chosen was that for the Wimpey-Elf structure (similar dimensions to the Cormorant A structure). The analysis produced a series of interaction curves of the ultimate axial load capacity versus the ultimate bending capacity of the section. These curves indicated that the tower may sustain substantial damage before collapse. For the design axial load coupled with the design wave (28 m), failure was predicted when the damage extended $110°$ of the circumference, representing a very large impact area. The report concluded that this analysis may be misleading for a number of reasons, which are listed in the report.

7.4.2 Department of Energy Review of Impact Damage Caused by Dropped Objects

Accidental damage to the domes of storage cells from dropped objects can lead to both loss of strength and seawater leakage. This type of damage was reviewed in a Department of Energy project (Department of Energy 1987c). This leakage could reduce the beneficial effects of drawdown as discussed above. Hence, it is likely that significant damage would need repair as has already happened in a few cases (see Section 8.4). In one case, damage to a 500 mm thick cell roof slab resulted from a dropped object, causing a deep hole 300 mm in depth, leading to water flowing though the slab. Repair was undertaken by pre-packing aggregates within the hole and covering the hole with a steel plate and injecting grout to restore the original concrete profile.

7.4.3 HSE Review of Durability of Prestressing Components

High-strength prestressing tendons are required to maintain the structural integrity of the concrete structure, particularly in the towers. These tendons are placed in steel ducts which after tensioning are grouted. The ducts are placed within the walls of the structure. The degree to which grouting has been fully effective, given the long length of the ducts and in some cases their horizontal orientation (which is more difficult to grout), has led to concerns that seawater can penetrate into the ducts. This could cause corrosion of the very high–strength steel tendons with a potentially serious local loss of prestress. A review of the durability of prestressing components (HSE1997) concluded that the first group of concrete offshore structures (pre-1978) was more vulnerable to poor grouting and hence corrosion of the prestressing tendons. Due to the length of the ducts, high pressure had to be used which caused grout in these early structures to bleed so that water collected at the tops of vertical ducts and the resulting voids had to be filled with fresh grout. It was known that later platforms benefited from improved grouting materials and procedures.

The HSE report (HSE 1997) investigated the potential for loss in strength of a typical offshore concrete structure due to local loss of prestressing caused by corrosion. The analyses undertaken showed that there would have to be a loss of 10% of the prestress to cause the reliability to fall below the target value and a loss of 40% before a leg could fail under loading from a design wave. In addition, it would be necessary for most of the tendons to fail in the same section, which is unlikely for grouted ducts. However, as noted in the report (HSE 1997), there has been evidence from land-based structures (e.g. bridges) where prestressing failures have occurred leading to overall failure. Nonetheless, the tendons do not play such a critical role in offshore concrete structures. The report concluded that locations in the post tensioning system most at risk are in the vertical ducts in the region of the splash zone, particularly at anchorages. The inspection of prestressing tendons in ducts is recognised as difficult, because of their length and because they are embedded within the concrete.

The report (HSE 1997) also identified the most vulnerable area in an offshore concrete structure as being the steel-to-concrete transition between the deck and platform legs. It is recommended in the report that inspection should be made of this location.

7.4.4 HSE Review of Major Hazards to Concrete Platforms

As part of the Concrete in the Oceans (CiO) programme, a review of major hazards to concrete platforms was commissioned by HSE and undertaken by Ove Arup and Partners (HSE 1994). A total of twenty-seven potential major hazards were identified and assessed, including all phases

Table 47 Risk Assessment Matrix for Offshore Concrete Structures.

Hazard	Frequency	Consequence	Risk
Ship impact	5	4	20
Internal pool and jet fire	5	4	20
External sea fire (pool fire)	5	4	20
Explosions	5	3	15
Oil storage systems	4	3	12
Implosion	2	5	10
Dropped objects	5	2	10
Excess hydrodynamic loading	3	3	9
Prestressing failure	2	3	6
Design errors	2	3	6
Fatigue	1	5	5
Scour	1	4	4

from design to abandonment. For each major hazard, the consequence of failure and the frequency of occurrence were assessed to create a risk assessment matrix indicating areas of uncertainty and research needs at that time. An extract from this table is shown in Table 47 for the hazards most relevant for in-service operation. The higher risk hazards are addressed in this book with the exception of fire and explosion damage to concrete structures.

7.4.5 Department of Energy Review of the Effects of Temperature Gradients

The storage of hot oil in the concrete tanks at the base of many concrete installations can lead to thermal stresses that can produce cracking of the concrete. Concrete is vulnerable to significant temperature differences, which in this particular case arise from hot oil on one side of a wall and cold seawater on the other side. Tests, as reported in the Department of Energy report OTH 87 234, have shown that temperature differences of up to 45° C can be sustained with the correct design details (Department of Energy 1987d), with some cracking occurring but within acceptable limits. The oil is cooled before storage, but if the coolers fail or unusual conditions occur, oil with temperatures of up to 90° C can be diverted into the storage cells, which could cause more significant cracking of the walls. Over a long period of operation these effects could accumulate. The increased stresses from these thermal effects could lead to cracking of the concrete and overstress of steelwork in and around the walls and roofs of the storage cells, including the critical junction with the legs.

7.4.6 Concrete in the Oceans Review of Corrosion Protection of Concrete Structures

As part of the Concrete in the Oceans (CiO) programme, a review of corrosion protection of concrete structures was undertaken by DNV and Wimpy Laboratories in the report OTH 87 248 (Department of Energy 1987e). The review together with exposure tests concluded that for a totally submerged structure, static cracks up to 0.6 mm wide at the surface on a 50–75 mm cover can be accepted.

However, in the splash zone, corrosion occurred in quite narrow cracks 0.1 mm wide. Overall, it is expected that local losses of the concrete cover to the reinforcement are likely to result in minimal loss of strength. However, many losses of cover, unless repaired, can lead to reinforcement corrosion which if extensive could affect strength.

7.4.7 Norwegian Road Administration Guideline V441

For Norwegian concrete offshore structures, the relevant parts of the Norwegian Road Administration guidelines (Statens vegvesen 2019) have been used in evaluation of damage and anomalies. These guidelines are similar to those used by other regional road administrations such as in the UK.

The guideline requires that a report shall be developed during inspection of concrete structure describing the damage types and anomalies observed. This should include a description of the level of damage and of the extent to which the damage will reduce the integrity of the structure. Standardised damage types are stipulated and shall be used in the reporting. The description of the level of damage and anomalies is required to include a verbal description specifying the type found, a clear description of where on the structure this is found and the extent and photos of the finding.

The level of damage or anomaly is required to be evaluated according to the categories minor, medium, major and critical. The level of damage can be exemplified by cracks as given in this guideline where a crack < 0.3 mm is regarded as minor, 0.3–1.0 mm is regarded as medium, 1.0–2.0 mm is regarded as major and > 2 mm is regarded as critical.

Further, the consequences of the damage or anomaly are required to be evaluated based on the following categories:

- minor consequences of damage or anomaly: no mitigation necessary;
- medium consequences of damage or anomaly: to be registered in the structural integrity management database for further assessment;
- large consequences of damage or anomaly: proposed mitigation to be registered in the structural integrity management database and inspection intervals should be re-evaluated; and
- critical consequences of damage or anomaly: owner or responsible structural integrity engineer are to be contacted immediately.

A damage or anomaly on a road bridge can affect the structural integrity, the serviceability of the bridge, future inspection and repair cost and aesthetics; the guideline indicates that these consequence types are listed in decreasing level of importance. The guideline requires the consequence type to be included in the evaluation. For the most relevant consequence type for offshore structures, the following description is provided for the evaluation:

- minor consequences for the structural integrity indicate that the strength of the structure is to a very small extent affected by the damage and the damage is regarded as stable (not developing);
- medium consequences for the structural integrity indicate that the strength of the structure is to some extent affected by the damage and there is a possibility that the damage is developing;
- large consequences for the structural integrity indicate that the strength of the structure is significantly affected by the damage and the structure is at immediate risk of collapsing;
- critical consequences for the structural integrity indicate that the strength of the structure is significantly affected and has been or will be overloaded. Owner or responsible structural integrity engineer are to be contacted immediately.

Table 48 Selected Types and Causes of Damage from Guideline V441.

Standardised damage types	Standardised damage causes
Material independent damage types	Errors in design and fabrication
• settlements;	• insufficient standards;
• rotation or movement of structural components;	• inappropriate use of standards;
• deformation of structural components;	• inappropriate choice of materials;
• leaks;	• errors in design calculations;
• discolouring.	• errors in fabrication.
Foundation damage	Material
• erosion and scour.	• inappropriate material strength;
Damage to concrete	• cracking.
• cracks;	Environmental
• damaged coating;	• frost;
• thaw;	• chloride;
• wear;	• carbonatization;
• delamination;	• bacteria;
• spalling;	• chemical.
• reinforcement corrosion.	Loading
	• service loads;
	• environmental loads;
	• accidental impact;
	• accidental fire and explosion.

Source: Modified from Statens vegvesen (2019). Håndbok V441 Bruinspeksjon, Vegdirektoratet. Oslo, Norway (in Norwegian).

The guideline also requires the cause of the damage to be identified as this may have significant impact on the decision of repair and mitigation and avoiding similar future damage on the structure. The inspection history shall be evaluated to check for any trends in damage development that needs to be corrected or mitigated.

The standardised types and causes of the damage listed in the guideline relevant for offshore concrete structures are shown in Table 48.

7.5 Practice of Evaluation and Assessment of Offshore Structures

7.5.1 General

A broad range of anomalies, damage and changes will be reported to the structural integrity engineer from the inspections and surveys that have been performed. The structural integrity engineer has to evaluate each of these and decide whether:

- the anomaly, damage or change can remain as it is based on the information available, often combined with a plan on monitoring further development or change going forward;
- more analysis and assessment are needed in order to decide on further action; and
- repair is needed and the urgency of the repair execution.

Simple rules may be used such as "Any developing crack in critical areas shall be assessed and shall have top priority for repair while all other cracks shall be repaired when possible" and "Any sign of corrosion shall be removed by abrasive blasting, cleaned and re-coated". However, many operators would prefer to optimise such guidelines with the aim of minimising repair costs.

DNVGL (2019b) indicates that a decision on whether to mitigate or repair should be made when damage such as fatigue cracks, significant corrosion, fabrication defects, overload and accidental damage are detected. Further, the need for mitigation should also be evaluated if changes in use or increase in loading are identified. Such changes in use or in loading may be due to subsidence, addition of new modules, addition of new risers or conductors and new information or knowledge about loading (e.g. environmental loading). Repair or mitigation should also be evaluated if new standards or regulatory requirements or results from new or more advanced analyses indicate that the structure has insufficient strength or fatigue capacity.

If the evaluation on a component basis indicates that the finding is significant and hence fails to demonstrate acceptable strength with the damage present, an assessment including global structural analysis may be required. MSL (2004) proposed an approach as presented in Table 49 as a way to perform the evaluation and assessment. Three levels of complexity were proposed for analysis and component checks. Basic analysis and standard code checks would be sufficient if these prove that the structure is fit-for-purpose with the damage present. More advanced analysis and component checks could be selected if needed.

DNVGL RP-C208 (DNVGL 2019b) highlights the importance of determining the extent of the damage, whether the damage occurs at several similar locations and the root cause of the damage taking into account the operational history and the age of the structure. In addition, the impact of the damage on the integrity of the structure should be determined in order to decide whether repair or

Table 49 Evaluation and Assessment Possibilities for Structures with Damage or Anomalies.

Evaluation	Structural analysis	Component check
Checking the effect of damage or anomaly on component strength and fatigue life. Significant damage or anomaly is often defined as causing a capacity reduction of 10% or more. If so, structural analysis and component checking is needed.	**Basic analysis** Simple structural model based on as-built drawings, including subsequent modifications. Damaged members are removed, or their stiffness is reduced according to damage level	**Standard code check** Strength formulae and variables such as yield stress and partial factors are used according to recognised standards. SN-check of remaining fatigue life.
	Complex analysis More detailed analysis model with, for example, local joint flexibilities included. Load data are updated and based on recent metocean specifications and joint occurrence data of loads (e.g. wind and wave).	**Refined checking** Yield stress can be based on mill certificates. The relevance and accuracy of variables such as effective lengths and SCFs can be checked and updated. Capacity of damaged elements can be included based on existing test data. Fracture mechanics crack growth analysis to establish remaining fatigue life.
	Advanced analysis Non-linear finite element analysis for a more complete understanding of structural behaviour and redistribution of loads in structure.	**Advanced checking** Finite element component checks, e.g. in accordance with DNVGL RP-C208 (strength) and BS7910 (SCFs and crack growth). Perform laboratory tests to establish strength or fatigue checks. Reliability analysis is also a possibility (but difficult to justify).

modification is needed and, if repair is needed, which repair method should be used (see Chapter 8 of this book). DNVGL (2019b) further highlights the importance of having access to quality information about the structure. This requires access to information such as inspection and analysis reports, weight reports, as-built drawings, any modifications to the structure and possible access to reanalysis models. A full description of the information needed is provided in, for example, NORSOK N-006 (Standard Norge 2015).

As indicated previously some structural items are uninspectable because of restricted access either externally or internally. Typical examples are the pile, the shear plates for pile-sleeve connectors, the grout between the pile and the pile sleeve and internal stiffeners in joints for fixed steel structures. The fatigue evaluation of these can make use of the availability of data from nearby inspectable components with similar loading and significantly lower fatigue lives. In reality, this approach has significant limitations for the uninspectable pile-related items but may be used for a joint without ROV access. For piles and pile-sleeve connections this leaves structural monitoring techniques as the primary method for providing the necessary information to evaluate their performance. The typical method used for this purpose is natural frequency monitoring as a significant reduction in pile capacity or pile sleeve connection will lead to change in the stiffness and behaviour of the platform.

As noted previously, ISO 19902 (ISO 2007) requires that *"all available data on the structure or group of structures shall be evaluated"* and indicates that the following structural considerations should be included in the evaluation:

- structure age, condition, original design situations and criteria, and comparison with current design situations and criteria;
- analysis results and assumptions for original design or subsequent assessments;
- structure reserve strength and structural redundancy;
- fatigue sensitivity;
- degree of uncertainty in specified environmental conditions;
- extent of inspection during fabrication and after transportation and installation;
- fabrication quality and occurrences of any rework or rewelding;
- damage (including fatigue damage) during transportation or installation;
- operational experience, including previous in-service inspection results and lessons from the performance of other structures;
- modifications, additions and repairs or strengthening;
- occurrence of accidental and severe environmental events;
- criticality of structure to other operations;
- structure location (geographical area, water depth);
- debris;
- structural monitoring data, if available; and
- potential reuse or removal intents.

7.5.2 Fixed and Floating Steel Structures

Several standards are as described in Chapter 2 available for the evaluation and assessment of fixed structures, including ISO 19902 (ISO 2007), ISO 19901-9 (ISO 2019), API RP-2A (API 2014a), API RP-2SIM (API 2014b) and NORSOK N-006 (Standard Norge 2015). Formulae for evaluation of dented, corroded and cracked tubular members are provided in, for example, ISO 19902 (ISO 2007). For floating structures standards such as API RP-2FSIM (API 2019a),

API RP 2MIM (API 2019b), NORSOK N-006 (Standard Norge 2015) and classification society guidelines such as DNVGL-OS-C102 (DNVGL 2015) provides codified capacity checks for many relevant situations. In addition, the work by organisations such as the Ship Structure Committee (SSC) and the Tanker Structure Cooperative Forum (TSCF) provides valuable data and methods. Such information can also be found in Paik and Thayamballi (2007) and Paik and Melchers (2008).

A first evaluation of corrosion damage will often be made visually based on some form of classification. Often a "good, fair and bad" classification is used in DNVGL-CG-0288 (DNVGL 2017) where less than 3% breakdown of coating or area rusted is considered good, 3–20 % is considered fair and more than 20% is considered poor and particularly requiring further assessment and most likely repair (see Figure 79 for illustration of amount of corrosion per area). For hard rust scale, more than 10% by area would be regarded as poor condition requiring attention while for the edges and weld lines more than 20% is regarded as fair and more than 50% is regarded as poor (DNVGL 2017). However, corroded welds will be fatigue sensitive and the SN-curve will need to be modified to allow for free corrosion which often will reduce the fatigue life by a factor in the order of 3.

Finite element analysis is often used for detailed checks of specific damage and anomalies. However, these need to be calibrated to actual physical tests of similar situations before they are used. This calibration implies that finite element analysis of specimens from the laboratory tests described earlier in this chapter are performed to check that the finite element analysis produces similar results. If the finite element analysis does not provide similar results, physical correct adjustments should be made to the finite element analysis in order to better represent the physics of the structural behaviour.

The amount of further crack growth from an identified fatigue crack is an important parameter that needs to be studied. The evaluation and assessment should determine the rate of which the crack will be growing and predict when the crack would reach a size where unstable propagation could occur. A few standards for such assessment exist but BS7910 (BSI 2019) is to a large extent the industry standard.

Fatigue crack growth is in BS7910 (BSI 2019) predicted by fracture mechanics using a fatigue crack growth law where the rate of fatigue crack growth under cyclic loading is related to the range of the stress intensity factor, ΔK, as shown in Figure 80. There are three stages of crack growth:

- Stage I: crack propagation is considered to occur only when the stress intensity factor range exceeds the threshold stress intensity factor range, ΔK_{th}.
- Stage II: at intermediate values of ΔK, there is an approximate linear relationship between the crack growth rate and ΔK on a log-log scale.

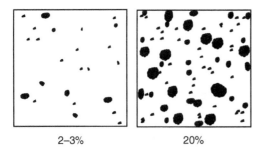

2–3%   20%

Figure 79 Illustrations of percentage of corrosion per area.

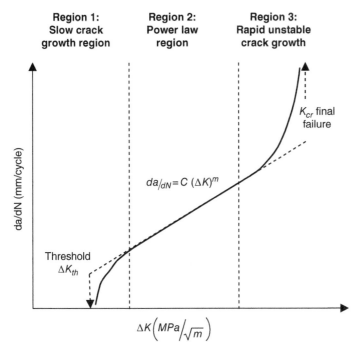

| Region 1: Slow crack growth region | Region 2: Power law region | Region 3: Rapid unstable crack growth |

K_{cr} final failure

$da/_{dN} = C\,(\Delta K)^m$

Threshold ΔK_{th}

da/dN (mm/cycle)

$\Delta K \left(MPa / \sqrt{m} \right)$

Figure 80 Fatigue crack growth rate curve. *Source:* Based on BSI (2013), BS 7910:2013 + A1:2015, 'Guide to methods for assessing the acceptability of flaws in metallic structures. British Standards Institution, 2013.

- Stage III: this is characterised by accelerated crack growth which becomes unstable and results in fracture when the maximum stress intensity factor attains a critical level, K_{cr}.

The often-used Paris and Erdogan (1963) equation is applicable to the Stage II region only. Various other fatigue crack growth laws have been proposed to take account of Stage I crack growth, which can represent a significant proportion of the total fatigue life. The crack growth rate under cyclic loading is a function of the loading conditions, geometry, material and the environment in which the material is placed. Further information can be found in Anderson (2017), Ersdal et al. (2019) and particularly in BS 7910 (BSI 1999).

The evaluation of the stability of a fatigue crack can be performed by using a failure assessment diagram (FAD). The failure assessment diagrams represent the interaction between the failure modes of brittle or ductile fracture and plastic limit state or collapse dependent on the fracture toughness properties and yield behaviour, respectively. The FAD method is described in standards such as BS 7910 (BSI 1999) and API 579 (API 2016).

Figure 81 shows an FAD based on BS 7910 where the vertical axis K_r is the ratio of the linear elastic stress intensity factor of the crack to fracture toughness and the horizontal axis L_r is the ratio between the load and the plastic limit or collapse load with the crack present. If these coordinates calculated for a crack in a given geometry and stress field lie within the FAD boundary, then the crack can be considered to be stable. If the coordinates lie outside the boundary, then unstable crack propagation is a possibility.

Three levels of assessment are available depending on the data available and the sophistication of analysis required. The fracture assessment procedure for offshore structures is usually based on the use of the Option 2 failure assessment diagram for low work hardening materials which represents the behaviour of offshore structural steels. The BS 7910 (BSI 1999) procedure allows the collapse

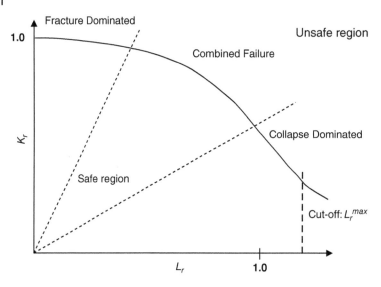

Figure 81 Failure Assessment Diagram. *Source:* Based on BSI (2013), BS 7910:2013 + A1:2015, 'Guide to methods for assessing the acceptability of flaws in metallic structures. British Standards Institution, 2013.

parameter L_r for tubular joints to be calculated using either local collapse or global collapse criteria. The local collapse approach is usually very conservative whilst the use of the global approach, which incorporates more realistic boundary conditions, tends to give more realistic predictions.

The evaluation and assessment of a structure for a damage or anomaly can include many different types of analysis and checks, the obvious examples being fatigue crack growth analysis and the static strength of a damaged member. An overview of possible evaluation methods for different types of damage and anomalies in steel structures is given in Table 50.

Table 50 Suggested Evaluation Methods for Steel Structures.

Cause of damage	Fixed steel and floating structures	Suggested evaluation method
Fatigue	Fatigue cracks particularly in welded structures with potential for through-thickness cracks, member severance (separation) and missing member	Crack growth rate and remaining fatigue life estimation; residual strength determination to determine consequence of cracking; static strength of component in accordance with appropriate standard (e.g. ISO 19902 for example tubulars in fixed steel structures); failure assessment diagram (FAD) if potential for brittle fracture is suspected
	Fatigue crack in chain link, link failure, loss of mooring line, drifting unit	Crack growth rate and remaining fatigue life estimation; static strength determination to determine need for replacement link or wire rope
Corrosion	Corrosion generally will result in reduced wall thickness and in some cases patch corrosion with potential for holes, flooding, member severance and missing member	Visual evaluation of the extent of corrosion; static strength of components and fatigue check by corrosion SN curves

Table 50 (Continued)

Cause of damage	Fixed steel and floating structures	Suggested evaluation method
	Corrosion pits leading to flooding and potential for fatigue initiation	Extent and depth of pitting is needed and can be used to determine loss of strength and local thickness and whether repair is needed
	Corrosion of chain link or wire strand, loss of mooring line, drifting unit	Static strength determination to determine need for replacement link or wire rope
Overload (hurricanes and major storms)	Buckles, dents, holes, cracks at welds and into chords, tears, out-of-plane bowing, severed members, tears, pancake leg severance, punch-through of brace, pull out of brace, conductor torn loose from guide, broken water caissons, broken riser standoffs	Residual strength determination to determine consequence of damage; crack growth rate and remaining fatigue life estimation may be required; static strength of component in accordance with appropriate standard (e.g. ISO 19902 for example tubulars in fixed steel structures)
	Damage to links, snapping of anchor chains, drifting units, dragging of anchors	Replacement of link or rope is normally required
	Drifting units and dragging of anchors may impact other structures and pipelines	Re-installing of anchors and mooring lines is normally required
Overload and accidental damage (vessel impact and dropped object)	Buckling in plates and members	Residual strength determination to determine consequence of damage; if cracks are present, crack growth rate and remaining fatigue life estimation may be required
	Dented and bowed member, member severance and missing member	Static strength of component; fatigue check including additional stresses and unintentional details
	Mechanical damage from mooring line seabed impact, wear and bending over the fairlead and handling of anchor chains and wire ropes	Static strength and fatigue check of mooring line (wire rope) or chain link—replace if significant
Wear and tear	Abrasion/wear of link or rope and possible loss of stud (chain)	Static strength and fatigue check of mooring line (wire rope) or chain link—replace if significant
Marine growth	Increased loading and potential overload damage	Take into account in load calculations and fatigue analysis
	Possibility for sulphate-reducing bacteria resulting in corrosion	
Settlement and subsidence	Platform settlement and subsidence may lead to change in loading pattern and exposure condition	Take into account in load calculations and fatigue analysis and estimate possibility of wave-in-deck loading
Scour and build-up of drill cuttings	Scour around the platform exposing the piles with the potential for pile failure	Estimate effect on pile and mud-plate strength; residual strength assessment (redundancy) may be required
	Drill-cuttings can bury or bear onto lower frames and result in damage.	
	Loss of holding power to anchors due to scour	Evaluate loss of holding power and need for rock dumping or anchor relocation

(Continued)

Table 50 (Continued)

Cause of damage	Fixed steel and floating structures	Suggested evaluation method
Material deterioration	Cracking from hydrogen embrittlement particularly for high strength steel (e.g. Jack-ups)	Estimate level of cracking and loss of static strength in addition to crack growth rate and remaining fatigue life estimation; the degree of overprotection from the CP system needs to be determined and adjusted if necessary
	Brittle fracture of chain links	Replacement of link is normally required
Fabrication fault	Lack of penetration or excessive undercutting that could lead to unexpected fatigue cracks, failure to provide vent holes for intended flooded members could lead to implosion, inferior material, incorrect member sizes, incorrect member positions, omissions.	Crack growth rate and remaining fatigue life estimation; residual strength determination to determine consequence of damage. Imploded members subjected to compression loads normally need to be replaced
Under-design	Member buckling, joint failure, tearing, fatigue cracking	Fatigue and crack growth analysis, strength analysis and residual strength assessment (redundancy) may be required to determine need for strengthening or repair
	Damage to links and failure of anchor chains possibly leading to drifting units. Under-designed anchors (soil strength) could lead to dragging of anchors.	

7.5.3 Concrete Structures

The international standards for concrete structures do not offer much information about evaluation of damage and anomalies to concrete structures. Previous studies have investigated the effect of impact damage on both towers and cells taking into account damage to the concrete, reinforcement and pre-stress. These studies have also provided some guidance on evaluation of this type of significant damage.

Table 51 Evaluation Scheme for Concrete Structures.

Consequence	Description	Repair need
Minor	The strength of the structure is to a very small extent affected by the damage and the damage is regarded as stable (not developing)	No immediate action needed
Medium	The strength of the structure is to some extent affected by the damage and there is a possibility that the damage is developing	Further monitoring should be implemented
Large	The strength of the structure is significantly affected by the damage, but the structure is not at immediate risk of collapsing	Repair or strengthening should be planned
Critical	The strength of the structure is significantly affected and has been or will be overloaded. Owner or responsible structural integrity engineer are to be contacted immediately	Repair or strengthening is immediately required

Source: Based on Statens vegvesen (2019). Håndbok V441 Bruinspeksjon, Vegdirektoratet. Oslo, Norway (in Norwegian).

The evaluation scheme shown in Table 51 is often used to determine the need for repair to concrete offshore structures.

With the lack of standardised evaluation methods for damage types to concrete structures, Table 52 has been developed to provide an overview of the most likely forms of significant damage and the suggested evaluation methods for each of these damage types.

Table 52 Anomalies and Possible Evaluation Methods for Concrete Structures.

Cause of damage	Types and description	Suggested evaluation method
Corrosion	Spalling and delamination (above water) as result of reinforcement corrosion as well as potential loss of reinforcement area	Determine the extent of spalling and delamination. The need for repair depends on the level and extent of spalling and delamination, typically in accordance with EN 1504-3:2005 (EN 2006). Monitor if low level.
	Seepage of corrosion products at cracks under water as result of reinforcement corrosion	Evaluate CP levels against design criteria and determine level of cracking. If CP levels are insufficient, consider replacement of anodes. Evaluation of the need for repair (filling epoxy in the crack) would depend on the consequence evaluation.
	Chloride ingress above water (reinforcement corrosion and potential loss of reinforcement area)	Determine level of chloride at reinforcement. However, as concrete is a porous material, chloride ingress is expected during life. The need for repair depends on the chloride level and the availability of oxygen at the reinforcement, potentially resulting in future corrosion. The need for repair is normally dependent on visual damage, such as spalling and delamination.
	Corrosion of steel attachments (e.g. pre-tension cable at anchorages, steel support attachments) and potential loss of prestressing capacity	Evaluate effect of local loss of prestress on performance and need for local concrete repair. If significant, consider additional prestress (external)
Overload and accidental damage (vessel impact and dropped object)	Local impact damage to towers resulting in reduction in concrete section area and possible water ingress	Calculation of local loss of tower strength and likelihood of water penetration. Evaluation of the extent of damage to concrete, reinforcement and prestress: <ul><li>if reinforcement is not damaged, patch repair will be sufficient;</li><li>if reinforcement or prestress are damaged, the effect on local strength needs to be determined and replacement of the reinforcement or prestress may be required.</li></ul> In the event of water penetration, repair methods designed to stop water flow are needed.
	Local impact damage to tops of storage cells from dropped objects resulting in reduction in concrete section area and possible water ingress	Evaluate the extent of damage to concrete and reinforcement and calculation of local loss of cell top strength. If strength loss or water ingress is significant, repair is needed. Repair should always be implemented if damage fully penetrates cell top.

(Continued)

Table 52 (Continued)

Cause of damage	Types and description	Suggested evaluation method
Wear and tear	Abrasion (local wear) and erosion resulting in minor reduction in concrete section area	Determine the remaining local reinforcement cover. If loss of cover is significant, local repair may be needed.
Marine growth	Increased loading	Evaluate the level of marine growth versus design basis and if in excess, take into account in load calculations or remove growth.
Settlement and subsidence	Settlement and subsidence	Evaluate the level of settlement and subsidence and their effect versus design basis. If settlement or subsidence is beyond design bases, an assessment of overall strength is required.
Scour and build-up of drill cuttings	Scour around the base of the platform possibly leading to instability	Evaluate effect on overall strength and need for, for example, rock dumping
Material deterioration	Loss of the concrete material due to aggressive agents (sulphate, chloride)	Evaluate loss of cover to reinforcement; evaluate need for local removal of concrete and replacement
	Cracking due to shrinkage (after construction)	Evaluate level of cracking against design criteria; evaluate effect on local cover to reinforcement
	Cracking, scaling and crumbling due to freeze and thaw (above water)	Evaluate degree of cracking against design criteria; evaluate remaining reinforcement cover
	Strength loss of the concrete material due to sulphate-producing bacteria (primarily in oil-storage tanks)	Often this anomaly is accepted partly due to the enormous difficulty in accessing interior of the tank; however, no real data exists on the status of such tanks after operation
Fabrication fault	Defective construction joints, minor cracking, surface blemishes, remaining metal attachments, patch repairs from construction, low cover to the reinforcement possibly leading to spalling in the splash zone and possible water ingress	Determine water leakage from defective construction joints. If significant, repair (epoxy filling) may be required. The other types of faults will be evaluated depending on the presence of corrosion or water leakage and may require repair in line with similar damage types mentioned above in this table.
Under-design	Cracking due to low reinforcement or prestressing, crushing of concrete in rare cases, failure in the junction between shafts and cells	Local loss of strength needs to be determined as a result of significant cracking, crushing or failures being detected. Repair may be needed if loss of strength is significant

References

Allan, J.D. and Marshall, J. (1992), *The Effect of Ship Impact on the Load Carrying Capacity of Steel Tubes, Department of Energy, Offshore Technology Report, OTH 90 317*, HMSO, London, UK.

Anderson, T.L. (2017), *Fracture Mechanics: Fundamentals and Applications*, CRC Press, 4th Edition.

API (2014a), API RP-2A *Recommended Practice for Planning, Design and Constructing Fixed Offshore Platforms*, API Recommended practice 2A, 22nd Edition, American Petroleum Institute, 2014.

API (2014b), API RP-2SIM *Recommended Practice for Structural Integrity Management of Fixed Offshore Structures,* American Petroleum Institute, 2014.

API (2016). API RP-579 *Fitness-For-Service*, Third Edition, American Petroleum Institute *(also published by ASME as ASME FFS-1–2016).*

API (2019a), API RP-2FSIM *Floating System Integrity Management*, American Petroleum Institute, 2019.

API (2019b), API RP-2MIM *Mooring Integrity Management*, American Petroleum Institute, 2019.

Atteya, M., Mikkelsen, O., Oma, N. and Ersdal, G. (2020), "Residual Strength of Tubular Columns with Localized Thickness Loss", In Proceedings of the 39th International OMAE2020 Conference, ASME.

Bolt, H.M. (1995), "Results from Large Scale Ultimate Strength Test of K-Braces Jacket Frame Structures", Presented at the 27th Annual Offshore Technology Conference (OTC) as paper OTC 7783.

Bruin, W.M. (1995), "Assessment of the Residual Strength and Repair of Dent-Damaged Offshore Platform Bracing", Master Thesis presented at Lehigh University, Bethlehem, Pennsylvania, US.

BSI (2019), *BS7910: Guide to Methods for Assessing the Acceptability of Flaws in Metallic Structures, British Standardisation Institute (BSI).*

Burdekin F.M. and Frodin J.G. (1987), "Ultimate Failure of Tubular Connections, Final Report Project DA709", Department of Civil and Structural Engineering, UMIST, November 1987.

Burdekin F.M. and Cowling M.J. (1998), "Defect Assessment in Offshore Structures—A Procedure", CMPT Publication 100/98, ISBN 1-870553-330.

Burdekin F.M., Thurlbeck S.D., Sanderson D., Kiwanuka F. and Hamour W. (1999), "Joint Industry Project on the Reliability of Flooded Member Detection as a Tool for Assurance of Integrity, Final Report", EQE Ltd. and UMIST, 1999.

BSI (1989), *BS 7191 Specification for Weldable Structural Steels for Fixed Offshore Structures*, British Standards Institution, 1989. Now withdrawn and superseded by *EN 10225:2001 Weldable Structural Steels for Fixed Offshore Structures. Technical delivery conditions.*

BSI (2013), *BS 7910:2013 + A1:2015, Guide to Methods for Assessing the Acceptability of Flaws in Metallic Structures*, British Standards Institution, 2013.

Cheaitani, M.J. and Burdekin, F.M., (1994), "Ultimate Strength of Cracked Tubular Joints," in *Tubular Structure VI*, Edited by P. Grundy, A. Holgate and B. Wong, A.A. Balkema, Rotterdam, pp. 607–616.

Cho, S-R. (1987), "Design Approximations for Offshore Tubulars against Collision", PhD thesis, Department of Naval Architecture and Ocean Engineering, University of Glasgow, Scotland.

Department of Energy (1987a), *Offshore Technology Report OTH 87 259 Remaining Life of Defective Tubular Joints—An Assessment Based on Data for Surface Crack Growth in Tubular Joints Tests*, Prepared by Tweed, J.H. and Freeman, J.H., 1987 for the Department of Energy, HMSO, London, UK.

Department of Energy (1987b), *Offshore Technology Report OTH 87 278 Remaining Life of Defective Tubular Joints: Det of Crack Growth in UKOSRP II and Implications*, Prepared by Tweed, J.H. for Department of Energy, HMSO, London, UK.

Department of Energy (1987c), *Offshore Technology Report OTH 87 240 The Assessment of Impact Damage Caused by Dropped Objects on Concrete Offshore Structures*, HMSO, London, UK.

Department of Energy (1987d), *Offshore Technology Report OTH 87 234 The Effects of Temperature Gradients on Walls of Concrete Oil Storage Structures*, HMSO, London, UK.

Department of Energy (1987e), *Offshore Technology Report OTH 87 248 Concrete in the Ocean Programme—Coordinating Report on the Whole Programme*, HMSO, London, UK.

Department of Energy (1988), *Offshore Technology Report OTH 87 250 Repair of Major Damage to the Prestressed Concrete Tower of Offshore Structures*, HMSO, London, UK.

DNV (1999), *ULTIGUIDE—Best Practice Guideline for Use of Non-Linear Analysis Methods in Documentation of Ultimate Limit States for Jacket Type Offshore Structures*, Det Norske Veritas, Høvik, Norway.

DNVGL (2015), *DNVGL-OS-C102 Structural Design of Offshore Ships*, DNVGL, Høvik, Norway.

DNVGL (2017), *DNVGL-CG-0288 Corrosion Protection of Ships*, DNVGL, Høvik, Norway.

DNVGL (2019), *DNVGLRPC208 Determination of Structural Capacity by Non-Linear Finite Element Analysis Methods*, DNVGL, Høvik, Norway.

DNVGL (2019b), "Reparasjonsmetoder for bærende konstruksjoner", Report for the Norwegian Petroleum Safety Authority, DNVGL, Høvik, Norway (in Norwegian).

Ellinas, C.P. (1984), "Ultimate Strength of Damaged Tubular Bracing Members", *Journal of Structural Engineering*, ASCE, Vol. 110, No. 2, pp. 245–259.

EN (2006), *EN 1504-3:2005 Products and Systems for the Protection and Repair of Concrete Structures—Definitions, Requirements, Quality Control and Evaluation of Conformity—Part 3: Structural and Non-Structural Repair*, European Standardisation Organisation.

Ersdal, G. (2005). *Assessment of existing offshore structures for life extension*. Doctorate thesis, University of Stavanger, Norway.

Ersdal, G., Sharp, J.V., Stacey, A. (2019), *Ageing and Life Extension of Offshore Structures*, John Wiley and Sons Ltd., Chichester, West Sussex, UK.

Frieze, P.A., Nichols, N.W., Sharp, J.V. and Stacey, A. (1996), "Detection of Damage to the Underwater Tubulars and its Effect on Strength", OMAE Conference 1996, Florence, Italy.

Garbatov, Y., Vodkadzhiev, I. and Guedes Soares, C. (2004), "Corrosion Wastage Assessment of Deck Structures of Bulk Carriers", Proceedings of the International Conference on Marine Science and Technology, 24–33.

Garbatov, Y., Guedes Soares, C. and Wang, G. (2005), "Non-Linear Time Dependent Corrosion Wastage of Deck Plates of Ballast and Cargo Tanks of Tankers", Proceedings of the 24th International Conference on Offshore Mechanics and Arctic Engineering (OMAE2005), OMAE2005-67579, Halkidiki, Greece.

Gibstein M.B. (1981), "Fatigue Strength of Welded Tubular Joints Tested at DNV Laboratories", International Conference on Steel in Marine Structures, Paper ST 8.5, October 1981, Paris, France.

Guedes Soares, G. and Garbatov, Y. (1999), "Reliability of Maintained, Corrosion Protected Plate Subjected to Non-Linear Corrosion and Compressive Loads", *Marine Structures*, 10, 629–653.

Hadley I., Dyer A.P., Booth G.S., Cheaitani M.J., Burdekin F.M. and Yang G.J. (1998), "Static Strength of Cracked Tubular Joints: New Data and Models", Proc. OMAE 1998, ASME 1998, Lisbon, Portugal.

Hebor, M.F. (1994), "Residual Strength and Epoxy-Based Grout Repair of Corroded Offshore Tubular Members", Lehigh University, Bethlehem, Pennsylvania, US.

Hebor, M., and Ricles, J. (1994), "Residual Strength and Repair of Corroded Marine Steel Tubulars", ATLSS Report No. 94-10, ATLSS Eng. Research Center, Lehigh University, Bethlehem, Pennsylvania, US.

Hellan, Ø. (1995), "Nonlinear Pushover and Cyclic Analysis in Ultimate Limit State Design and Reassessment of Tubular Steel Offshore Structures", PhD Thesis 1995:117, Norwegian Institute of Technology, University in Trondheim, Norway.

Hestholm, K. and Vo, T. (2019), "Buckling Capacity of Corroded Steel Tubular Members", BSc Thesis at the University of Stavanger, Norway.

HSE (1999), *Offshore Technology Report OTO 1999 054 Static Strength of Cracked tubular joints: New data and models,* Prepared by TWI for Health and Safety Executive (HSE), HSE Books, Sudbury, UK.

HSE (1994), "Major Hazards to Concrete Platforms", Project undertaken by Ove Arup and Partners (report is not in public domain).

HSE (1995), *Guidance on Design, Construction and Certification*, 4th Edition of January 1990 including amendment No.3 published in Feb. 1995, HMSO, London, UK.

HSE (1997), *Durability of Prestressing Components in Offshore Concrete Structures*, Gifford and Partners, OTO 97 053, HMSO, London, UK.

HSE (1999), *Offshore Technology Report OTO 1999 084 Detection of Damage to Underwater Tubulars and Its Effect on Strength*, Prepared by MSL Engineering for the Health and Safety Executive (HSE), HSE Books, Sudbury, UK.

HSE (2002a), *Offshore Technology Report 2000/077 Fracture Mechanics Assessment of Fatigue Cracks in Offshore Tubular Structures*, *Prepared by the University of Wales Swansea (D Bowness and M M K Lee) for the Health and Safety Executive (HSE)*, HSE Books, Sudbury, UK.

HSE (2002b), *Offshore Technology Report 2000/078 Static Strength of Cracked High Strength Steel Tubular Joints*, Prepared by University College London for the Health and Safety Executive (HSE). HSE Books, Sudbury, UK.

HSE (2002c), *Offshore Technology Report OTO 2001/080 The Static Strength of Cracked Joints in Tubular Members*, Prepared by UMIST (Professor F M Burdekin) for the Health and Safety Executive. HSE Books, Sudbury, UK.

HSE (2002d), *Offshore Technology Report OTO 2001/081 Experimental Validation of the Ultimate Strength of Brace Members with Circumferential Cracks*, Prepared by UMIST and University College London for the Health and Safety Executive (HSE), HSE Books, Sudbury, UK.

HSE (2004), *Growth of Through-Wall Fatigue Cracks in Brace Members*, Prepared by TWI Ltd (Dr Marcos Pereira) for the Health and Safety Executive (HSE) as Research Report 224, HSE Books, Sudbury, UK.

HSE (2005), *RR344 The Effect of Multiple Member Failure on the Risk of Gross Collapse over Typical Inspection Intervals*, Prepared by MMI Engineering Ltd (Andrew Nelson and David Sanderson) for the Health and Safety Executive (HSE), HSE Books, Sudbury, UK.

ISO (2007), *ISO 19902:2007 Petroleum and Natural Gas Industries—Fixed Steel Offshore Structures*, International Organization for Standardization.

ISO (2018), *ISO 15653:2018 Metallic Materials—Method of Test for the Determination of Quasistatic Fracture Toughness of Welds*, International Organization for Standardization.

ISO (2019), *ISO 19901-9 Petroleum and Natural Gas Industries—Specific Requirements for Offshore Structures—Part 9: Structural Integrity Management*, International Organization for Standardization.

Landet, E. and Lotsberg, I. (1992), "Laboratory Testing of Ultimate Capacity of Dented Tubular Members", ASCE, *Journal of Structural Engineering* 118 (4): 1071–1089.

Lutes, L.D., Kohutek, T.L., Ellison, B.K., Konen, K.F. (2001), "Assessing the Compressive Strength of Corroded Tubular Members", *Applied Ocean Research*, October 2001.

Machida S., Hagiwara Y., and Kajimoto K. (1987), "Evaluation of Brittle Fracture Strength of Tubular Joints of Offshore Structures", Proc. Sixth International OMAE Conference, 1987, Vol III, pp 231–237, ASME.

MSL (2003), *Guideline for the Definition and Reporting of Significant Damage to Fixed Steel Offshore Platforms*, MSL Services Corporation, Houston, Texas, US.

MSL (2004), "Assessment of Repair Techniques for Ageing or Damaged Structures", Project #502, MSL Services Corporation, Egham, Surrey, UK.

Nakai, T., Yamamoto, N. (2008), "Corroded Structures and Residual Strength", In *Condition Assessment of Aged Structures*, Edited by J.K. Paik and R.E. Melchers. Woodhead Publishing.

Nichols, N., Sharp, J.V. and Kam, J. (1994), "Benchmarking of Collapse Analysis of Large-Scale Ultimate Load Tests on Tubular Jacket Frame Structures", Proceedings of the 3rd International Conference Offshore Structure Design, Hazards, Safety and Engineering, ERA, London. UK.

Ocean Structures (2009), "Ageing of Offshore Concrete Structures—Report for Petroleum Safety Authority Norway", Laurencekirk, Scotland.

Ostapenko, A., Wood, B.A., Chowdhury, A. and Hebor, M.F. (1993), "Residual Strength of Damaged and Deteriorated Offshore Structures: Volume 1; Residual Strength of Damaged and Deteriorated Tubular Members in Offshore Structures", Lehigh University, Bethlehem, Pennsylvania, US.

Ostapenko, A., Berger, T.W., Chambers, S.L. and Hebor, M.F. (1996), "Corrosion Damage—Effect on Strength of Tubular Columns with Patch Corrosion", ATLSS Report No. 96.01. ATLSS Engineering Research Center, Fritz Engineering Laboratory Report No. 508.5, Lehigh University, Bethlehem, Pennsylvania, US (also published as MMS report 101).

Padula, J.A. and Ostapenko, A. (1987), "Indentation and Axial Tests of Two Large-Diameter Tubular Columns" (also published as MMS TAP project report 101).

Padula, J.A. and Ostapenko, A. (1989), "Axial Behavior of Damaged Tubular Columns", Sponsored by Minerals Management Service of the US Department of the Interior (Contract No. 14-12-0001-30288) and American Iron and Steel Institute (Project No. 338). DOIIAISI COOPERATIVE RESEARCH PROGRAM. Fritz Engineering Laboratory Report No. 508.11, Lehigh University, Bethlehem, Pennsylvania, US.

Paik, J.K., Kim, S.K. and Lee, S.K. (1998), "Probabilistic Corrosion Rate Estimation Model for Longitudinal Strength Members of Bulk Carriers", *Ocean Engineering*, 25, 837–860.

Paik, J.K., Lee, J.M., Hwang, J.S. and Park, Y.I. (2003a), "A Time-Dependent Corrosion Wastage Model for the Structures of Single and Double Hull Tankers and FSOs and FPSOs", *Marine Technology*, 40, 201–217.

Paik, J.K., Lee, J.M., Park, Y.I., Hwang, J.S. and Kim, C.W. (2003b), "Time-Variant Ultimate Longitudinal Strength of Corroded Bulk Carriers", *Marine Structures*, 16, 567–600.

Paik, J.K. and Melchers, R.E. (2008), *Condition Assessment of Aged Structures*, Woodhead Publishing.

Paik, J.K. 2004, "Corrosion Analysis of Seawater Ballast Tank Structures", *International Journal of Maritime Engineering*, 146(A1), 1–12.

Paik, J.K., Thayamballi, A.K., Park, Y.I., Hwang, J.S. and Kim, C.W. (2004a), "A Time-Dependent Corrosion Wastage Model for Seawater Ballast Tank Structures of Ships", *Corrosion Science*, 46, 471–486.

Paik, J.K. and Thayamballi, A.K. (2007), *Ship-Shaped Offshore Installations—Design, Building and Operation*, Cambridge University Press, Cambridge, UK.

Paris, P.C. and Erdogan, F. (1963), "A Critical Analysis of Crack Propagation Laws", *Journal of Basic Engineering*, 1963, 85, 528–533.

PMB (1987), *AIM II Assessment Inspection Maintenance*, Published as MMS TAP report 106ak.

PMB (1988), *AIM III Assessment Inspection Maintenance*, Published as MMS TAP report 106ap.

PMB (1990), *AIM IV Assessment Inspection Maintenance*, Published as MMS TAP report 106as.

Qin, S. and Cui, W. (2003), "Effect of Corrosion Models on the Time-Dependent Reliability of Steel Plated Elements", *Marine Structures*, 16, 15–34.

Ricles, J., Gillum, T. and Lamport, W. (1992), "Residual Strength and Grout Repair of Dented Offshore Tubular Bracing", ATLSS Reports, Lehigh University, Bethlehem, Pennsylvania, US.

Ricles, J.M., Bruin, W.M. and Sooi, T.K. (1997), "Repair of Dented Tubular Columns—Whole Column Approach", ATLSS Reports, Lehigh University, Bethlehem, Pennsylvania, US (also published as MMS TAP report 101).

Ricles, J. M. and M. F. Hebor (1994), "Residual Strength and Epoxy-Based Grout Repair of Corroded Offshore Tubulars", International Conference on the Behaviour of Offshore Structures, BOSS, Cambridge, Massachusetts, Pergamon Press, 3: 307–322.

Sharp, J.V., Stacey A., Wignall C.M. (1998), "Structural Integrity Management of Offshore Installations Based on Inspection for Through-Thickness Cracking", OMAE Conference 1998.

SINTEF (1988), *USFOS—A Computer Program for Progressive Collapse Analysis of Steel Offshore Structures, User's Manual*, Report No. STF71 F88039. SINTEF, Trondheim, Norway.

Skallerud B., Eide O.I., and Berge S. (1994), "Ultimate Strength of Cracked Tubular Joints: Comparison between Numerical Simulations and Experiments", Proc. Seventh BOSS Conference Boston 1994, Volume III, pp 241–260, July 1994.

Skallerud, B. and Amdahl J. (2002), *Nonlinear Analysis of Offshore Structures*, Research Studies Press Ltd, Baldock, Hertfordshire, UK.

Smith, C.S., Kirkwood, W., Swan, J.W. (1979), "Buckling Strength and Post-Collapse Behaviour of Tubular Bracing Members Including Damage Effects", Proc. 2 Offshore Structures (BOSS 79), London, Aug 1979 (also appeared as Smith, C.S., Kirkwood, W., Swan, J.W., "Buckling Strength and Post-Collapse Behaviour of Tubular Bracing Members Including Damage Effects", Department of Energy, Offshore Technology Report, OT-R-7837, HSMO, 1979, London, UK.

Smith, C.S., Somerville, W.L. and Swan, J.W. (1980), *Compression Tests on Full-Scale and Small-Scale Tubular Bracing Members Including Damage Effects,* Department of Energy, Offshore Technology Report, OT-R-8079, HMSO, 1980, London, UK.

Smith, C.S. (1983), "Assessment of Damage in Offshore Steel Platforms", Proc. of International Conference on Marine Safety, September 1983, Glasgow, Scotland.

Smith, C.S., Swan, J.W. and Kirkwood, W. (1984), *Compressive Strength of Damaged Tubulars,* Department of Energy, Offshore Technology Report, OTO-8315, HMSO, 1984, London, UK.

SSC (1995), "SSC-381 Residual Strength of Damaged Marine Structures", Ship Structure Committee, Washington, DC, US.

SSC (2000), "SSC-416 Risk-Based Life Cycle Management of Ship Structures", Ship Structure Committee Report no 416, 2000.

Stacey A., Sharp J.V., and Nichols N.W. (1996a), "The Influence of Cracks on the Static Strength of Tubular Joints", Proc. OMAE Florence 1996, Volume III, Materials Engineering, Florence, Italy.

Stacey A., Sharp J.V., and Nichols N.W. (1996b), "Static Strength Assessment of Cracked Tubular Joints", Proc. OMAE Florence 1996, Volume III, Materials Engineering, ASME 1996, Florence, Italy.

Standard Norge (2004), *NORSOK N-004 Design of Steel Structures,* Standard Norge, Lysaker. Norway.

Standard Norge (2015), *NORSOK N-006 Assessment of Structural Integrity for Existing Offshore Load-Bearing Structures,* Edition1, March 2009, Standard Norge, Lysaker, Norway.

Statens vegvesen (2019), *Håndbok V441 Bruinspeksjon, Vegdirektoratet*, Oslo, Norway (in Norwegian).

Taby, J., Moan, T. Rashed, S.M.H. (1981), "Theoretical and Experimental Study on the Behaviour of Damaged Tubular Members in Offshore Structures", *Norwegian Maritime Research,* No 2, 1981, 26–33.

Taby J. and Moan T. (1985), "Collapse and Residual Strength of Damaged Tubular Members", 4th Int. Conf. on Behav. of Offshore Struct. 395–409, Delft, Netherlands.

Taby, J. (1986), "Ultimate and Post-Ultimate Strength of Dented Tubular Members", *Norges Tekniske Høgskole*, Univ. Trondheim, Rep. No. UR-86-50, Oct. 1986, Trondheim, Norway.

Taby J. and Moan T. (1987), "Ultimate Behaviour of Circular Tubular Members with Large Initial Imperfections Annu.", Tech. Session, Struct. Stab. Res. Counc.

Talei-Faz B., Dover W.D., and Brennan F.P. (1999), "Static Strength of Cracked High-Strength Steel Tubular Joints", UCL NDE Centre, February 1999.

TSCF (1986), "Guidance Manual for the Inspections and Condition Assessment of Tanker Structure", International Chamber of Shipping Oil Companies, International Marine Forum.

TSCF (1995), *"Guidance Manual for the Inspection and Maintenance of Double Hull Tanker Structure"*, London, Witherby & Co. and Tanker Structure Cooperative Forum.

TSCF (1997), "Tanker Structure Cooperative Forum Guidance Manual for Tanker Structures", Issued by Tanker Structure Co-operative Forum in Association with International Association of Classification Societies, Witherby & Co. Ltd.

Ueda, Y. and Rashed, S.M.H. (1985), "Behavior of Damaged Tubular Structural Members", *J. Energy Res. Technol.* 107: 342–349.

Vo, T., Hestholm, K., Ersdal, G., Oma, N. and Siversvik, M. (2019), "Buckling Capacity of Simulated Patch Corroded Tubular Columns—Laboratory Tests", Second Conference of Computational Methods in Offshore Technology (COTech2019), IOP Publishing.

Zhang, Y. H. and Wintle, J. B. (2004), "Review and Assessment of Fatigue Data for Offshore Structural Components Containing Through-Thickness Cracks", TWI Report for HSE, April 2004 (unpublished).

Zhang, Y.H. and Stacey, A. (2008), "Review and Assessment of Fatigue Data for Offshore Structural Components Containing Through-Thickness Cracks", OMAE2008-57503.

Øyasæter, F.H., Aeran, A., Siriwardane, S.C. and Mikkelsen, O. (2017), "Effect of Corrosion on the Buckling Capacity of Tubular Members", First Conference of Computational Methods in Offshore Technology (COTech2017), IOP Conf. Series: Materials Science and Engineering 276 (2017). IOP Publishing.

8

Repair and Mitigation of Offshore Structures

> *The major difference between a thing that might go wrong and a thing that cannot possibly go wrong is that when a thing that cannot possibly go wrong goes wrong, it usually turns out to be impossible to get at and repair.*[1]
>
> —Douglas Adams

8.1 Introduction to Underwater Repair

Repair and mitigation are often needed when inspection has revealed some damage or anomalies that have affected the integrity of the structure. In addition, such actions may also be necessary when a structure is found by analysis to have for example insufficient strength or fatigue life even without any damage or anomalies present. In the course of the long development of offshore production, history clearly indicates that repairs are necessary to maintain structural safety. Many structures are now being used beyond their original design life and life extension is common. Some of these older platforms in life extension show clear signs of an increasing rate of damage with a growing need for repair.

The first offshore platforms based in the Gulf of Mexico experienced corrosion and accidental damage which required repair in many cases. When the offshore industry moved to the North Sea, many of the designs were based on Gulf of Mexico practice. Due to the more hostile environment in the North Sea, these structures suffered early fatigue damage requiring repair. Later, fixed steel platforms have been designed for fatigue and this has improved the problem. However, continued operation of these platforms has led to some issues of corrosion, fatigue damage and permanent deformation due to accidental events. In addition, insufficient strength and fatigue life are common issues for older but also for newer platforms. Fixed steel, floating and concrete platforms have also proved to need a significant number of repairs due to similar causes. Floating steel platforms are, in particular, experiencing significant amounts of corrosion and fatigue cracking.

The decision whether a repair is needed is normally based on an evaluation and assessment of the damage and anomalies and their effect on structural integrity, as discussed in Chapter 7. The

---

1 *Source:* Douglas Adams.

Underwater Inspection and Repair for Offshore Structures, First Edition.
John V. Sharp and Gerhard Ersdal.
© 2021 John Wiley & Sons Ltd. Published 2021 by John Wiley & Sons Ltd.

need for repairs can vary for different types of structures, depending on the material and design configuration. As noted above, repair of steel structures is primarily needed due to corrosion, cracks and accidental damage, and the extent of repairs increases with age. In comparison, repair of concrete structures is dominated by reinforcement corrosion due to chloride ingress in and above the splash zone followed by possible spalling and delamination. Underwater, the main types of damage on concrete structures are due to impacts from vessels or dropped objects. For both steel and concrete structures, maintaining the corrosion protection system is important to keep the structures safe.

The design and fabrication of repairs requires particular care and attention so that the repair itself does not introduce defects, degradation and other features that could undermine the integrity of the structure. Examples include the introduction of dissimilar metals which have led to corrosion cells, excess loading being introduced by the presence of the repair (e.g. a large clamp) and new defects introduced by a weld repair. In addition, any repair should be included in the inspection plan to verify that the repair is performing as intended and not creating further issues.

Five basic approaches to repair and mitigation work are proposed, based on the work of MSL (2004) as reviewed in Section 8.3.13. These are arranged as shown in Table 53.

Major repairs (e.g. strengthening and component addition) of structures are often costly and complicated and involve the use of expensive equipment and vessels. In addition, the damage experienced is often particular to the specific structure and location, which requires new design solutions for each individual case. The planning and preparation of major repairs can take several months and the repair may be further delayed in the installation phase as the team will often have to wait for a suitable weather window. In many cases the operators will choose to perform a trial installation, involving execution of the repair on a mock-up model of the structure, either on land or in shallow water. This for obvious reasons involves increased duration of the preparation stage. This lengthy period with the structure in a damaged condition has implications for managing its integrity, and a temporary repair or other mitigating measures may be required.

Taking into account that a major repair is often costly and time consuming, its design and execution need to be properly undertaken to ensure an optimal installation and a sufficiently safe

Table 53 Repair and Mitigation Approaches and Examples.

Repair and mitigation approaches	Examples
Detail and weld improvement	Hammer peening, grinding and softening of details (reducing stress concentration factor), replacement or addition of coatings for corrosion protection
Local repair	Welding of steel joints, spraying and re-casting of concrete cover
Load reduction and new load paths	Removal of marine growth, reduction in topside weight
Strengthening	Clamping and grout filling of damaged and understrength members in fixed steel platforms
Structural component removal or addition	Removal of redundant members on fixed offshore structures, addition of new bracing in fixed steel structures and adding stiffeners for floating structures, anode replacement

structure. In addition, the design of the repair needs to ensure that the repair itself is durable and avoids future deterioration. A good understanding of the root cause of the damage is important in achieving this.

An extensive expertise has been accumulated on repair through the many years of experience of operating offshore structures, and this knowledge, to a large extent captured in reports from Joint Industry Projects needs to be retained and made available for future use. In addition, the authors recognise that the inclusion of a review of these reports in this book will hopefully avoid repetition of research work.

The majority of these reviews were undertaken and published in the period 1983 to 2000, and for ship-shaped structures earlier studies also exist. With this large amount of reviews, it is difficult to provide an extensive overview of them all. Projects where major governmental organisations including MMS, BOEME, BSEE, UK Department of Energy, UK HSE, Norwegian Petroleum Directorate and Petroleum Safety Authority Norway are included in the overview provided here. At the time of writing this book many of these reports are not readily available and a summary of these is provided in the following sections as indicated in Table 54.

Table 54 Overview of Previous Repair Studies Relating to Different Forms of Damage and Anomalies (F fatigue, C corrosion, MD mechanical damage, IC insufficient capacity).

		Steel				Concrete			
Report	Sect.	F	C	MD	IC	F	C	MD	IC
UEG Repair to North Sea Offshore Structures	8.2.1	X	X	X	X	X	X	X	X
MTD Repair to North Sea Offshore Structures	8.2.2	X	X	X	X	X	X	X	X
Department of Energy 4th Guidance Note	8.2.3								
DNVGL Repair Methods for Offshore Structures	8.2.4	X	X	X	X		X	X	X
Grout Repairs to Offshore Structures OTH 84 202	8.3.1	X	X	X	X				
Repairs Research Project (JIRRP)	8.3.2	X	X	X	X				
Fatigue Performance of Repaired Tubular Joints OTH 89 307	8.3.3	X							
Bonded Repairs OTH 88 283	8.3.4	X	X	X	X				
Adhesive Bonding for Offshore Structures OTO 96 030	8.3.4	X	X	X	X				
Strength of Grout-Filled Damaged Tubular Members OTH 89 314	8.3.5	X	X	X					
Fatigue Life Enhancement of Tubular Joints by Grout Injection OTH 92 368	8.3.6	X							
ATLSS Reports on Repair to Dent Damaged Tubular Members	8.3.7			X					
ATLSS Reports on Repair to Corrosion Damaged Tubulars	8.3.8		X						
MSL Strengthening, Modification and Repair of Offshore Installations	8.3.9	X	X	X	X				
MSL Underwater Structural Repairs Using Composite Materials	8.3.10								
Experience from the Use of Clamps OTO 00 057 & 058	8.3.11				X				

(Continued)

Table 54 (Continued)

Report	Sect.	Steel				Concrete			
		F	C	MD	IC	F	C	MD	IC
MSL Study on Neoprene Lined Clamps	8.3.12								
MSL Repair Techniques for Ageing and Damaged Structures	8.3.13	X	X	X	X				
KBR Energo Hurricane Damage to Platforms	8.3.14	X	X	X	X				
Wet Weld Repairs and Modifications	8.3.15	X	X	X	X				
Repair of Major Damage to Concrete Offshore Structures OTH 87 250	8.4.2							X	
Scaling of Underwater Concrete Repairs OTH 89 298	8.4.3							X	
Assessment of Materials for Repair of Damaged Concrete Underwater OTH 90 318	8.4.4							X	
Effectiveness of Concrete Repairs HSE RR 175, 184, 185 and 186	8.4.5					X	X	X	X
SSC Reports on Repair of Plated Structures	8.5	X	X	X	X				

8.2 Previous Generic Studies on Repair of Structures

8.2.1 UEG Report on Repair to North Sea Offshore Structures

In 1983 UEG undertook a study of repairs of North Sea structures, steel and concrete structures. This was the first comprehensive study of causes of damage to North Sea offshore structures. It considered the several techniques used to repair structural damage. The experience of using these techniques for some 50 underwater repairs used up to 1982 was also described. The results from this study were incorporated in the later report undertaken by MTD (MTD 1994).

For steel structures, UEG (1983) reports that four methods had been used for strengthening and repair to enable a section to have its geometric properties increased or to have additional members to be fixed to the structure to reduce loading in the area. The methods mentioned included welding (habitat welding, coffer-dam welding and wet welding), clamps (friction, grouted and stressed grouted), grout filling of existing tubular members and bolts to connect plates to an existing member. However, the report states that "No operator has used wet welding for structural components on any repair covered by this review".

For concrete structures, the UEG report (1983) indicates that only five repairs had been reported. The repair methods mentioned are surface application or injection of cementitious or resinous materials. These repair methods are reported to be used both for spalling and more serious damage to concrete sections.

8.2.2 MTD Study on Repairs of Offshore Structures

A very useful analysis of repairs to fixed steel and concrete structures was undertaken by the Underwater Engineering Group (UEG) for the UK Marine Technology Directorate (MTD 1994) as a follow-up project to the UEG report (1983). The study covered the sub-sea strengthening and repair of both steel and concrete offshore installations as well as pipelines. Repairs to topsides were

Table 55 Information within the Repairs Database.

	Steel structures	Concrete structures	Total
Total number of recorded repairs to early 1992	158	14	172
• of which the date the repair was carried out is known	141	13	154
• of which the date the damage was discovered is known	129	12	141
• of which the date at which the damage was caused is known	69	7	76
Total exposure to the end of 1991 in structure years[1]	3846	266	4112

[1] Structure years is the number of repairs divided by the total exposure of similar platforms.

excluded although splash zone damage was included, even when the repair was above the waterline.

The project was overseen by a steering group led by J.V. Sharp of UK HSE and consisting of representatives from the Phillips Petroleum, Elf UK, Amoco, Texaco, Total, Technomare SpA, British Gas and the Norwegian Petroleum Directorate (now the Petroleum Safety Authority Norway).

In total the review included 172 repairs. At the time of the survey there were 352 steel and 23 concrete structures on the North West European Continental shelf, which was the basis for the survey of repairs. Details of the information in the UEG database are shown in Table 55. Although this survey only covered details of repairs undertaken up to 1991, it provides a valuable source of information on repairs as a result of the limited availability of more recent data. It is not expected that recent damage is different and repairs are not undertaken in a significantly different manner.

Figure 82 shows the cumulative total of platforms (both steel and concrete) versus installation year and the number of repairs per repair year in this review.

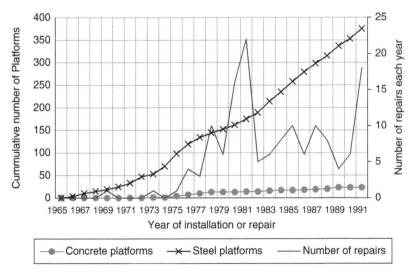

Figure 82 Cumulative total of platforms by material versus installation year and number of repairs versus repair year. *Source:* Based on MTD (1994). Review of repairs to of structures and pipelines. MTD report no. 94–102, Marine Technology Directorate, London, UK.

It can be seen that the majority of repairs were undertaken in the 1980s, but this is linked to the growing number of installed platforms. There are two peaks shown for which the review could find no explanation, other than possible association with the date of the surveys assuming it's easier to recall more recent events (the results of the UEG survey undertaken in 1983 described in Section 8.2.1 were included in the this survey).

Figure 83 charts the relative performance of steel and concrete structures. The annual number of repairs to steel structures reviewed in this study remained static during the 1980s in proportion to the total population. From the early to mid-1980s the number of repairs to concrete and steel structures related to the respective populations of these two types of structure. From 1987 to 1991 (the last year included in this study) concrete structures were observed to require a higher rate of repair.

The frequency of repairs per structure-year versus water depth was analysed, which showed an increasing frequency of repairs per structure-year with increasing depth. This is believed to be the result of structures in deeper waters having more structural members than shallow water structures and hence a greater chance of incurring damage. However, the age of the structures was also believed to be of relevance to the number of damage types.

Figure 84 shows the frequency of repairs per structure year versus water depth. This shows an increasing frequency of repairs per structure year with increasing depth.

Table 56 shows an analysis of structural elements requiring repair or strengthening for steel platforms. It can be seen that the number of repairs to primary structural elements had declined, but the repairs to appurtenances had more than tripled over the time period analysed. The number of repairs to secondary structures such as conductor guide frames remained more or less constant over the time period.

Table 57 shows that most damage was discovered as it happened (e.g. ship impact, dropped object) or by routine or non-routine inspection (typically chance discoveries during inspections for other purposes). However, the table shows that chance discovery of damage did occur, such as

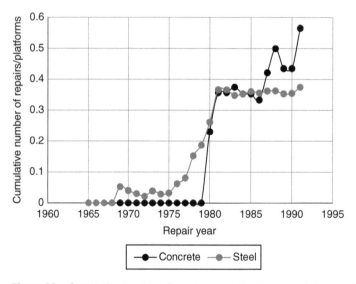

Figure 83 Cumulative number of repairs normalised by material population versus repair year. *Source:* Based on MTD (1994). Review of repairs to of structures and pipelines. MTD report no. 94–102, Marine Technology Directorate, London, UK.

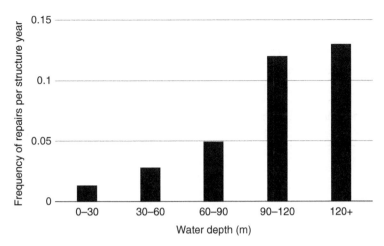

Figure 84 Frequency of repairs per structure year versus water depth. *Source:* Based on MTD (1994). Review of repairs to of structures and pipelines. MTD report no. 94-102, Marine Technology Directorate, London, UK.

Table 56 Analysis of structural elements requiring repair or strengthening.

		Percentage of total for each timeband				
Date of repair	Number of incidents	Primary structure	Secondary structure	Appurtenances	Fabrication aids	Sub-sea structures
1974–79	24	75	17	8	0	0
1980–82	38	63	24	13	0	0
1983–85	23	61	26	9	4	0
1986–88	20	50	25	25	0	0
1989-early 92	33	36	24	27	3	9
Total	138					

Source: MTD (1994). Review of repairs to of structures and pipelines. MTD report no. 94–102, Marine Technology Directorate, London, UK. © 1994, Marine Technology Directorate Limited.

Table 57 Analysis of Method of Discovery, over Time.

		Percentage of total for each timeband				
Date on which damage occurred	Number of incidents	As incident occurred	Routine inspection	Non-routine inspection	Chance	Unknown
1969–76	13	54	23	0	8	15
1977–79	14	79	21	0	0	0
1980–82	16	75	6	0	0	19
1983–85	9	67	22	11	0	0
1986–88	12	50	42	0	0	8
1989–early 1992	11	46	27	9	9	9
Total	75					

Source: MTD (1994). Review of repairs of structures and pipelines. MTD report no. 94–102, Marine Technology Directorate, London, UK. © 1994, Marine Technology Directorate Limited.

Table 58   Analysis of Method of Discovery, by Cause.

Cause of damage	Number of incidents	Percentage of total for each timeband				
		As incident occurred	Routine inspection	Non-routine inspection	Chance	Unknown
Corrosion	13	0	62	8	8	23
Fatigue	40	0	63	3	8	28
Vessel impact	37	68	19	0	5	8
Dropped objects	16	75	13	0	0	13
Total	106					

Source: MTD (1994). Review of repairs of structures and pipelines. MTD report no. 94–102, Marine Technology Directorate, London, UK. © 1994, Marine Technology Directorate Limited

during visual inspections of other structural elements. This clearly has implications for management of structural integrity depending on the severity of the damage.

Table 58 shows that not surprisingly, damage from accidents such as vessel impact and dropped objects were more frequently found soon after the incident occurred, whereas corrosion and fatigue were more usually discovered by routine inspection. It also shows that there was a significant percentage of damage being discovered by chance for corrosion and fatigue, with implications for structural integrity.

The MTD review of repairs (MTD 1994) provided details of ~140 repairs to offshore fixed steel platforms with 14 in water depths greater than 100 m, using a range of methods including clamps and grouted members. Table 59 shows that mechanical and grouted clamps dominated the repair methods. However, welding techniques including hyperbaric also played an important part in repair and strengthening. Grouting of members had also been used on several occasions. It should be noted that wet welding was reported as being in use after 1986, possibly for non-structural purposes.

Table 59   Analysis of Repair Types over Time.

Date of repair	Number of incidents	Percentage of total for each timeband							
		Mechanical clamps	Grouted clamps	Air weld	Cofferdam weld	Hyperbaric weld	Wet weld	Grouted members	Bolts, plates
1974–76	8	38	12	8	0	12	0	0	0
1977–79	21	10	19	24	0	24	0	10	14
1980–82	38	34	21	18	8	11	0	8	0
1983–85	20	15	20	15	10	15	0	15	10
1986–88	17	18	18	24	6	18	12	6	0
1989–91	21	29	10	38	0	10	10	5	10
Total	125								

Source: MTD (1994). Review of repairs of structures and pipelines. MTD report no. 94–102, Marine Technology Directorate, London, UK. © 1994, Marine Technology Directorate Limited.

Table 60 Analysis of Vessel Impact Leading to Repairs by
Vessel Type.

Type of vessel	No. of incidents
Errant	4
Operational	19
Work	5
Unknown	9
Total no. of incidents	37

Source: MTD (1994). Review of repairs of structures and pipelines.
MTD report no. 94–102, Marine Technology Directorate, London,
UK. © 1994, Marine Technology Directorate Limited.

MTD (1994) assessed the number of fatigue repairs versus installation year. The number was shown to be concentrated in the years from 1969 to 1980 representing the problem that the early designs of North Sea structures did not adequately cover the fatigue problem in the more aggressive North Sea wave climate. Later designs had fewer fatigue problems requiring repair.

Vessel impact causing damage was also reviewed as shown in Table 60. Not surprisingly impact damage from operational vessels dominates the list. However, the more serious impact from errant vessels is shown; these are vessels en route to another location colliding with the offshore installation. Errant vessels have caused serious damage; one of the incidents was a vessel travelling through a field at speed in fog; a second was a vessel on autopilot colliding with an installation.

It is recognised that the above analysis of vessel impact is dated and generic. More recent and platform-specific data often needs to be obtained from recognised risk analysis consultants. Some more recent data can be found in HSE RR053 (HSE 2003e).

8.2.3 UK Department of Energy Fourth Edition Guidance Notes

HSE has published separate reports on different topics from the Fourth Edition Guidance Notes (HSE 1994). The section on repairs to offshore installations was issued in the report OTO 96 057 (HSE 1997). This section was quite extensive, covering both steel and concrete, and probably the most detailed description of these topics at the time. The Guidance Notes were based on data that had been acquired as a result of research work as described later, supported by the Department of Energy and the offshore industry. The main sections were:

- procedure for dealing with damage and repair;
- general considerations for repair work;
- welded repairs to steel structures;
- repairs to steel structures other than by welding; and
- repair of concrete structures.

A specific requirement for assessing bowing damage was stated (bowed members for which the maximum measured deviation from the nominal centreline exceeded 1/1200 of the nominal length should be regarded as significantly damaged). The Guidance Notes also referenced OTH 87 259 (Department of Energy 1987a) and OTH 87 278 (Department of Energy 1987b), which gave some guidance on the remaining fatigue life to be expected at cracked tubular joints. These reports are discussed in Chapter 7.

Welded repairs to steel structures were described with emphasis on the issue of the quality of welding in a repair situation recognising the difficulty of achieving high-quality welds in the field, particularly underwater. It was also noted that if the cause of damage was solely due to fatigue, repair welding to an equivalent standard would only be likely to reinstate the original life. Weld improvement techniques, such as controlled weld toe grinding of tubular joints, could then be necessary to improve the fatigue life. The Guidance Notes also listed underwater welding problems and limitations, as discussed in Section 8.6.8. These include depth limitations of certain methods and problems of wet welding particularly for repairs to the primary structure.

Repairs to steel structures other than by welding were described including the following techniques:

- Bolted friction clamps; these included those relying on metal to metal contact, those using either a grout or resin filler and those using elastomeric linings. The Guidance Notes also gave information on acceptable coefficients of friction that could be used in different circumstances, with a maximum value of 0.25 between steel surfaces. This value was also recommended to apply to a grout to steel interface.
- Grout filling: this addressed the filling of members with grout to improve the strength of the member or joint. Advice was provided on the strengths of grouted members/joints as given in BS 5400-5 (BSI 2005), OTH 88 283 (Department of Energy 1988) and OTH 88 299 (Department of Energy 1989b).
- Grinding: the Guidance Notes indicated that surface cracks could be removed by carefully controlled grinding and this should be confirmed by MPI.

Repairs to Concrete Structures

The Fourth Edition Guidance Notes (HSE 1994) stated that the accepted materials for repair offshore were concrete, cement grouts, mortars and epoxy resins (as discussed in more detail later). The importance of the ability of the repair material to bond to concrete, reinforcement or prestressing ducts needed to be considered. The Guidance Notes also noted that when selecting a repair material for protection of the reinforcement against corrosion, the properties of the material to be considered should include issues such as permeability to water, presence of chloride ions and resistivity. An important factor was the durability of the repair material in the marine environment.

Repair of fire damage to concrete structures was also considered, mainly relevant to above water structures. It was noted that following a fire, the damage should be inspected to estimate the temperatures reached by the fire (hydrocarbon fire temperatures may be as high as ~1100°C). From this, the relaxation of prestressing cables and loss of strength of the concrete should be assessed to consider repair options.

Repairs to steel reinforcement were considered where the reinforcement has been damaged or cut away, as addressed later in this chapter.

Repairs to prestress were also addressed making the point that it is very difficult to establish the effectiveness of any damaged prestress, as this is to a large extent dependent on the bond with the grout in the cable duct. The effectiveness of any repairs to tendons which have been broken would require being re-stressed.

8.2.4 DNV GL Study on Repair Methods for PSA

A very recent study of repair methods for both steel and concrete offshore structures was performed by DNV GL for the Petroleum Safety Authority Norway (DNVGL 2019). Structural types included ship-shaped structures, semi-submersible structures, fixed steel structures and gravity-based concrete structures.

Repair was in this report defined in a broad manner, matching the definition of SMR (strengthening, modification and repair) by MSL (2004) and included measures to restore a damaged detail or area to its original condition, strengthening and also measures implemented for intact structures with insufficient strength. The report focusses on experience of repairs on the Norwegian Continental Shelf (NCS).

Triggers for when repairs are necessary included:

- damage such as fatigue cracks, corrosion, fabrication errors, overloading, accidental damage including impact from collisions and damage from installation;
- changes in use or increase in loading from subsidence, addition of new modules, risers, conductors, etc. and life extension; and
- new information such as improved knowledge on, e.g. environmental loading, improved standards and regulatory requirements and results from a more advanced analysis.

Any of these triggers may start a process for evaluating the need for a method of repair. The following steps in this process are recommended by DNVGL (2019) by determining:

- the amount of damage and whether the damage occurs at several similar locations;
- the root cause of the damage taking into account the operational history and the age of the structure;
- the consequences of the damage on the integrity of the structure. This may often imply the need for a non-linear finite element and fracture mechanics analysis and capacity checks of the structure in the as-is situation;
- whether repair or modification is needed or whether the structure has sufficient strength in the as-is condition with the damage present; and
- which repair method is to be used.

DNVGL (2019) states that it is crucial in this process to have access to quality information about the structure. This requires access to information such as inspection reports, analysis reports, weight reports, as-built drawings, any modifications to the structure and possible access to reanalysis models. A full description of the information needed is provided in NORSOK N-006 (Standard Norge 2015).

An appropriate method to execute the work is needed if it is concluded that the damage needs to be repaired. There are often several methods that may be relevant to a specific case. Although a specific solution may be perceived as an obvious choice, alternative methods are recommended to be considered to ensure that the final choice is optimal in each case. Important criteria to evaluate in this decision-making process are:

- the effect of the repair on the structural strength;
- reliability of the method (with what confidence the prescribed result can be achieved);
- the cost of the repair;
- operational consequences of the chosen method (e.g. if shut down is needed);
- requirements for competence;
- requirements for execution;
- tolerances in designing and fitting new components (e.g. clamps and new members); and
- operators own preferences.

The repair methods described for steel structures are well known and have been used for decades on both the UK and Norwegian continental shelves. The effect and long-term properties of these repair methods are also understood. A few relatively new methods are also described by DNVGL (2019) where knowledge about the long-term properties are more limited. In addition, a description of repair methods using adhesives is provided.

If the structure is overloaded, DNVGL (2019) states that it will be necessary to determine the magnitude of permanent deformations and any plastic strains. Structures may experience dents and permanent deflections from ship impact, dropped objects or local buckling due to overload. Normally, a check would be made that the damage is sufficiently small to allow the structure to still meet the strength requirements without repair, see also Chapter 7. For example, rules for checking braces damaged by buckling are given in ISO 19902. For plated structures, analyses can be performed with reduced or no stiffness in the damaged area to get a quick indication of the consequence of the damage. The structure should be checked for strength taking into account any permanent deformations. If the structure has experienced plastic strain, material tests may be required to determine whether the structure has been weakened.

Corrosion is a major challenge for steel structures in the marine environment. However, to what extent corrosion is an issue varies with the type of installation and when they were built. The early fixed steel structures on the Norwegian continental shelf were protected with both paint and cathodic protection. Although many of these structures have been in operation far beyond their original design life, corrosion has not been reported to be a major problem. This is believed to be a result of well-designed and maintained cathodic protection systems. These fixed steel structures have proved to be durable as a result of the CP systems in combination with high-quality coatings. Later designs have followed the same design method, and the experience related to corrosion on fixed steel structure on the Norwegian continental shelf is regarded as good at the time of writing.

The use of ship-shaped production vessels was introduced in the late 1980s and their use has increased since that date. The early versions of ship-shaped units were often designed without any corrosion allowance. Paint and cathodic protection systems were used, for example, in ballast tanks, but experience indicates that these have not been of a sufficient quality. Consequently, ship-shaped production vessels have experienced major corrosion problems in ballast tanks. Storage tanks have been less of an issue, but this is sensitive to temperature and the concentration of corrosive agents.

These have typically been painted at the bottom and top (to minimise condensation problems). Large variations in corrosion problems for floating units have been observed which is believed to be due to differences in requirements for paint systems in the NORSOK standards, compared to class rules. Presently, FPSOs are designed with a corrosion allowance in hulls, tanks and decks. If corrosion is observed, it is less of a problem in these FPSOs, thus reducing the need for early and urgent repairs. Accordingly, it should be sufficient to prevent further corrosion by, for example, cleaning, abrasive blasting and the application of protective coatings such as paint.

The choice of repair methods has been given considerable attention for concrete structures in DNVGL (2019), including the various activities and roles essential in this process (inspection plan, documentation and assessment of findings, choice of repair method, choice of material, method for testing, quality control and documentation during the execution). However, concrete repairs above water have primarily been described with minimal information on underwater repair.

8.3 Previous Studies on Repair of Tubular Structures

8.3.1 Grout Repairs to Steel Offshore Structures

The UK Department of Energy in 1984 funded a test programme, OTH 84 202, to investigate the strength of grouted connections between steel tubular members (Department of Energy 1985). These were initiated as a result of increasing interest in design information for repairs to North Sea structures. The tests included 14 pipe-to-pipe connections at quarter scale with no mechanical shear connectors (weld beads), using a split sleeve with bolted flanges to represent the fixing of a sleeve around an existing member. A separate programme involved 14 pipe-to-pipe connections with welded shear connectors to represent the end-to-end connection of new members (also at quarter scale). In addition, there were supplementary tests to extend the range of geometries of the pipe-to-pipe connections. There were also 3 tests on T-shaped repair systems for tubular joints.

The report also addressed a number of practical issues. These included tolerances, making the point that centralisers are normally required between a sleeve and the parent member to ensure that there is a minimum grout thickness at all points (recommended 25 mm). Another issue was detailing of the flange connection to ensure that there are effective seals at the ends of each sleeve to contain the grout, where rubber seals were recommended. Regarding the addition of shear connectors to the parent member, underwater welding methods were reviewed. These included wet welding and although this technique was known to produce reduced mechanical properties, the report concluded that for the addition of weld beads, this could be acceptable. Welding in a water-tight box was also reviewed but concluded to be more expensive and to have limitations.

The test results showed that in pure bending, the strength of a grouted connection was much greater than that of the steel member. For combined axial and bending loads, a small reduction in strength of the connection was reported. The strengths of the specimens with shear connectors were much greater than those of plain pipe connections. With only a small amount of shear connectors the strength was increased by a factor of three. An optimal value for shear connector spacing (D_p/s) of 8 was found where D_p is the member diameter and s the connector spacing. The use of more closely spaced weld beads led to no further increases in strength.

Overall, the results from the pipe-to-pipe tests showed that for the geometries tested, the strength of the grouted connections was adequately described by the design rules for grouted pile-to-sleeve connections as reported in the Department of Energy Guidance Notes at that time (Department of Energy 1977 and 1984). The tests also indicated that the bond strength was a linear function of the shear connector ratio (h/s, where h is the height of the weld bead and s its spacing) for the range of h/s up to 0.06.

The results from the three T joint tests demonstrated that it was possible to design a grouted repair system for a simple T joint against static failure.

Later, in 1988, Department of Energy published the report OTH 89 289 on grouts and grouting for construction and repair of offshore structures, providing a summary of a university-led programme managed by the London Centre for Marine Technology (Department Energy 1988b). The project was funded by the Department of Energy and industry partners. The programme examined problems encountered in offshore grouting applications and generated information which can be used to design grouts for specific tasks. The ten projects in the programme included research into the three aspects of grouting, which were materials, techniques and structural applications.

In the materials part of the programme, grout formulations were investigated to provide the required properties for different offshore applications. These included, for example, the need to develop strength rapidly at low temperatures for grouts to be used for clamp repairs. Eight of the projects extended the understanding of how to develop specific formulations, using four different types of cement, including Portland cement with ash, slag or silica fume replacement. Seawater was used for mixing the grouts and they were cured at 8°C, typical of North Sea temperatures. Other projects also investigated the placing of grout. The water volume needed to maintain fluidity is greater than that required to fill the voids between the particulate components. The volume occupied by a settled grout may be only 85% of the mix volume. This potential bleed (or settlement) can cause voids leading to structural problems. To identify the critical stages in settlement, the theory of bleed was investigated using laboratory tests (with grouts in vertical and inclined ducts).

Methods to reduce the bleed, such as high shear mixing and use of admixtures, were also investigated.

An investigation of the residual strength of dented tubular members showed that filling the member with grout could restore strength. A series of tests examined the effects of geometry and grout age on the ultimate load capacity. It was concluded that for an isolated member, a grout infill restored the strength lost by denting in most cases. In addition, it was found that fatigue was not a problem with a high-strength grout, the ultimate load increased with the age of the grout and full-scale tests showed proportionally higher failure loads than small-scale tests. Another project improved the understanding of the failure of a grouted pile-sleeve connection which had weld beads on the tubulars to improve the key between grout and metal. Failure of the grout beneath the weld-beads was shown to follow circumferential yielding of the pile. A finite element (FE) program to predict the strength of weld-beaded connections was developed and showed good agreement with experimental results.

8.3.2   UK Joint Industry Repairs Research Project

The OTH 88 283 report (Department of Energy 1988) is the outcome of the Joint Industry Repairs Research Project funded by the UK Department of Energy and nine oil companies. The steering committee for this Joint Industry Project was led by J.V. Sharp. The research carried out investigated the static strength and fatigue performance of grouted and mechanical connections and clamps of the types used to strengthen or repair underwater steel members. Data from the research and from other sources was assimilated and used in the compilation of Part 1 of this report which was entitled the "Designers Manual". The designer's manual was intended to provide engineers with basic information and design formulae from laboratory tests on repairs by grouting and mechanically strengthening carried out between 1982 and 1984. It was in the review recognised that the use of grouted and mechanical repair systems underwater and in the splash zone had increased dramatically up to date. This "Designers Manual" is now considered dated, but many of the elements in this manual have been transferred into new standards, recommended practices and guidance with the required updates.

In the review it was differentiated between a clamp and a connection. A clamp was defined as a repair or strengthening of a tubular joint by providing an alternative load path through the clamp whilst a connection was defined as a device for joining concentric tubular members together. The primary causes for repair were in the review identified as fatigue, increased code requirements (wave loading), ship impacts and impact by dropped objects.

The different grouted and mechanical repair schemes that had been employed to date were in the review defined as shown in Table 61 (Department of Energy 1988).

Part II of the document (Engineering Assessment of Test Data) provides more detailed information, including a detailed discussion of the factors affecting the strength and behaviour of each type of clamp or connection.

For each type of underwater repair, the analysis includes its applications and factors affecting its strength, ranges of application, permissible working loads, safety considerations, applied loads and fatigue.

The concluding chapters of the manual describe grouting materials and procedures and other relevant considerations in the design of repairs, such as bolting and sealing systems, cathodic protection, inspection and wet welding.

There are two main types of grout material used for repairs, which are (Department of Energy 1988):

Table 61 Repair Methods.

Repair method	Description	Comments
Grouted connection	A connection between two concentric tubulars formed by the injection of a cementitious material into the annular space between the tubulars	Grouted connections were reported to be regularly used to connect piles to fixed steel structures (still the case) but also for repairs such as sleeving understrength members or providing an alternative load path across defective areas.
Stressed grouted connection	A connection formed between two concentric tubulars. The outer tubular is formed in two or more segments. Cementitious material is placed into the annular space between the tubulars and allowed to reach a predefined strength prior to the application of an external stressing force normal to the steel-grout interface.	Stressed grouted connections were reported to be used as a repair technique for the Viking Field. General formulae for static strength and guidance on fatigue analysis of grouted connection were provided.
Mechanical connection	A connection formed between two concentric tubulars relying for load transfer on the friction capacity of the interface between the two tubulars. The outer tubular will be formed from two or more segments which are stressed together to generate a force normal to the friction surface.	Applications were reported to include connecting two parts of a member together and for connecting a new member into a structure. Formulae for static strength were provided.
Mechanical clamp	A clamp in which the outer sleeve is formed in two or more segments which are placed around an existing tubular joint. The clamp body will be formed from two or more segments which are stressed together to provide the load path in the clamp.	Applications were reported to include strengthening of one or more brace members at a tubular joint (static strength and fatigue) and connecting members at an inclined angle into an existing structure. Guidance for static strength and fatigue analysis were provided.
Grouted clamp	A clamp in which the outer sleeve is formed in two or more segments which are placed around an existing tubular joint. The splits are closed by pre-tightened bolts prior to the injection of a cementitious material into the annular space between the clamp and the existing tubular joints.	Applications were reported to include strengthening of one or more brace members at tubular joints and connecting a new brace member into the structure. Formulae for static strength and stress concentration factors were provided.
Stressed grouted clamp	A clamp in which the outer sleeve is formed in two or more segments which are placed around an existing tubular joint. Cementitious material is placed into the annular space between the clamp and the existing tubular joint and allowed to reach a predefined strength prior to the application of an external stressing force normal to the steel-grout interface.	Applications were reported to include strengthening of one or more brace members at tubular joints against static or fatigue loading and connecting a new brace member into at an inclined angle into an existing structure. Formulae for static strength and stress concentration factors were provided.
Grout-filled tubular	A tubular which has been filled with a cementitious material	The major uses were reported to be increase of capacity of understrength, impact prone, dented or buckled members.

Source: Modified from Department of Energy (1988). Offshore Technology Report OTH 88 283 Grouted and mechanical strengthening and repair of tubular steel offshore structures. Prepared by Wimpey Offshore Engineers and Constructors LTD (R.G. Harwood and E.P. Shuttleworth) for the Department of Energy. HMSO, London, UK

- Portland cement grouts with or without inert fillers, mixed preferably with fresh water. Seawater can be used; however, for repairs particularly in the splash zone, it is not recommended because of potential corrosion issues due to the presence of chloride and the availability of oxygen.
- High alumina cement (HAC) grouts mixed with fresh water for faster curing times. The water-cement ratio should not exceed 0.4 as HAC is prone to conversion reducing its strength. It is also vulnerable to chemical attack when exposed to water for long periods and hence limited for underwater applications.

Admixtures are available to improve the properties of either the slurry or hardened grout, but tests need to be undertaken to demonstrate that these have no harmful effects on the performance of a connection.

8.3.3 UK Department of Energy and TWI Study on Repair Methods for Fixed Offshore Structures

The report OTH 89 307 (Department of Energy 1989) describes a series of fatigue tests at the Welding Institute on welded tubular T joints in which fatigue cracks were repaired by a number of alternative methods. The main objectives were to establish a ranking for the repair methods in terms of residual fatigue performance and to establish whether it is necessary to repair the entire joint or only the cracked region. The joints were tested in out-of-plane bending.

The repair methods investigated were:

- repair welding (cracks were removed by grinding and the resulting excavation filled by manual metal arc welding);
- repair welding and burr grinding (repair welds, made as described above, were fully burr ground to remove the repair toe weld and to achieve a smooth finish on the repair weld face);
- hole drilling and cold expansion (the fatigue crack tips were removed by drilling through-thickness holes, which were then cold expanded to include compressive residual stresses in the hole circumference); and
- grinding alone (part wall cracks were removed by burr grinding and the resulting excavations left unrepaired).

Repair welding included removing the crack by grinding and welding the resulting excavation by manual arc welding. Various factors were seen to play a part in the fatigue life of a repair welded detail, for example the hot spot increased marginally but the length of the weld increased reducing the stress at the weld toe. Repair welding was carried out in the air at atmospheric pressure under shop conditions, rather than trying to reproduce conditions likely to be met in an underwater repair of an offshore structure. In a subsidiary investigation fatigue crack growth rates were measured in weld metal deposited under hyperbaric conditions to establish whether hyperbaric repair welds would give similar results to those for one-atmosphere repairs. All in all, the conclusion was that the remaining life was similar to or possibly marginally greater than the fatigue life of the original detail.

Burr grinding to remove the repair weld toe and to achieve a smooth finish on the repair weld face significantly improved the fatigue life of a repaired weld. If the applied loading was unchanged, the fatigue life was indicated to be in excess of five times that before repair. It was noted that burr grinding of the unrepaired weld toes on both the chord and brace side was necessary to avoid premature failure.

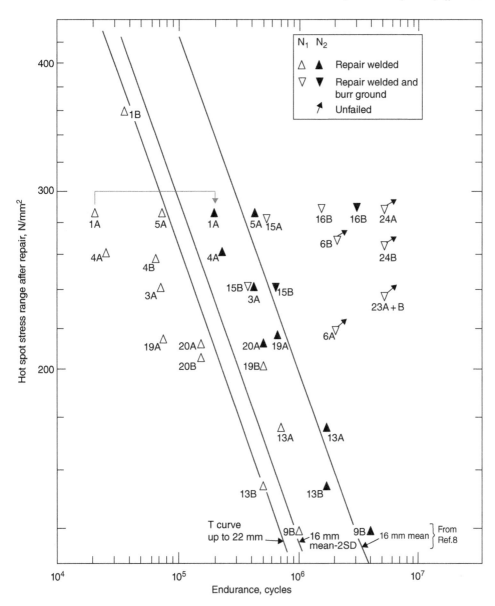

Figure 85 Fatigue test results for repair welded tubular joints plotted in terms of the measured hot spot stress range after repair. White dots indicate fatigue life prior to repair and black dots after repair. *Source:* Department of Energy (1989). OTH 89 307 Fatigue performance of repaired tubular joints. Prepared by the Welding Institute (P.J. Tubby) for the Department of Energy. HMSO, London, UK.

Results of the fatigue tests of repair welded specimens with and without burr grinding are shown in Figure 85.

In hole drilling the fatigue crack tip was removed by drilling a through-thickness hole which was cold expanded to induce compressive residual stresses in the hole circumference. The report concluded that cold expanded holes at the crack tips were not an effective way of delaying crack propagation; see Figure 86. However, later work on the use of hole drilling indicates that this is not necessarily the case (Atteya et al. 2020).

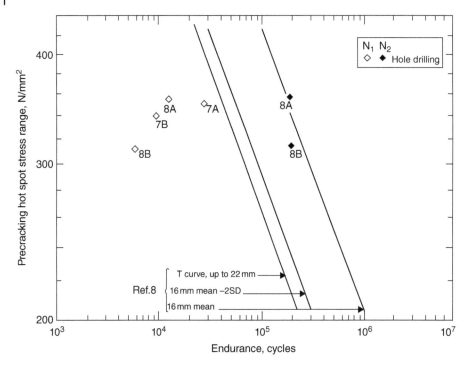

Figure 86 Fatigue test results obtained for tubular joints repaired by hole drilling and cold expansion plotted in terms of the hot spot stress range during pre-cracking. White dots indicate fatigue life prior to repair and black dots after repair. *Source:* Department of Energy (1989). OTH 89 307 Fatigue performance of repaired tubular joints. Prepared by the Welding Institute (P.J. Tubby) for the Department of Energy. HMSO, London, UK.

Grinding alone, as shown in Figure 87, may be used for part wall cracks and includes that the crack is removed by burr grinding and the resulting excavations are left unrepaired. This study indicates that this was an effective repair method giving fatigue lives up to four times greater than that of the unrepaired weld. It was further noted in this project that it is necessary to burr grind the unrepaired weld toes on both the chord and the brace side to avoid fatigue failure at an early stage.

The main findings of the review were the following:

- Expressed in terms of the hot spot stress range, the fatigue strength of as-welded repair welds was marginally lower than that of unrepaired welds. However, this was compensated by the fact that in making the repair, the overall weld length was increased, which led to a reduction in the stress range of the repair weld toe. As a result, if the applied loading was kept the same before and after repair, the fatigue endurance to through-thickness cracking after repair was on average similar or marginally greater than that before repair.
- Burr grinding the repair weld was found to considerably improve the fatigue strength, although a smooth surface finish should be achieved to avoid premature crack initiation on the weld face. Where the applied loading continues unchanged after the repair, the fatigue life to through-thickness cracking may be in excess of five times that before repair.
- Through-thickness and part-wall weld repairs behaved similarly. Under the mode of loading investigated, a relatively large penetration of defects could be tolerated at the root in through-thickness repairs.
- Cold expanded holes at the crack tips were not effective as a means of delaying crack propagation.

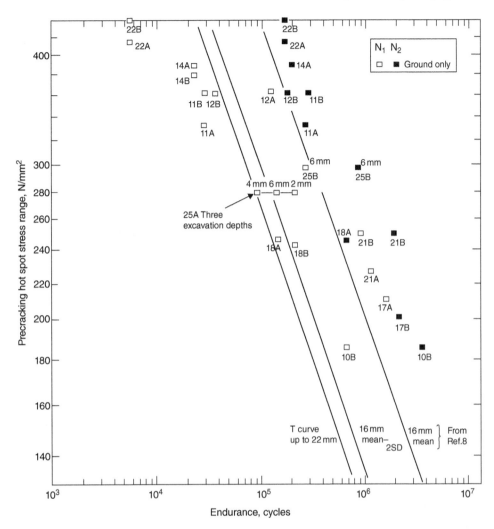

Figure 87 Fatigue test results for tubular joints repaired by grinding alone plotted in terms of the hot spot stress range during pre-cracking. White dots indicate fatigue life prior to repair and black dots after repair. *Source:* Department of Energy (1989). OTH 89 307 Fatigue performance of repaired tubular joints. Prepared by the Welding Institute (P.J. Tubby) for the Department of Energy. HMSO, London, UK.

- Removal of part-wall flaws by grinding was an effective repair method giving endurances after repair up to four times greater than the mean for unrepaired joints. Data were presented which allowed the likely residual life to be estimated for a given excavation depth.
- The report indicated that crack growth rates in hyperbaric repair welds would be expected to be similar to those in the one-atmosphere repairs.

8.3.4   UK Department of Energy–Funded Work on Adhesive Repairs

The UK Department of Energy funded a programme of work to develop repair methods for application to underwater sections of an offshore steel platform (Clarke et al. 1986). Typical underwater repair systems for fatigue cracking are based on grouted connections. Alternatives to these would have benefits if they could provide equivalent or improved mechanical properties and cost advantages. Previous work for the UK Ministry of Defence had developed resins for repair of ships where the resin was capable of bonding to a wet steel surface. The Department of Energy programme was aimed at exploiting these resins for offshore use for repair of platforms (Clarke et al. 1986).

In selecting a suitable resin, several physical properties needed consideration. These included bond strength to a wet surface, cure time at temperatures typical of the North Sea, viscosity as it is needed to penetrate the annular space in a connection. Another consideration was the exotherm generated during curing. Thick sections of resin could lead to high temperatures, which would degrade the properties of the resin.

Based on the Ministry of Defence work, the adhesive selected was an epoxide-based, cold curing formulation called UW 45. It was selected as having a good blend of chemical and physical properties, with the ability to cure when immersed in water at temperatures as low as 5°C. It also flowed well at these temperatures. In addition, several other commercially available resins at the time were tested for comparison purposes.

The programme involved work on the adhesive at the Admiralty Research Establishment at Holton Heath, together with structural analyses of proposed repair systems at the Admiralty Research Establishment at Dunfermline.

The materials-focussed research involved first grit blasting the steel specimens underwater. This led to a surface where water was strongly absorbed onto the freshly exposed steel. In this condition it was found that the adhesive failed to "wet" the surface leading to a poor bond. It was established that the application of a sacrificial pre-treatment technique (SPT) led to the displacement of the water and to the deposition of a hydrophobic film over the surface. This film was found to be compatible with the adhesive and use of the SPT method led to conditions where the adhesive would spontaneously spread over the surface and provide a good bond. A technique was developed for applying the SPT using a hood where bulk water was displaced from the hood by compressed air; an atomised spray of the SPT was then applied, which displaced the residual water film. This left behind a useful hydrophobic surface to which the adhesive could bond.

Many small steel specimens were prepared to investigate the basic properties of the adhesive and the effect of different surface treatments. These included both tensile butt joints and lap shear joints. Joints made using the SPT and without it were compared, and it was found that use of the SPT led to strengths 3 to 4 times higher and with a much smaller variation in properties. Joints made using the SPT failed mainly within the adhesive layer, not at the steel-adhesive interface.

To test the durability of adhesive-bonded butt joints they were exposed to seawater at room temperature for a period of up to 8 years. These showed good performance over the long exposure period. To test the performance in real seawater, a number of single overlap tensile joints were exposed just below the waterline on a raft in a harbour on the south coast of England. These showed no loss in strength after 3 years' exposure.

Stress rupture tests were also carried out to establish the effect of continuous loading. Specimens were exposed at 20°C under seawater at loads representing between 10–40% of their short-term failure stress. Although failure of specimens initially loaded at the 40% level occurred within a few hours, the specimens subjected to the 20% loading level survived about 3 years before failing.

In addition to the small-scale specimens, tube-to-tube joints were selected as an example of a repair required for offshore structures. Typical dimensions of these were an outside diameter of 110 mm for the tube and 127 mm for the sleeve. Resin thicknesses of between 1.5 and 5.5 mm were tested. All the specimens were bonded in simulated seawater at a temperature of 6°C. Grit blasting of the steel surfaces was done both in air and underwater using the SPT treatment. For the latter specimens, an adhesive pre-coat was applied as it was found that the resin bonded well to itself, and the pre-coat provided a protective layer to prevent corrosion.

The tube specimens were tested in tension at a temperature of about 10°C. Good strengths were found for those specimens grit-blasted in air and with the resin applied above water. Specimens cleaned underwater and using the SPT treatment provided strengths in excess of 450 kN corresponding to the yield strength of the tube.

A comparison of these resin-bonded joints was undertaken in OTH 88 283 using results for similar grouted tube-to-tube joints from a separate programme (Department of Energy 1988). The grouted tests included both specimens with and without shear connectors (weld beads applied to the surface). The results for the resin-bonded joints were encouraging, with the strength of the resin-bonded joint approximately twice that of the grouted specimen with weld beads and 6 times that of the plain grouted connections (see Table 62).

The practical aspects of an adhesive-based repair to steel underwater were examined. This involved several stages starting with an initial grit blasting followed by application of the SPT. For preparation of larger specimens, grit blasting and the SPT deposition needed to be mounted in one unit in order to keep to a minimum the time before the SPT is applied. It was found that the useful life of the SPT film varied depending on the exposure conditions with current and wave action having an effect. The key stage was to apply the adhesive pre-coat as quickly as possible to avoid damage to the SPT film.

The research programme demonstrated a repair technique for using an epoxy resin adhesive to steel underwater. This included the use of a surface preparation technique which was shown to be essential to achieve good bonds. The use of the resin to connect tube-to-tube connections also demonstrated good properties, with strengths up to a factor of two compared to similar tube-to-tube connections using grout. Patents were applied for the process, which were granted.

A company (Wessex Resins) was commissioned by the Department of Energy to commercialise the resin, pre-treatment and general process. Wessex Resins and Adhesives Limited is a manufacturer and formulator of epoxy resin systems for the marine, transport, construction and general engineering industries. At the time of writing, Wessex Resins offered a modified adhesive, UW 4701, which is an improvement on UW45. Its claim is that it can cure at low temperatures and is highly tolerant to oil and water. The improved formulation provides a good bond to wet steel without the use of the surface pre-treatment tested in the research programme. This has the advantage of simplifying the process underwater. At present, adhesive repair using the above-mentioned resins have been reported to be used for an underwater caisson repair (Clarke 2020).

Table 62 Comparison of Grouted and Bonded Repairs.

Type of repair	D/t			Tube OD	Average shear stress MPa
	Tube	Sleeve	Adhesive/ grout		
Adhesive	27–33	32	20–25	110	6.6–7.6
Grout	36.3	50–83	32.5	230	0.6–1.3
Grout with weld beads	36.3	50–83	32.5	230	3.1–3.9

Source: Clarke, J.D., Sharp, J.V. and Bowditch M. (1986). An Underwater adhesive based repair method for offshore structures. Conference on inspection, maintenance and repair, Aberdeen, 1986.

Review of Adhesive Bonding for Offshore Structures OTO 96 030

A review was commissioned by the Department of Energy (HSE 1997) on the applications of adhesives in offshore structures. It was limited to dry applications relevant to topsides. The review addressed health and safety issues in handling adhesives, as well as inspection and repair. On occupational health the review concluded that adhesives at that time posed no particular hazards to operatives, provided established handling procedures were used.

On inspection the comment was that it was inherently difficult, and for that reason, quality assurance should be built into the bonding process and post-bonding coupon testing. It also commented that in most situations adhesively bonded connections were significantly defect tolerant. It was concluded that at the time there were a very limited number of inspection methods available.

The review also addressed repair and identified three situations, which were:

i) to completely remove the existing joint and continue with surface preparation as for a new joint;
ii) to attempt to introduce more or different adhesives into a partially bonded connection; and
iii) to use some other form of joining such as mechanical to provide security for the joint.

Option (ii) requires introduction of adhesive into the joint either from the edge or from holes drilled into the adherend. This clearly has many limitations. Option (iii) is particularly relevant if the defect is at the edge of an attachment or at the end of a stiffener where failure is due to excessive cleavage forces. A mechanical attachment (or possibly welding) can help reduce these cleavage forces. Welding is likely to produce local damage to the joint. These repairs would be difficult underwater.

8.3.5 Residual and Fatigue Strength of Grout-Filled Damaged Tubular Members

Report OTH 89 314 (Department of Energy 1989b) presented details and results of a test programme carried out on tubular members to determine the effectiveness of grout filling as a means of repairing damage.

The tests involved a number of indented specimens to be grout-filled under controlled conditions. After denting, the tubulars were tested and then after failure cut open for the condition of the grout to be examined. Comparisons were made with the unfilled damaged and undamaged parent tubulars. A comparison was also made with the experimental work and theoretical analysis undertaken separately.

It was shown that for the number and dimensions tested, the presence of grout enhanced the strength of the member when compared with an identical but unfilled damaged tubular. The results showed that this increase in strength was dependent upon grout strength, dent size, and L/r and D/t ratios. At higher grout strengths, increases in ultimate strength ranging from 45% to 125% were achieved. The size of dent influenced the strength in grout-filled members with an increase of approx. 70% for small dents ($d/D = 0.04$) and a smaller increase of 55% for larger dents ($d/D = 0.16$). It was shown that the L/r ratio had little effect at low values of the D/t ratio, but a reduction was observed at higher D/t ratios. The strength increased with the D/t ratio for lower values of the slenderness ratio (L/r), but there was little change at higher ratios. Except for the most severe case of damage tested ($d/D = 0.16$), the presence of high-strength grout increased the strength of a damaged member beyond that of the same unfilled undamaged member. The increase in strength, however, resulted in post-collapse losses and rapid unloading.

Scale effects would appear to have been present, but these effects did not change the main conclusion of enhanced strength from grout filling. From the limited test results, it would appear that fatigue is not a significant problem at acceptable grout strengths. Since the ultimate load was found to be influenced by grout strength in the static case, it might be that fatigue is an important consideration at lower grout strengths.

It was noted that the results in this report were based upon tests carried out on specimens that have been completely filled with grout. Losses due to voids or partial filling had not been investigated in this study.

The grout used for these tests was a standard Oilwell B cement grout with 0.36 water/cement ratio by weight of cement used. This is typical of the grouts used for structural grouting of North Sea platforms.

8.3.6 Fatigue Life Enhancement of Tubular Joints by Grout Injection

Report OTH 92 368 (Department of Energy 1992) described static stress analysis and fatigue tests on repaired and fully internally grouted tubular welded T joints. Two fatigue damaged 914 mm diameter tubular T joints, originating from the UKOSRP programme, were repaired by internally grouting. Extensive stress analysis was carried out on each specimen to determine the influence of full internal grout on the stress and strain concentrations factors generated under 3-point loading, axial loading through the brace member, in-plane bending and out-of-plane bending.

Each specimen was then subjected to axial fatigue loading through the brace members to evaluate the effect on fatigue performance and failure modes due to the introduction of this type of stiffening. Throughout each test, crack initiation sites and crack propagation data were recorded along with joint flexibility data.

The results, which were presented in the form of stress concentration factors and stress endurance curves, indicated that the technique may be applied to existing nodes to extend their fatigue lives by reducing the hot spot stresses around the chord to brace intersection due to the loads to which the nodes are subject in-service. In addition to the possible extension in fatigue lives, crack propagation and local joint flexibility indicated that failure modes had not changed from those exhibited by conventional nodes.

Introduction of grout considerably reduced the stress generated under axial and out-of-plane bending conditions, thereby reducing the SCFs around the brace/chord intersection weld.

The fatigue lives of two grout-stiffened nodes were less than that predicted by the mean T curve but were within the scatter band of that displayed by conventional nodes. Failure modes of grout-stiffened nodes of the geometry investigated in this programme had not been altered from that of conventional nodes.

It was concluded that fatigue lives of existing undamaged nodes could be extended by the introduction of grout provided the dominant load condition was axial. It was also concluded that the existing S-N curve approach to structural analysis was still applicable for grout-stiffened nodes.

8.3.7 ATLSS Projects on Repair to Dent-Damaged Tubular Members

As part of the ATLSS projects, Rickles et al. (1992) in the project "Residual Strength and Grout Repair of Dented Offshore Tubular Bracing" performed laboratory tests of thirteen steel tubular braces of various diameter-to-thickness ratios. These tests were performed to examine the effect of:

- a dent damage of 0.1 of the diameter D on the residual strength;
- internal grout repair; and
- grouted clamp repair (sleeve).

The braces were subjected to axial loading, and in some specimens, this was combined with load eccentricity of 0.2 of D to simulate a bending moment. The main results of this study included the following conclusions:

- A dent of 0.1 D gave a significant loss of strength (up to 50% reduction);
- The effect of load eccentricity of 0.2 D gave a pronounced effect on the strength (up to 50% reduction);
- Internal grout repair of a 0.1 D dent-damaged brace successfully reinstated the original un-damaged strength (limited by the D/t ratio of the damaged member due to local buckling); and
- Grouted steel clamp repair of a 0.1 D dent-damaged brace successfully reinstated the original un-damaged strength.

Several strength formulations, computer programs and general finite element analysis were used to compare with the test results. It was found that non-linear finite element analysis was effective in modelling the experimental results, if care was taken in the material modelling and mesh refinement.

Later, Ricles et al. (1997) investigated the effect of grout repairs to larger dented specimens up to 0.5 D. The review confirmed previous experimental studies that internal grout repair could reinstate the capacity of a tubular member with up to a dent of 0.15 D. A series of laboratory tests were performed to investigate tubulars with dents > 0.25 D and investigate the effect of grouting for tubulars with D/t of 33.8 and 67.9. The project used these tests in verification of analytical methods and computer programs relevant at that time. However, the test results as shown in Figure 88

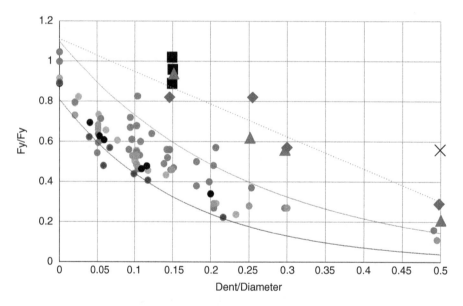

Figure 88 Capacity of repaired and unrepaired dented tubulars based on Ricles et al. (1992), Bruin (1995) and Ricles et al. (1997). Circles indicate unrepaired tests, triangles represent tests results of internal grout-repaired tubulars with D/t~34, squares with D/t~45, diamonds with D/t~67 and the X indicates one external grouted sleeve repair for a tubular with D/t~67. *Sources:* Based on Ricles, J., Gillum, T. and Lamport, W (1992). Residual strength and grout repair of dented offshore tubular bracing. ATLSS Report No. 92-14, Lehigh University, Bethlehem, PA, US ; Bruin, W.M. (1995). Assessment of the residual strength and repair of dent-damaged offshore platform bracing. Master Thesis presented at Lehigh University, Bethlehem, PA, USA ; Ricles, J.M., Bruin, W.M. and Sooi, T.K. (1997). Repair of dented tubular columns – whole column approach. ATLSS Reports, Lehigh University, Bethlehem, PA, USA. Also published as MMS TAP report 101.

clearly indicate that grouting is not able to reinstate the capacity of dented tubular members with dents of 0.25 D or 0.5 D. Approximately 65–90% of strength was restored (depending on D/t ratio) by internal grouting for a tubular member with a 0.25 D dent and as low as 20–35% of strength was restored by internal grouting for a tubular member with a 0.5 D dent. Interestingly, more capacity was restored by external grouting (clamp or external sleeve).

The formulae of Loh et al. (1992) and Ellinas (1984) were recommended for calculating the as-dented capacity (Loh for dents of less than 0.3 D and Ellinas for dents between 0.3 D and 0.5 D). Finite element analysis of such members was recommended to include the denting process. For grouted members Loh's formulae (Loh 1991) were recommended for dent depths up to 0.5 D. Finite element analysis was not recommended due to cost, effort and uncertainty.

Ellinas' (1984) estimation of the ultimate member stress σ_{ud} of the dented tubular member was found by solving the following quadratic equation including second-order effects:

$$\frac{\sigma_{ud}^2}{\sigma_e} - \left[1 + \alpha_0 \cdot \lambda_d + \frac{A_d \cdot e_d}{S_d} + \frac{f_e}{\sigma_e}\right] \cdot \sigma_{ud} + f_y + \sigma_{pd} \cdot \left[\frac{A_d \cdot e_d}{S_d}\right] = 0$$

where e_d is the eccentricity introduced by the dent, f_y is an equivalent squash stress of the dented section (not the yield stress σ_y), σ_e is the Euler buckling stress and σ_{pd} is equal to the initial plastification stress in the saddle of the dent. Geometric parameters for the imperfection α_0 and column slenderness λ_d are further discussed in Ellinas (1984).

The moment and axial load capacity of grout-filled dented tubes was studied by Ostapenko et al. (1996) as part of a Joint Industry Project also sponsored by US MMS. The aim of this project was to derive formulae to predict the deformation as a result of axial and bending loads. The formulae were developed based on an assumption that both steel and concrete materials had a bi-linear elastic-plastic stress-strain relationship and that the grout had no strength in tension. By varying the location of the neutral axis and the value of curvature, the relationship between the axial load and curvature could be defined. The method was compared to laboratory tests of grouted tubulars where 3 specimens were dented and 3 specimens were undented. The formulae derived in this study were concluded to provide somewhat higher ultimate loads compared to the laboratory test specimens.

8.3.8 ATLSS Projects on Repair to Corrosion Damaged Tubulars

Repair options should be reviewed if the evaluation has found the residual strength of a corroded tubular member to be less than needed. Repair options include total member replacement, but this is often neither feasible nor necessary. A better option is indicated by Hebor (1994) to be the addition of a sleeve or partial or full internal grouting. For dent damage, internal grouting is effective as it prevents the dent from growing and the original strength can be restored for members with smaller dents.

However, as stated by Hebor (1994), repair of corrosion damage is influenced by load transfer across the corroded patch and an external steel sleeve was considered to provide a better load transfer. A cement-based grout between sleeve and member was an option, but Hebor (1994) used an epoxy-based grout to obtain higher bond strength in order to develop improved stress transfer across the corrosion patch. It was also indicated by Hebor (1994) that an epoxy-based grout restores the lost gross section and minimises the corrosion-induced internal moment by reducing the eccentricity in the corroded cross section. High-strength epoxy-based grouts will also allow for shorter steel sleeves to be used.

The choice of repair method may also depend on the size of the corrosion patch. Hebor (1994) reports that an unrepaired corrosion patch had a tendency to buckle outward if the aspect ratio of the corrosion patch (c/h) was greater than one and inward if the ratio was close to one. An outer sleeve would constrain outward buckling. An inward buckle would require both hoop restraint by the sleeve and radial tensile stresses in the grout to maintain compatibility between the tubular and the sleeve.

The required sleeve length is provided by Hebor (1994) as:

$$s = \left(\frac{P}{D \cdot \pi \cdot \sigma_b} \right) \cdot 2 + h_{critical}$$

where s is the sleeve length, P is the load needed to be transferred to the sleeve, D is the outer diameter of the tubular, σ_b is the bond strength of the grout and $h_{critical}$ is the span of the corrosion patch itself.

The tests performed by Hebor (1994) indicated that patch corrosion could be successfully repaired with a grouted external sleeve. Further, it was concluded that the bond strength of the grout influenced the behaviour of the grouted sleeve repair by preserving the compatibility between the tubular and the sleeve both in the longitudinal and radial directions. A grout with higher bond strength preserved the compatibility more effectively. Particularly when the aspect ratio of the corrosion patch was close to 1.0, higher bond strength grouts were recommended to avoid an inward local buckle.

The work in the ATLSS project on residual strength and repair of dented and corroded tubulars in offshore structures was summarised in Ricles et al. (1995). It was concluded that:

- The ultimate strength capacity of tubular members was significantly decreased by dent damage and corrosion. The reduction of the capacity of the patch-corroded steel tubulars was reported to be due to local buckling in the corrosion patch (reduced wall thickness in combination with an amplification of the stress due to a shift in the centroid of the cross section). Further, the local buckling strength was found to be independent of the height of the patch for heights equal to or greater than half the diameter. Capacity formulae based on simplified elastic analysis were developed and showed reasonable approximations of the residual strength.
- An exterior grouted sleeve repair was found to successfully reinstate the capacity of a corroded tubular member (low bound strength grouts were found to be acceptable, but epoxy-based grouts were found to be more capable of preventing yielding and buckling of the damaged cross section).
- Non-linear finite element analysis and moment-thrust-curvature analysis both were able to predict the behaviour of non-repaired dented tubular columns. The studies indicated that the interaction of dent-depth and out-of-straightness can have a significant effect on a member's ultimate capacity.
- An internal grout repair was shown to be able to reinstate the strength of the member for low dent depts and out-of-straightness ($<0.15D$ and $<0.002L$ respectively).

Patterson and Ricles (2001) reviewed the residual strength and repair of patch-corroded offshore steel tubular braces subjected to inelastic cyclic axial loading to determine whether existing strength prediction methods for monotonic loading conditions could predict cyclic strength. As part of this study, the effectiveness of using an epoxy grouted repair sleeve to restore member resistance was investigated. An experimental programme consisting of the testing of twelve specimens was undertaken, consisting of two non-damaged specimens, eight patch-corroded specimens and two repaired specimens with patch corrosion. Steel tubulars with diameter-to-thickness ratios (D/t) of 27 and 40 and a slenderness ratio (L/r) of 57 were used and mechanical removal of portions of the wall thickness over a controlled area of the surface was used to simulate the patch-corrosion. The repairs studied included both a steel sleeve and a carbon fibre composite sleeve. The results of the experimental programme demonstrated that corrosion damage can severely limit the ductility and strength of a tubular member. The presence of patch corrosion resulted in local buckling which led to a through-thickness crack in the cross section due to low cycle fatigue. The results from the repaired test specimen demonstrated that a patch-corroded tubular member could be restored to its original design strength. Comparison of the test results to design formulae were performed and the average error was found to be 2.6%, with a maximum error of 9.2% and a coefficient of variation of 5.8%.

Epoxy grout was found to be effective in transferring forces to the repair sleeve and eliminating the high stress concentrations around the patch corrosion and was as such an effective tool in restoring a corroded tube to its full undamaged axial capacity.

8.3.9 MSL Strengthening, Modification and Repair of Offshore Installations

A Joint Industry Project in two phases sponsored by several oil and gas companies and governmental organisations (HSE UK and MMS US) was performed by MSL on strengthening, modification and repair (SMR) of offshore installations (MSL 1995, MSL 1997a and MSL 1997b). The purpose of the project was stated to be to "bring together in one document all the information necessary for the planning and execution of offshore SMR work". Building upon these two reports, MMS commissioned MSL to revisit strengthening, modification and repair to review and update these reports in MSL (2004), further described in Section 8.3.13.

It should be noted that these four reports in combination form an extensive (close to 2000 pages) source of information on repair of fixed steel offshore structures. The reports are available at the US Bureau of Safety and Environmental Enforcement (BSSE 2020) as part of their Technology Assessment Programs (TAP). It is highly recommended to visit these reports for significantly more information than what can be provided in this book.

Phase one of this project (MSL 1995) consisted of six parts:

- Part I–Summary, which included structure and usage of the document, project description and outline of the remaining parts;
- Part II–Assessment engineering and repair technique selection, which included a guide to the choices to be made relating to strengthening or repairs.
- Part III–Design recommendations, which included a detailed recommended practice for the design of various strengthening, modification and repair (SMR) techniques. The recommended practices involved design formulae based on the data from Part IV.
- Part IV–Background data and assessment, which included research, databases and appraisals which form the supporting evidence to the design recommendations in Part III.
- Part V–Clamp stud-bolt load variations, which included new data generated in the project (experimental and numerical) in order to provide a rationale for clamp design.
- Part VI–Diver-less implementation studies, which included various feasibility studies and an actual case history on implementing repairs without the use of divers.

Part II included guidance on the use of non-linear structural analysis methods, the combination of fracture mechanics and fatigue calculations, how to deal with overstressed elements and how to integrate physical research into a design method. Further, the report gave guidance on the selection of repair including removal of damage, reducing of the loading and how to undertake a localised or global strengthening or repair.

An overview giving the full recognition to this project cannot be provided in a few pages. However, many of the recommendations are included into the review of repair methods in Section 8.6.

Phase two of this project (MSL 1997a), extending the work in Part VI of phase one included demonstration trials of diver-less strengthening and repair techniques for offshore installations. The objective of the project was to demonstrate that strengthening and repair systems could be implemented using remote intervention rather than the traditional diver intervention. The objective was met through in-water demonstration trials and experimental assessments for the following strengthening and repair scenarios:

- Repair of a T-joint with a stressed grouted clamp using an atmospheric diving system (ADS) intervention;
- Repair of a T-joint with a stressed grouted clamp using a work-class remotely operated vehicle (ROV) intervention; and
- Placement of an additional brace member into a structure, utilising an elastomer-lined clamp and a tube-to-tube stressed grouted clamp using ROV intervention. This scenario represented both the repair of an existing damaged member and introduction of a new brace member.

A significant number of innovative designs were reported to be introduced in this project for strengthening and repair systems to make the repairs ROV-friendly, including the following:

- A clamp manifold was provided on each strengthening and repair system to allow the ROV to interface with and provide power to the clamp hydraulic system and grouting system.
- A clamp closure system was provided to clamp two halves together via structural hinges.
- Stud-bolt restraint and engagement were developed.
- A direct interface self-centralising sealing system was developed.
- A modified grouting system was developed to allow the ROV to control grouting operations from the clamp manifold.
- Specialised stud-bolt tensioners powered from the clamp manifold were developed.

The trials were reported to provide valuable lessons learnt. Dry "fit-up" trials were found to be an invaluable part of the preparation for deployment, and it was recommended that this should be a part of any strengthening and repair operation before attempting to implement these offshore. It was further found that the ROV should be specified with 7-function manipulator arms as a minimum. In a harsh environment a separate station-holding device such as a hydraulic docking device or a suction "foot" were seen as preferable. Suitable lighting and camera configurations were seen as essential and an additional simple manipulator arm should be available for adjustable positioning of these. An "eyeball" observation ROV should be specified to assist the work class ROV pilot with a second visual perspective of both the work site and the work-class ROV itself. This ROV should carry a dedicated work sled containing all required tools and fittings for the completion of the task. Visual coloured and graduated indicators were recommended for all tasks requiring a work-class ROV observation.

8.3.10 MSL Underwater Structural Repairs Using Composite Materials

MSL initiated a Joint Industry Project (MSL 1999) to undertake a study for a number of participants to develop an underwater technique for the strengthening or repair of offshore steel structures, in which carbon fibre reinforced plastic (CFRP) was bonded to an existing structure. In this method dry fibres were pre-formed in a workshop. As part of the project, a resin specification was sent out to major resin manufacturers including the following required properties. The resin should:

- be able to cure in the presence of water;
- be capable of bonding to steel underwater;
- be capable of curing in temperatures as low as 3°C;
- have low viscosity and long pot life (1–3 hours); and
- have high initial Young's modulus and ultimate elongation.

Thirteen candidate resins were identified for further investigation (of which 11 were epoxy resins) and coupon tests were carried out on these. The coupons were prepared with the adherends saturated with seawater and lap shear tests were undertaken. The three most suitable resin systems were then used to manufacture test specimens to provide static material test data and long-term material properties. A single resin was selected (Kobe R10) as being the most suitable for the repair process and bonding to the steel. This resin was then evaluated for long-term durability, and it was found that it would deteriorate underwater over a period of time and would not therefore be suitable for permanent repairs. Further work (outside this JIP) was then undertaken to provide an improved resin, but to the knowledge of the authors it is unclear if this was achieved.

The JIP developed a manufacturing process for resin infusion under flexible tooling (RIFT). In this method, dry fibres are pre-formed in a workshop to match the profile of the intended repair and the necessary process materials and ancillary equipment are attached to the pre-form before packaging. Using divers, the laminate was cast in-situ. On-site the pre-form is unwrapped and attached to the substrate. A vacuum line was connected to remove the bulk water out of the fibre pre-form. After the initial de-watering, air was drawn through the pack to further eliminate any liquid residues. Finally, the resin was infused into the fibre preform and the laminate allowed to consolidate and cure. Fifty-five flat plate trials were undertaken to develop the process and the resin. Finite element analysis was used to design the preform and this was shown to be a valuable tool.

This technique was further demonstrated underwater for bonding laminates onto pipes and tubular joints. The two pipe specimens (diameter 610 mm, thickness 12.7 mm) had simulated defects (dent, crack, corrosion patch). The laminate wrap was intended to restore the full integrity of the pipe over the region containing the defects. The four T joints were designed with a brace-to-chord bolted connection, the bolts being removed before testing to simulate complete severance. The performance of the T joints tested under out-of-plane bending (OPB) was variable, with tests underwater in the tank showing values between 33 and 74% of the target strength. The design of the preform using FE analysis was shown to be important in achieving good performance. For the two pipe tests one failed at a low pressure, the second using a hand lay-up process passed the proof loading test and was then subjected to a fatigue pressure cycling regime. The specimen eventually failed during a subsequent burst test at a pressure close to that produced by the analysis.

Overall, the repair method showed some limited success for the bonding of laminates onto damaged pipes and also to damaged tubular joints. However, the main limitation was that the selected resin did not have long-term durability underwater and hence the technique as developed would only be suitable as a temporary repair.

8.3.11 HSE Experience from the Use of Clamps Offshore

The Viking AD platform was decommissioned in 1998 and one of the strengthening clamps introduced after installation was obtained by HSE for assessment in the offshore technology report 2000/057 and 2000/058 (HSE 2002a and HSE 2002b). This clamp was subject to careful examination following its twelve years in service by the Health and Safety Laboratory at Buxton. The first phase (HSE 2002a) of the project included the tasks listed below, which were undertaken after the clamp had been out of the water for approximately one year. In addition, it had been transported between sites on two occasions and had been lifted several times.

- compilation of a visual record of the clamp and its dimensions at the end of service;
- assessment of marine growth, in terms of thickness, density and distribution;
- corrosion survey;
- non-destructive assessment of the grout layer between the clamp shell and the jacket tubulars;
- crack detection of all welds;
- measurement of residual stress in the vicinity of the welds in the tubulars; and
- disassembly of one half of the clamp and measurement of stress in the stud-bolts.

Overall, the clamp was approximately 8 m long, including the tubular ends. It was approximately 1.7 m wide, i.e. across the strong-backs, and each clamp box was 0.75 m deep. The clamp box flanges were 1.0 m across—see Figure 89. The two halves of the clamp were held together by 12 stud-bolts made from medium carbon, low alloy steel, probably grade L7. On testing, the tensile strength of the stud-bolt material was found to be 930 MPa, equivalent to those of a grade 8.8 bolt.

After cleaning, a visual examination of the strengthening clamp was carried out to estimate the degree of corrosion that had occurred and to assess the effectiveness of the impressed current protection. This examination revealed that there had been no significant localised pitting or general corrosion of any of the plate material used to fabricate the clamp boxes or the strong backs. The only components that appeared to have suffered any noticeable corrosion were the threaded stud-bolts themselves. These exhibited signs of minor general corrosive attack where they passed through the pockets on the upper surface only. These appear to have suffered corrosion only where they had been covered by sand and grit, molluscs, and the like. Other areas of the threaded stud-bolts were in very good condition. The grout between the clamp shell and the jacket tubulars was found to be uniform in thickness (45–48 mm) using an ultrasonic instrument (PUNDIT) and thus the clamp and tube were concentric. Following strain gauging of the stud-bolts, it was found that all the stud-bolts had been correctly tensioned in accordance with the drawing requirements and

Figure 89 Bolted clamp used on the Viking AD platform after decommissioning. *Source:* John V. Sharp.

it was concluded that the clamp had been installed satisfactorily. The welding of the node in the centre of the clamp was checked and found to be satisfactory with no weld defects being found. The residual stress pattern in the vicinity of the node welds was measured and found to be complex and it appeared that complete stress relief had not been carried out. In addition, installation of the clamp appeared to have induced compressive stresses around the node welds.

The second phase of the project (HSE 2002b) consisted of the following tasks:

- dismantle one-half of the clamp;
- carry out compression tests on samples of grout;
- assess the residual fatigue life of the stud-bolts;
- measure the fracture toughness and impact toughness of parent plate, weld metal and heat-affected zone; and
- measure the fatigue crack growth rate and threshold stress intensity in the parent plate.

One-half of the clamp was dismantled successfully and it was found there was no indication of any significant deterioration. Visual examination and mechanical testing of the grouted annulus between the clamp body and the tubular showed that the annulus was complete, relatively uniform in thickness and there were no large voids present. The bond between the clamp body and the grout material was considered very good. Tests showed that the compressive strength of the grout material met the design requirement but that the shear properties were anisotropic, being influenced by the observed layered structure such that the shear strength along a layer was approximately one-third of that across a layer. It was concluded that these results could have had implications for the strength of the clamp, which is dependent on the bonding and the mechanical properties of the grout material.

Test samples, each approximately 0.5 m in length, were cut from the stud-bolts removed from the upper half of the clamp. Fatigue tests were conducted in air at a frequency of approximately 2 Hz. An S-N curve for the axial tensile loading of the stud-bolts was developed and compared with that in BS 7608. It was concluded that the residual fatigue life of the stud-bolts was consistent with what would be expected from new bolts and thus the fatigue life had not been significantly reduced by the period in service. Following further testing, it was concluded that the fatigue crack growth rate and the fatigue crack threshold stress intensity were consistent with values reported previously for this material. In addition, it was found that the fracture toughness and impact toughness of the tubular material were within the range normally expected for 50D steel.

Overall, twelve years in service had not led to significant corrosion and the measured material properties were similar to those for new materials.

8.3.12 MSL Study on Neoprene-Lined Clamps

Neoprene-lined clamps contain a liner that lies between the clamp steelwork and the enclosed member. This liner provides tolerance against lack of fit of the clamp saddle around the tubular brace. The neoprene liner is usually plain for structural connections designed to transmit axial or rotational loads, although ribbed linings are sometimes used to accommodate potentially large lack of fit tolerances. Stressed neoprene-lined clamps rely on applied stud bolt pre-loads to generate compressive forces normal to the interface between the clamp liner and the surface of the clamped brace. The strength is considered to be dependent on the magnitude of the normal force, the relative stiffness of the steel and liner and the effective coefficient of friction at the liner-brace interface.

A Joint Industry Project funded by HSE and two North Sea operators was undertaken by MSL in HSE RR 031 (HSE 2002c). At that time there was only limited data available on the slip capacity of this type of clamp, despite the widespread use of neoprene-lined clamps throughout the world. The test programme consisted of two phases. In the first phase 16 full-scale neoprene-lined clamps were subjected to axial and torsional loading, including parameters such as the bolt load, neoprene thickness, pipe surface condition, clamp length-to-diameter ratio and pipe radial stiffness. The primary Phase I finding was that the coefficient of friction for neoprene-lined clamps was substantially below the range of values that had been adopted in practice. One factor was the unexpected slip behaviour of the clamp, particularly with regard to the relationship between applied bolt load and clamp capacity. There was also insufficient test data to permit a proper clarification of the role of bolt loads.

As a result of the findings in Phase I, further tests were undertaken in Phase II, consisting of six axial slip tests on neoprene-lined clamps. This programme consisted of either quasi-static loadings or cyclic loadings that simulated wave action in the UK Southern North Sea. The parameters that were investigated in this second phase included bolt pre-load and neoprene hardness.

The tests with clamps having different neoprene hardness confirmed that this hardness did affect the capacity of the clamp. Cyclic loading indicated that at the design capacity, the relative displacement of the clamp and member that occurred was recoverable as the displacement was largely due to neoprene shear deformation as opposed to true slip. It was found that time-dependent phenomena such as creep occurred only when the loads were applied statically.

Design guidance was formulated based on the results of both Phase I and Phase II test programmes (HSE 2002c). In this guidance it was recommended that the factor of safety should be adjusted depending on whether quasi-static or dynamic loading was being considered.

8.3.13 MSL Repair Techniques for Ageing and Damaged Structures

MSL (2004) evaluated repairs and remediation in a review for MMS (project number 502), and this report can be seen as the third report by MSL on repair techniques for fixed offshore structures building upon (MSL 1995 and MSL 1997a and b). The review was prepared for the US MMS as an assessment of the current status of repair techniques and a guide for selecting an optimal SMR scheme and individual SMR techniques. The review indicates that developments in repair techniques since the earlier work by MSL in 1995 and 1997 primarily consisted of a rise in composite material repairs and developments in welding consumables and grout materials. The report states that the objective was to build on the earlier reports and to bring these up to date with recent developments in the field. The report included:

- past, present and future perspectives on strengthening, modification and repair (SMR) techniques;
- triggers (initiators) for carrying out an assessment possibly leading to SMR requirements;
- an assessment process to decide whether SMR is needed and if needed how extensive the SMR scheme needs to be;
- a guide for the possibilities that exist for various SMR schemes; and
- detailed information on each of the SMR techniques and their strengths and weaknesses.

As shown in Figure 90, the initiators for strengthening, mitigation and repair included situations where damage was found on the structure during inspections and where the structure was found to have insufficient strength. These were then recommended to be a part of a structural assessment similar to the process described in Chapter 7 in this book. The possible results of the structural

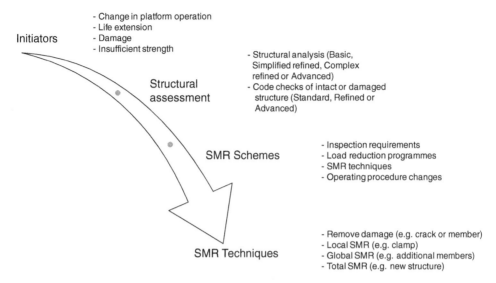

Figure 90 Illustration of the process leading up to SMR. *Source:* Based on MSL (2004). Assessment of repair techniques for ageing or damaged structures. Project #502. MSL Services Corporation, Egham, Surrey, UK.

assessment included updates of the inspection requirement, load reduction programmes and operational procedure changes, in addition to strengthening, mitigation and repairs.

MSL proposed four basic approaches to strengthening, modification and repair (SMR) work, which included the following:

- remove damage (e.g., grinding out of cracks or removal of bent or bowed members);
- local SMR (where no change in the load path of the structure occurs as a result of using an SMR scheme, e.g. employing a clamping mechanism around a joint or member);
- global SMR by provision of new members (a change of system load path occurs, e.g. by the addition of a new member); and
- total SMR by tying into a new adjacent structure.

MSL stated that the full extent of any remediation can be only determined after the assessment phase of the SIM process. Initially when considering the SMR options it is necessary to determine whether a local SMR option is viable or if a more detailed global SMR action is required. It was noted that local SMR options generally tend to be less costly and less complex to install than global SMR actions.

MSL (2004) showed the links between SMR actions for both undamaged structures where there is insufficient static or fatigue strength and those for damaged structures as a result of denting and bowing, corrosion or fatigue as shown in Figure 91.

Local and global repair for the various situations were recommended as shown in Table 63.

MSL (1995 and 2004) discussed the various techniques and their applicability to address commonly occurring damage scenarios as indicated in Table 64. The footnotes in this table provide additional information that the authors to some extent find questionable, for example, the note that dry welding is usually performed in conjunction with additional strengthening measures.

MSL (2004) included a useful summary of the SMR techniques as further discussed in Section 8.6. Composites were included as they have been used as a containment formwork in the case of repair work to corroded conductors as described earlier for tubular repairs.

Nichols and Khan (2017), as a follow-up of MSL (2004), listed the different repair options with some details on each one. This included member removal, welding, weld improvement,

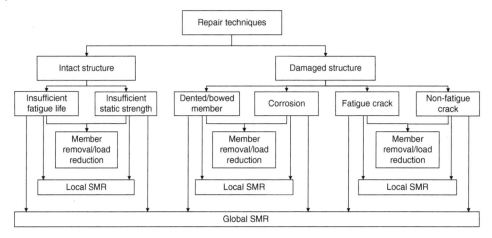

Figure 91 Interrelationship between scenarios, SMR schemes and SMR techniques (Simplified from MSL 2004). *Source:* Modified from MSL (2004). Assessment of repair techniques for ageing or damaged structures. Project #502. MSL Services Corporation, Egham, Surrey, UK.

Table 63 Local and Global Repair as Recommended by MSL (2004).

		Toe grinding	Hammer peening	Weld repair	Welded doubler plates	Grouting of joint	Grouting of member	Grouted sleeve	Nodal clamp	Brace member	Weld in new member	Clamp in new member
Intact structure	Insufficient fatigue life	X	X			X			X		X	X
	Insufficient static strength					X	X	X	X		X	X
Damaged structure	Dented / bowed member				X		X			X	X	X
	Corrosion				X		X			X	X	X
	Fatigue crack	X			X			X	X		X	X
	Non-fatigue crack	X	X	X	X			X	X		X	X

Source: Modified from MSL (2004). Assessment of repair techniques for ageing or damaged structures. Project #502. MSL Services Corporation, Egham, Surrey, UK.

clamp technology, grout filling and bolted connections. Nichols and Khan (2017) also included several examples of the use of SMR techniques such as damage to a drilling platform located at 68 m depth in the Gulf of Mexico with a stress grouted clamp attached with sleeves to new piles as part of the SMR scheme. In the latter case post-hurricane inspection revealed that a pile was exposed and severed about 6 m below the mud line. The repair work involved the installation of stress grouted clamps to the existing piles.

Nichols and Harif (2014) advocated using an online monitoring (OLM) process to monitor the effectiveness of an SMR scheme. They added that inspection and monitoring techniques should

Table 64 Applicability of Repair Techniques (MSL 1995 and MSL 2004). *Sources:* MSL (1995). Strengthening, Modification and Repair of offshore installations – Final report for a joint industry project. MSL Document No. C11100R243. MSL Engineering Limited, MSL House, Sunninghill, Ascot UK. Also published as MMS TAP project report no. 189. ; MSL (2004). Assessment of repair techniques for ageing or damaged structures. Project #502. MSL Services Corporation, Egham, Surrey, UK.

	Fatigue crack	Non-fatigue crack	Dent	Corrosion	Inadequate static strength		Inadequate fatigue strength	
					Member	Joint	High loads	Fabrication defect
Dry welding	Y^1	Y	Y^3	Y^3	Y^1	Y^1	N	Y
Wet welding	N^2	Y	Y^3	Y^3	Y^1	Y^1	N	Y
Toe grinding	N	N	N	N	N	N	Y	N
Remedial grinding	Y	Y^1	N	N	N	N	N	N
Hammer peening	N	N	N	N	N	N	Y	N
Stressed mechanical clamps	Y	Y	N	Y	Y	N	Y	Y
Unstressed grouted connections	Y	Y	Y	Y	Y	Y	Y	Y
Unstressed grouted clamps	Y	Y	Y	Y	Y	Y	Y	Y
Stressed grouted clamps	Y	Y	Y	Y	Y	Y	Y	Y
Elastomer lined clamps	N	Y	N	Y	Y^4	N	N	N
Pressurised connections	Y	Y	Y	Y	Y	Y	Y	Y
Grout filling	N	N	Y	N	Y	Y^4	Y^4	N
Bolting	N	Y	N	N	N	N	N	N
Member removal	Y^5	Y^5	Y^5	Y^5	N	N	Y^5	Y^5
Adhesives	Y^6	Y^6	$Y^{3,6}$	$Y^{3,6}$	$Y^{3,6}$	Y^6	$Y^{3,6}$	$Y^{3,6}$

[1] Usually in conjunction with additional strengthening measures
[2] Except to apply weld beads in unstressed grouted connection and clamp repairs
[3] To apply patch plates
[4] Applicability depends on type and sense of loading
[5] If member is redundant
[6] Used as epoxy grout in clamps

be part of an operator's inspection plan to ensure that SMR schemes continue to perform over time. They also noted that modern monitoring technology had proven quite effective in monitoring structural behaviour and performance of jacket structures.

8.3.14 MMS Studies on Hurricane Damage and Repair

Energo performed several reviews to summarise the damage to offshore structures in the GoM in the hurricanes during the 2000s (Energo 2006, Energo 2007, Energo 2010). These also included examples of the repairs that had been done. The report provides a list of typical repair options in order of least complex to more complex:

- stop hole drilled into material at tip of crack;
- grinding surface cracks to smooth surface to prevent further growth;
- grout interior of member to increase strength;
- wet welding to weld on doublers, shear pups and other secondary structure to reinforce member in region of damage. Key concerns with wet welding are that the weld is not as strong as an above-water dry weld;
- stressed clamping using contact stress of member to clamp and repair damage;
- dry habitat welding used to enclose the damage in an underwater habitat and dewater to perform dry weld; and
- stress grouted clamping to account for deformations and fit-up tolerance.

Photos of several of these repairs are included in Section 8.6.

8.3.15   BOEME Report on Wet Weld Repairs to US Structures

The service history of several underwater wet welded repairs and modifications carried out in the period 1970 to 2010 in the US Gulf of Mexico and at several international locations were summarised as case studies in Reynolds (2010). Components repaired and modified consisted of primary structural members (e.g. jacket legs, vertical diagonals and horizontal members) and secondary structural members, such as conductor guide framing and control umbilical support structures. Specific details of repairs are given for several structures and are summarised in Table 65.

In most of the case studies, primary members were either repaired or replaced by wet welding. These primary members consisted of jacket leg repairs (100% separation of legs), vertical diagonal replacements, vertical X brace replacements, horizontal member replacements, horizontal X brace repairs and a few conductor guide framing repairs.

Table 65   Brief Details of Damage and Repairs to Platforms.

Structure	Type	Water depth m	Cause of damage	Type of damage	Repair type
Ship Shoal 169A	8 piled jacket	17.3	Hurricane Andrew	Crack on horizontal diagonal	Installation of doubler plate by wet welding
Ship Shoal 291 A	8 piled jacket	72	Fatigue due to movement of conductors in guide frame from wave loading	Separated member	New horizontal stub members were prepared and attached to doubler plates Clamps were installed on four selected conductors
Ship Shoal 246A	8 piled jacket	52	Fatigue due to movement of conductors in guide frame from wave loading	Cracks identified at six locations on members that supported the conductor bay framing	Repair procedure involved locating the ends of the cracks and drilling crack arresting holes, then designing and fabricating three T, one K and two X type cruciform clamps
Eugene Island, 231CB	4 leg design	34	Hurricane Andrew	Crack located at intersection of a vertical diagonal member and ring stiffener	Crack removed with an arc/water gouge and hydraulic grinder The resulting groove preparation area was filled in by wet welding

Table 65 (Continued)

Structure	Type	Water depth m	Cause of damage	Type of damage	Repair type
South Timbalier, 52	4 leg design	20	Hurricane Andrew	Two damaged, critical midpoint K brace nodes	New K brace nodes designed and installed involving wet welding
Arabian Gulf – Khazzan Offshore Storage and Loading Facility K1	Storage and loading	47	Installation damage	Several anodes were torn loose during launching	Anodes were repaired at site by re-welding the 100 mm standoffs to the tank wall
Arabian Gulf – Khazzan Offshore Storage and Loading Facility K2, K3	Storage and loading	47	Pile driving damage	K-2 and K-3 tanks were damaged from dropped piles during pile driving Structural tee stiffeners crushed due to falling piles	Damaged structural tee stiffeners were removed by chipping and oxy-arc burning and then new tee stiffeners were fitted in place and installed by wet welding
Atlantic Ocean – Trinidad, Samaan A Structure	8 leg design	51	Storm damage	Detached 406 mm diameter horizontal member at 38 m elevation	The existing damaged horizontal member was removed, the surface of the legs ground smooth and then the new assembly installed by wet welding
UK North Sea –Montrose Structure	8 leg jacket	91	Vessel collision	Impact damage including cracking in weld joining vertical diagonal member to a wall stub	Repair consisted of a modified scallop sleeve that stabbed over the existing undamaged, underwater stub to leg connection The single scallop sleeve was attached to the stub with a continuous fillet weld (wet welding)
Garden Banks 426 A, Auger Tension Leg Platform	TLP	305	Construction upgrade	Upgrade requirement to attach umbilical I tubes to columns	The below-waterline assembly had five doubler plates. Tube clamp assemblies were wet welded to the hull column. All clamps were attached to the column by wet welded fillets.

Source: Modified from Reynolds, T.J. (2010). Service History of Wet Welded Repairs and Modifications. In International Workshop on the State-of-the-Art Science and Reliability of Underwater Welding and Inspection Technology November 17–19, 2010 Houston, Texas, US ed. by Stephen Liu and David L. Olson.

The paper also includes additional repairs included in Table 66, however, with minimal detailed descriptions compared to those mentioned in Table 65. It can be seen that vertical diagonal members have received the most damage and subsequent repairs by wet welding.

The paper is for obvious reasons focussed on US practice using wet welding. However, in the North Sea wet welding is normally not accepted for repair of structural damage. Rather, hyperbaric dry welding techniques are preferred as a welding technique to restore the structure to its original design. Nevertheless, in the case of the Montrose platform repair as shown in Table 65, the

Table 66 Summary of Repair Types.

Repair types	No. of repairs
Vertical diagonal brace repairs	25
Vertical brace repairs	3
T, K and Y node replacements	2
Horizontal brace repairs	5
Horizontal X brace repairs	2
Conductor guide framing repairs	4
100 leg break repairs	2
Total	43

Source: Modified from Reynolds, T.J. (2010). Service History of Wet Welded Repairs and Modifications. In International Workshop on the State-of-the-Art Science and Reliability of Underwater Welding and Inspection Technology November 17–19, 2010 Houston, Texas, US Edited by Stephen Liu and David L. Olson.

operator chose to investigate wet welding as an alternate technique for reinstalling a replacement brace.

The report also assessed the performance of structures after repair. All of the structures described in the case studies in Table 65 performed satisfactorily for many years after the repairs; some were still supporting production and storage in 2010 when the report was prepared. This was seen as validation of wet welding methods used offshore.

8.4 Previous Studies on Repair of Concrete Structures

8.4.1 Introduction

There are several books and articles on concrete repair, mostly dealing with concrete above water. Some of these include parts dealing with underwater repair, such as the book *Repair of Concrete Structures* by Allen, Edwards and Shaw, which includes a chapter on underwater repair by R.D. Browne (1993).

Another book entitled *Materials, Maintenance and Repair* by Campbell-Allen and Roper (1991) also includes limited material on underwater concrete repair. This book provides background and understanding of how to achieve sound concrete construction and or remedying defects in damaged concrete structures. The sections on repair provide useful information for repair of structures above water with a few examples of repairs in the splash zone. It further notes that polymer-modified mortars are useful for placement underwater for local patch-type repairs. In addition, the book notes that pre-packed aggregate concrete (pre-pack) made by forcing grout into the voids of a compacted mass of coarse aggregate is particularly useful for underwater construction and repair. In the examples of repairs to structures, the reinstatement of concrete piles in the splash zone on an oil refinery jetty was addressed. This splash-zone repair has relevance for similar structural repairs for an offshore structure.

A further relevant book by El-Reedy (2019) includes a section on concrete repair. However, this mainly addresses concrete above water, with examples of repairs to beams and plate structures typically found in bridge structures and buildings. The book stresses the importance of removing the damaged material and cleaning the steel reinforcement prior to repair using, for example, sandblasting and coating the repaired area after the concrete has been added.

In addition, Hamakareem (2020) addresses several underwater repair methods. These include:

- Surface spalling repair in the splash zone:
  The boundary of the spalled area should be cut back (removed) to enable cementitious mortars to be used to fill the damaged area. A bonding coating can then be applied. If the damaged area is small, water tolerant epoxy mortars can be suitable for this repair. However, for larger areas requiring repair, formwork should be used to hold the repair material in position whilst it is setting and grout or cement injected into the gap.
- Large-scale repair of underwater structural concrete:
  The damaged area needs to be prepared, including cleaning the reinforcement. Suitable formwork is prepared and placed on site. Aggregate is then placed and compacted in the formwork. Grout is then injected into the base of the compacted aggregate, expelling water. The formwork is filled, allowing for shrinkage of the grout while setting.
- Injection techniques for restoring underwater concrete structures:
  Cementitious grout or resins can be injected to repair cracks or to fill voids. Epoxy resins are suitable for narrow cracks, while epoxy grout can be used for cracks up to 3 mm and cement grout appropriate for wider cracks. There are two methods of injection, pressure and gravity feed. Injection nipples need to be placed along the whole length of the crack and then the surface should be sealed along the whole length of the crack. Injection should commence at one end of the crack using the nipples. Following injection, the concrete surface is finally prepared along its length.
- Steel sleeve repairing technique for underwater concrete:

  This method involves placing a steel sleeve around a column or pile. The repaired sleeve should be able to take the load on the pile if the reinforcement has been damaged due to corrosion. Loose concrete and marine growth are removed from the damaged pile. The sleeve is then fitted in two semicircular sections. The space between the sleeve and the pile or column is then filled with mortar or concrete, pumped from the base. Any temporary supports are removed after setting and corrosion protection is then added for the steel sleeve.

8.4.2 Repair of Major Damage to Concrete Offshore Structures

The Department of Energy funded Wimpey Laboratories to undertake a study on the repair of major damage to prestressed towers of concrete offshore structures (Department of Energy 1988c). This report, OTH 87 250, reviewed damage from ship collisions, how this damage could be assessed and then proposed five different repair techniques, which were:

- repair of damage where broken prestressing cables are re-joined (two different methods);
- repair of damage where broken prestressing cables are not re-joined (two different methods); and
- repair of damage where prestressing cables are unbroken.

In all the repair methods proposed, a standard sized square hole of 3 m × 3 m was selected as being typical of the type which might result from an incident of moderate severity. The tower dimensions were taken as those investigated in a previous study with an inside diameter of 8 m with a wall thickness tapering from 400 mm to 650 mm near the waterline. The repairs were detailed for the lower wall thickness (400 mm) as it was recognised that this thin wall section would cause the greatest problems. It was noted that the location of the impact damage in the tower would depend on several factors including the tidal level, the type of ship involved in the collision and the sea state at the time of the incident. If the damage is below the waterline, the problem of repair becomes more difficult and a steel jacket would be placed and sealed around the relevant part of the tower to enable most of the repair work to be carried out in the dry.

The initial procedure for all the repair operations involved:

- mooring a small crane barge adjacent to the damaged zone of the tower;
- for damage below water level, if possible, fabricating and positioning a steel jacket around the part of the tower where damage has occurred;
- breaking out all severely damaged pieces of concrete and reinforcement using jack hammers and burning equipment;
- clearing marine growth from the concrete surface of the hole by water jetting; this should expose the extent of the smaller cracks;
- sealing the smaller cracks with epoxy putty; and
- injecting the larger cracks with epoxy injection grout.

In the case of methods involving repair and re-stressing of existing tendons, new lengths of tendon can be attached to the broken ends by means of a tendon coupler (Department of Energy 1988c). These tendons can be overlapped, anchored and stressed using normal anchorages. An alternative is that they can be curved outside the wall, joined together and stressed using new anchorages. This stage is undertaken after the damaged area has been infilled with epoxy resin.

There are three different methods where fractured prestressing tendons are not re-joined (Department of Energy 1988c):

- The original role of the prestressing can be replaced by using "flat jacks", which produce a similar effect in maintaining compressive stresses in the concrete.
- A second method utilises Macalloy bars, which can accommodate the tensile stresses in the ultimate load condition. The design needs to ensure that at ultimate strain the total force in the Macalloy bars will be equal or greater than the sum of the forces that the broken tendons would have taken.
- Another repair solution uses a steel "splint" which is capable of taking tensile loads. It is stated that this should be designed to have the same stiffness characteristics as the concrete that has been removed. Shear forces are transmitted through an epoxy adhesive layer between the steel plate and the concrete wall.

The study showed that it was possible to restore the structural integrity of a damaged tower with a number of different techniques depending on the level of damage. More detail on these repair methods including drawings can be found in the Department of Energy OTH 87 250 report (Department of Energy 1988c).

8.4.3   Scaling of Underwater Concrete Repairs

A study undertaken for the Department of Energy in the report OTH 89 298 (Department of Energy 1992b) noted that any materials used in concrete repair underwater must be compatible with underwater placement, be able to bond to a wet surface and develop full strength at low temperatures. As a result, four different test methods were investigated of which two at the time were recommended in BS 6319:1984 (BSI 1984) and the remaining two developed specifically for this test programme. These included two slant shear tests, which is a standardised test method for concrete repair. One of these was designed for cementitious grouts and one which was an adaptation of the BS slant shear test to be suitable for testing epoxy resins. In addition, two flexural tests were developed, one suitable for cementitious grouts, the other for epoxy resins. Eight different repair materials were tested (not named), all of which were proprietary formulations and with previous use as an underwater repair material. Concrete slabs were cast in pairs, one to be used for the repair, the second as a control. The concrete used was based on the standard grade mix tested in the CiO programme reported in OTH 87 248 (Department of Energy 1989c). The slabs were cured in seawater for 28 days before testing. The seawater used in the tanks was based on an artificial mix, similar to that used in the CiO programme and cooled to typical North Sea temperatures for the repair and subsequent curing.

The slant shear test used for resin repairs was a resin-bonded scarf-jointed prism tested in compression. The test specimens (typical specimen size was $500 \times 100 \times 100$ mm^3) were prisms for the epoxy repairs and prepared by fracturing the slab under compression, producing two halves which were then mounted in a jig to create a 5 mm crack width for the repair. The cracked slabs were held overnight in seawater at the North Sea temperature before being repaired. A control slab was also immersed in the tank for the same time. Resin was injected into the crack slowly to ensure displacement of the water in the crack. The repaired slab was then left for 28 days for curing before being removed from the tank for testing. The repaired slabs were then tested in compression to record an ultimate load and the mode of failure was also recorded.

The test procedure for the cementitious repairs was similar to that for the epoxy repairs, except the specimens were slightly smaller ($400 \times 100 \times 100$ mm^3) and the gap left between the two cracked sections in the jig was 100 mm. Cementitious grout was introduced for the repair and cured for a similar period before testing as shown in Figure 92.

Concrete test specimens were also prepared for flexural testing (bending), each $400 \times 100 \times 100$ mm^3 in size. These were prepared for repair both by resins and cementitious materials, by loading in a two-point flexural test. The two halves were then mounted in a jig for repair similar to the compression tested specimens as shown in Figure 92. After repair and curing in cold seawater, the specimen and control were loaded in flexure, on a two-point loading machine to test the strength of both the repaired specimen and the control.

The results for the resin bonded and cementitious repairs as shown in Table 67 clearly demonstrate that for repairs tested in compression, cementitious grout was more effective while for flexural testing, the epoxy resin repair was clearly better.

It is worth noting that three of the cementitious repairs shown in Table 67 tested in compression had strengths in the range of 83–90% of the control strengths.

In the resin-bonded slant shear tests it was found that failure was usually in the concrete adjacent to the bond, understood to be due to the stress concentrations caused by differences in the

Figure 92   Repaired prism using grout for compression testing (left) and repaired slab with grout for flexural testing (right). In both cases the repair material can be seen as a slightly darker band in the centre (Department of Energy 1992b). *Source:* OTH 89 298 Scaling of underwater concrete repair materials. Prepared by University of Dublin (S.H. Perry) and Imperial College of Science and Technology (J.M. Holmyard) for the Health and Safety Executive (HSE). HMSO, London, UK.

Table 67   Results of Repairs to Concrete Specimens.

Repair method	Compression (percentage of control strength)	Flexural (bending) (percentage of control strength)
Epoxy resin	35–54%	Up to 28%
Cementitious grout	62–90%	5–12%

Source: Modified from Department of Energy (1992b). OTH 89 298 Scaling of underwater concrete repair materials. Prepared by University of Dublin (S.H. Perry) and Imperial College of Science and Technology (J.M. Holmyard) for the Health and Safety Executive (HSE). HMSO, London, UK.

elastic moduli of the resin and the concrete. In the grout-bonded slant shear tests the repaired prisms always failed by vertical cracking in the grout or the concrete or by a failure of the whole prism. There was no evidence of the failure of the bond itself. In the flexural tests all specimens failed at the repair-concrete interface at relatively low strengths. This means that the strength of the bond was less than the tensile strength of the concrete.

In conclusion the report questioned the test procedures at the time, particularly those in BS 6319:1984 (BSI 1984). Since these tests were carried out, a new version of the BS standard has been published in BS 6319:1990 (BSI 1990). The report also made recommendations about improved test procedures. An important outcome of the tests series was that cementitious grout was considerably more effective in compressive testing whereas the strength of resin repairs was higher than cementitious repairs in bending. This information is relevant to the selected repair method depending on the loading situation.

8.4.4   Assessment of Materials for Repair of Damaged Concrete Underwater

Study OTH 90 318 was undertaken to assess the efficiency of different materials for repairing damage to the simulated domes of the roofs of underwater concrete storage tanks (Department of Energy 1990). The roofs of these storage tanks are vulnerable to damage from accidentally dropped objects and there have been several occurrences of such damage followed by repair (see Section 4.8).

Several concrete slabs and domes simulating roof members of storage tanks were subjected to controlled punching shear damage and then repaired underwater in a tank of artificial seawater cooled to a typical North Sea temperature of 7°C. After suitable storage time in the tank (usually 28 days), each member was tested to destruction by static loading to assess the performance of the different materials.

The set of domes and slabs comprised:

- ten circular reinforced concrete slabs (diameter 1500 mm, thicknesses of either 60 or 80 mm);
- three circular reinforced concrete domes (1500 mm in diameter, 40 mm thickness at the crown);
- two circular reinforced concrete domes (2300 mm in diameter, 800 mm thick with a 250 mm square section outer ring beam); and
- one circular reinforced concrete dome (without a ring beam, 2300 mm in diameter and 800 mm thick).

As part of an earlier study, part of this set of concrete elements had been subjected to hard impact from a dropped steel billet (100 mm in diameter), representing a dropped object. Different degrees of damage as shown in Figure 93 and Figure 94 had occurred through varying the mass

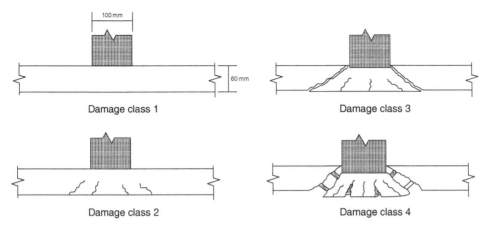

Figure 93 First four levels of impact damage to concrete slabs. *Source:* Department of Energy (1990). Offshore Technology Report OTH 90 318 Assessment of materials for repair of damaged concrete underwater. Prepared by University of Dublin (S.H. Perry) and Imperial College of Science and Technology (J.M. Holmyard) for the Health and Safety Executive (HSE). HMSO, London, UK.

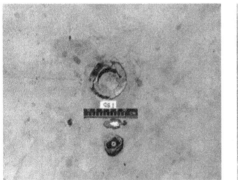
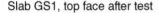

Slab GS1, top face after test   Slab GS1, back face after test

Figure 94 Slab GS1 after test, front and back-face (Department of Energy 1990). *Source:* Offshore Technology Report OTH 90 318 Assessment of materials for repair of damaged concrete underwater. Prepared by University of Dublin (S.H. Perry) and Imperial College of Science and Technology (J.M. Holmyard) for the Health and Safety Executive (HSE). HMSO, London, UK.

and terminal velocity (Department of Energy 1990). The three larger domes had been subject to static loading by means of a steel loading platen of either 150 or 300 mm diameter. Five classes of damage were considered, which were:

- no damage;
- cracking on the back of the element but no evidence of shear plug formation;
- full-depth cracking and small shear plug formation (less than a tenth of the thickness) but with no significant loss of concrete from the back face;
- severe full-depth cracking, distinct displacements of the shear plug and visible deformation of the steel; and
- hole in the elements (total loss of concrete from the shear plug); the reinforcing steel may be present but damaged.

The repair of these slabs consisted of sealing the cracks and injecting repair materials into these with the aim of re-establishing the strength and water tightness. These included both epoxy and cementitious materials selected to be suitable for bonding underwater. These were:

- Epoxy resin
  - Sikadur 53LV, two-part solvent free epoxy resin formulated for the repair of water-immersed concrete structures by crack injection;
  - Colebrand CXL878R, also a two-pack aromatic amine-cured solvent-free epoxy resin formulated for repair of damage concrete both in dry and wet conditions;
- Cementitious
  - Conbextra HF, high-fluidity non-shrink cementitious grout;
  - Conbextra UW, non-shrink cementitious grout especially formulated for underwater application. Additives contained in the grout gave resistance to wash-out of the cement when placed underwater;
  - Armorex L2, high flow. This was a high fluidity, non-shrink cementitious grout.

All three cementitious grouts were supplied as a pre-packaged dry powder requiring only the addition of water. It is uncertain whether all of these products are still available, but alternative products with possibly improved properties are on the market (Wessex Ressins 2020).

Each damaged slab and dome was prepared for repair before it was immersed in seawater by sealing the cracks on the back face with a siliconised rubber sealant. Care was taken to ensure that the sealant did not penetrate into the cracked area. The authors claimed that the sealant could be regarded as a form of shuttering and did not contribute to the strength of the repaired slab or dome. For the planned epoxy repairs, injection nipples were added to the cracked area using a special cement. After this preparation, each specimen was immersed in the chilled seawater tank for seven days before the actual repair took place.

For the epoxy repairs, plastic tubes were connected to the installed nipples. The damaged area was first flushed out with clean seawater. The resin was then pumped into the damaged area using an injection gun. Pumping was continued until clean resin came out the other injection nipples.

For the cementitious repairs, the grout was mixed with water as prescribed by the manufacturer and injected into the crack system, relying on the properties of the grout to displace water from the cracked region. Pumping continued until the crack appeared to be full and a pool of grout was left on the surface.

After repair, each of the specimens was left in the seawater tank to cure at 7°C. For most specimens, the period in the tank was 28 days. However, four slabs were left in the tank for longer periods up to 330 days before being tested.

Testing of the repaired specimens recovered from the test tank was performed using a specially made test rig, enabling the load to be applied to the centre of each specimen. The 2.3 m diameter dome specimens were supported by a steel I-section beam rolled into a complete circle, which was a similar diameter to the large dome specimens. The bearing surface was 145 mm in width. The specimens were instrumented using linearly variable displacement transducers to measure displacement. For testing the repaired specimens, circular section steel billets with diameters ranging from 63 to 190 mm and thicknesses ranging from 60 to 140 mm were used. The loading platens were placed directly on the surface concentric with the original damage.

In addition to the above tests, slant shear tests were also carried out in accordance to BS 6319 (BSI 1984) to measure the bond strength of the different repair materials. All repairs to the slant shear prisms were carried out underwater in the seawater tank. The results from the slant shear tests are given in Table 68.

The results of the tests are given in Table 69.

Table 68 Slant Shear Test Results.

Material	Average compressive strength concrete, (N/mm^2)	Average compressive strength, grout (N/mm^2)	Slant shear bond strength (N/mm^2)
Conbextra HF (G)	53.9	60.0	31.6 (58.6%)
ConbextrA UW (G)	54.0	44.0	26.4(48.9%)
Armorex L2 (G)	46.7	42.2	27.1 (58%)
Colebrand CXL878R (E)	45.8	44.4	36.3 (79.3%)
Sikadur 53LV (E)	No results, as proved difficult to inject the resin into the cracked plaque		

Note: Figures in brackets in column 4 are the percentage of the concrete strength

Table 69 Results of Tests.

Type of test	Class of damage	Type of repair material	Strength (kN)	Comments
Slab – GS3	4	Epoxy (S)	43 (116%)	Concrete failed along the old lines of cracking. Penetration of resin into crack system good. Adhesion of resin to concrete good.
Slab – GS5	4	Grout (CoH)	33 (89%)	New shear cone formed. Penetration and adhesion of repair grout was good.
Slab – GS7	4	Grout (A)	37 (100%)	New shear cone formed within old one, with considerable deflection of reinforcement.
Slab – GS1	3	Epoxy (Col)	18 (50%)	New shear cone formed within the old one. Failure mainly in the concrete, but some debonding of resin noted. Resin had penetrated full depth of slab. Damage shown in Figure 93.
Slab- GS8	3	Epoxy (Col)	25 (68%)	Resin had penetrated and adhered well but concrete within shear cone was in poor condition. Failure partly in the resin and partly in the concrete.
Slab – GS16	4	Epoxy (S)	33 (90%)	Resin penetrated into crack system well. Failure in the parent concrete and also by debonding the resin in part of the failure.
Slab- GS2	3	Epoxy (Col)	30 (82%)	Penetration of resin complete, with all visible cracks filled. Failure mainly at resin to concrete interface.
Slab – GS12	4	Grout (A)	31 (85%)	Penetration not good, but where penetrated, bond was good. Failure not at the grout-concrete interface.
Slab GS13	3	Grout (CoH)	58 (159%)	Penetration of repair grout good. Failure by old shear plug.
Small Dome – D08	4	Grout (CoU)	17	Grout not penetrated, voids present. Failure by original shear plug.
Small dome – D05	4	Grout (CoH)	23	Failure by original shear cone and at interface between parent concrete and repair material.

(Continued)

Table 69 (Continued)

Type of test	Class of damage	Type of repair material	Strength (kN)	Comments
Small dome – D010	4	Grout (CoH) followed by Epoxy (S)	17	Good penetration of the grout with remaining fine cracks filled with resin. However, resin had failed to bond with repair grout or parent concrete. Good penetration of the grout, remaining fine cracks filled with resin. However, resin had failed to bond with repair grout or parent concrete due to a soft coating being formed because of the time between the two stages of repair.
Large dome – D/86/1	4	Grout (CoH)	68 (13%)	Failure in parent concrete and grout, also failure of grout-concrete bond observed. Poor penetration of grout
Large dome – D/86/2	4	Grout (CoU)	71 (24%)	Very good penetration of grout, reaching full depth of repair. Failure by original shear cone. Repair failure through body of grout.
Large dome – D/86/4	3	Epoxy (S)	53 (21%)	Failure occurred via new shear cone. Failure in both concrete and resin. Resin de-bonded from parent concrete, probably due to difficulty of flushing out part of the repair.

Note: Grout (CoH) is Conbextra HF, Grout (CoU) is Conbextra UW, Epoxy (S) is Sikadur 53LV, Epoxy (Col) is Colebrand CXL 878R, A is Amorex L2. The figures in brackets in column 4 are the percentage regain in strength. * Indicates longer-term tests, in the range 200–330 days.

The authors noted that the project had been about testing repair materials and not the method of repair. It was noted that the selection of repair materials was dependant on the width of the cracks through which the repair materials had to penetrate. Cementitious grouts had difficulty in penetrating cracks of surface width of 10 mm and below 5 mm it was impossible. Hence for cracks less than 10 mm in width, epoxy resins would be the repair material of choice.

The authors also noted the risks to divers from making repairs underwater using these materials. The lighter epoxy resin (Colebrand CXL78R) tended to break into small globules which floated when coming out of an exit port. There was less problem with the heavier resin (Sikadur 53LV) where any material tended to sink. When the cementitious grouts came into contact with water, some clouding was observed, reducing visibility.

Regarding bond strength, which is an important parameter, the tests demonstrated that surface preparation was a key factor. Flushing out the damaged area with seawater was undertaken in each case but because of the nature of the cracking and that the damaged area could not drain easily, surface preparation was uncertain.

Results of the slab, dome and slant shear tests showed that in most cases the epoxy resin bonded well and bond strengths were high. However, where there was evidence that there had been a problem of surface precondition (e.g. dome D010), the bond was weak.

As noted earlier, four repaired slabs were stored for a period of up to 330 days in cold seawater before testing. Although there was some variation in specimen properties between the 28- and 300-day tests, there appeared to be no detrimental effect with time on the Colebrand CXL878R resin or the Conbextra HF grout, both of which showed an increase in strength over the slabs tested at 28 days.

Overall, the report concluded that there was a large variation in the performance of the epoxy resins whilst the cementitious grouts behaved more consistently. It was also noted that the results obtained from the slant shear tests were not directly comparable to the results obtained on large-scale repairs. The project has shown from the slab tests that underwater injection of crack systems with either epoxy resins or cement grouts could result in the recovery of a large proportion of the original strength. It was also shown that repairs of the type undertaken could satisfactorily seal cracked elements, limiting water ingress. The authors also commented that for those undertaking full-scale repairs, there were many obstacles to overcome and that some additional strengthening such as a cap or fender layer might be necessary to guarantee the full structural integrity of repaired structures.

As mentioned earlier, three methods involving repair and restressing of existing tendons are presented (Department of Energy 1988c). These include methods where broken prestressing tendons are not re-joined and three different methods described in more detail with drawings, which are:

- a method where the original role of the prestressing can be replaced by using "flat jacks", which produce a similar effect in maintaining compressive stresses in the concrete;
- a method that utilises Macalloy bars which can take the tensile stresses in the ultimate load condition; the design requirement for this method is that at ultimate strain, the total force in the Macalloy bars will be equal or greater than the sum of the forces that the broken tendons would have taken; and
- a method where an alternative repair solution uses a steel "splint" which is capable of taking tensile loads. It should be designed to have the same stiffness characteristics as the concrete that has been removed, providing the same stiffness characteristics to allow the plate to take some of the stresses in the elastic range. Shear forces are transmitted through an epoxy adhesive layer between the steel plate and the concrete wall.

New lengths of tendon can be attached to the broken ends by means of a tendon coupler. The tendons can be overlapped, anchored and stressed using conventional anchorages or they can be curved outside the wall, joined together and stressed using new anchorages. This stage is after the damaged area has been infilled with epoxy resin.

8.4.5 Effectiveness of Concrete Repairs

A study on the effectiveness of concrete repairs was undertaken by Mott MacDonald Ltd for HSE with a focus on land-based repairs, such as bridges (HSE 2003a, b, c and d). Hence, it has limited value to offshore structures, other than regarding the general principles of placing and monitoring repairs. The first phase included a literature review; the third and fourth phases addressed field studies of the effectiveness of concrete repairs including a number of case studies mainly from bridges, car parks and power stations.

As part of the first phase, an expert group was set up from industry and academia for consultation. In addition, a thorough literature study was undertaken to identify the state of the art at the time with regard to repair materials, methods for selecting and specifying repairs, in addition to the measurement of the effectiveness of different repair procedures. Degradation mechanisms were also identified similar to those already identified in this book. A particular case was the chloride ingress into reinforced concrete due to the presence of de-icing salts (which is not directly relevant for offshore structures, but chloride ingress from seawater may react similarly). Durability models for these issues were put forward for use in maintenance planning.

The classification of repair materials and techniques was reviewed. Repair materials considered included cement-only systems, polymer-modified cementitious systems, resins and fibre fillers. The placement methods reviewed included hand trowelling, hand packed, poured in shuttering and spraying. The report concluded that there was little published information describing or comparing the long-term performance of different repair types.

A framework for measuring the performance of a repair in situ was developed. This included factors such as the preparation and application, as well as long-term performance. Methods for testing the effectiveness of repairs were similar to those already discussed in this book.

The report also reviewed relevant standards and guidance relevant to UK repairs. It concluded that at the time there was no current British Standard for concrete repair. It noted that a forthcoming Eurocode would provide a framework for identifying the causes of deterioration in a structure and selecting appropriate methods of repair. The code would also standardise the measurement of properties of repair materials to provide some comparison of relative performance. This standard is now published as EN 1504:2005 (EN 2005).

8.5   Previous Studies on Repair of Plated Structures

The Ship Structure Committee (SSC) has published several reports on the repair of ship structures. A brief summary of some of these reports is provided in this section. It should be noted that most of these reports are concerned with internal repair in dry conditions as would be expected for floating structures. The exception is report SSC 370 on underwater welding (SSC 1993).

In 1969 the Committee issued a "Recommended Emergency Welding Procedure for Temporary Repairs of Ship Steels" (SSC 1969), acknowledging that the steel quality of the time required close control of the welding procedures as well as special techniques to assure serviceability.

Further in 1993, SSC published report "SSC 370 Underwater Repair Procedures for Ship Hulls—Fatigue and Ductility of Underwater Wet Welds" (SSC 1993). Wet welding was seen as a method with a potential to avoid dry docking of ships. The report saw fatigue performance and low tensile elongation properties of wet welds as a potential problem. However, the S-N tests for the underwater wet welds without backing bars indicated fatigue strength levels comparable to dry surface welds. The results for welds with backing bars were less good but those results were based on more limited data. Further, the study indicated that wet welds did not appear to have adequate weld metal ductility for areas where tensile strains exceeded 6%.

In 2000, the SSC published report "SSC 416 Risk-Based Life Cycle Management of Ship Structures" (SSC 2000) including a review of repairs in Appendix B. It was reported that there was no reasonable consensus on what, how and when to repair and that "the general lack of readily retrievable and analysable information on repairs and maintenance makes repair and maintenance tracking very difficult". Further, it was noted that that many common crack repairs appeared to be ineffective, such as veeing (arc-gouging, grinding or cutting a V-shaped slot) and welding as cracks quickly develop again. Replacement of the cracked plate and modification of the design by adding a bracket or a lug were reported as longer lasting, although more costly. In addition, crack repairs, steel renewal and pitting and grooving repairs were reported as possibilities. A summary of the suggested repair methods in this report is presented in Table 70.

Table 70 Repair Options for Ship Structures as Provided by SSC (2000).

Damage type:	Repair option
Cracks	No repair and monitor Temporary fix and monitor: • drill hole at crack tip • drill hole at crack tip, tighten lug to impose compressive stresses at crack front • add doubler plate • cover crack with cold patch Permanent fix with same design: • gouge out crack and re-weld • cut out section and butt weld • apply post weld improvement techniques Permanent fix with modified design: • gouge out crack, re-weld and add, remove or modify scantlings, brackets, stiffeners, lugs or collar plates • cut out section, reweld, and add, remove or modify scantlings, brackets, stiffeners, lugs or collar plates • apply post weld improvement techniques

Corrosion	Minor coating breakdown:
	• General corrosion
	  – no repair and monitor
	  – spot blast and patch coat
	  – add and maintain anodes
	• Pitting corrosion (small shallow pits less than 50% plate thickness in depth)
	  – no repair and monitor
	  – spot blast and patch coat
	  – add and maintain anodes
	Major coating breakdown
	• General corrosion
	  – no repair and monitor
	  – spot blast and patch coat
	  – re-blast and recoat
	  – Add and maintain anodes
	• Pitting corrosion (large, deep pits greater than 50% plate thickness in depth, small in number)
	  – no repair and monitor
	  – spot blast, weld fill and patch coat
	  – add and maintain anodes
	• Pitting corrosion (large, deep pits greater than 50% plate thickness in depth, large in number)
	  – no repair and monitor
	  – spot blast, weld cover plate and patch coat (temporary repair)
	  – cut out, weld new plate, blast, coat (permanent repair)
	  – add and maintain anodes

Source: Modified from SSC (2000). SSC-416 Risk-based Life Cycle Management of Ship Structures, Ship Structure Committee report no 416. Washington, DC, US.

A four-step repair strategy was described in SSC (2000):

- Inspection of structural failure (locate failure and describe the basic properties such as location, orientation, length and percentage wastage).
- Determination of mode of structural failure (such as fatigue damage, corrosion fatigue damage, fracture buckling and stress corrosion cracking).
- Determination of cause of structural failure (design issues such as analysis and material selection, insufficient quality control and welding procedures, overloading due to collisions or poor seamanship in extreme weather, environmental factors that cause corrosion and combined effects).
- Evaluation of repair alternatives and selection.

In 2003, SSC prepared report 425 on "Fatigue Strength and Adequacy of Weld Repairs" (SSC 2003). In this report tests were conducted on large-scale structural details typical for tankers and bulk-carriers (which to a large extent are also relevant for ship-shaped offshore platforms). These included girders, tee joints and longitudinal-to–transverse web frame connections. The repair methods tested included arc-gouging and welding (veeing and welding), adding doubler plates, hole drilling and hammer peening (alone or in some combination with other repair methods). The specimen types were fatigue-tested until cracking occurred and then had their cracks repaired by different techniques. After repair, the specimens were retested under fatigue loading until cracking occurred again. The process was either continued until the specimens could not be further repaired due to multiple cracks or the repair technique resulted in a fatigue life significantly better than that of the original detail. The results of the tests are briefly outlined in Table 71.

In 2005 SSC undertook a review (SSC 443) to establish design guidelines for doubler plate repairs in ship structures (SSC 2005). The review noticed that the use of doubler plates had become routine for temporary ship repairs due to their relative ease and low cost. However, as the review indicated there was a lack of performance data and engineering design guidance. As a result, repairs with doubler plates were considered only temporary. Consequently, the review was performed to develop a set of guidelines for designing and applying doubler plate repairs to ship structures, including criteria for stress analyses, buckling strength, primary stress assessment, corrosion types and rates, weld types and doubler plate fatigue and fracture assessment.

SSC report 469 (SSC 2015) reviewed the use of composite patches for preventing crack growth, hence extending the service life of ship plating. The primary aim of the report was to review the strength and fatigue capacity of composite patch repairs to cracked steel plates. The report indicated that cracks in ship plates were most often repaired by welding, but more permanent repairs required the plating to be cropped and renewed. However, the welding might induce changes to the material properties and stress concentrations and hence not provide the required fatigue life. Hence, the study examined the repair of cracked ship plating using composite patches as the use of adhesively bonded composite patches for repairing cracked or corroded structural components had been seen to experience a significant increase in repair of ship structures.

Numerical simulations of the cracked plates were undertaken and indicated approximately two orders of magnitude increase in service life for the specimens tested. However, the test results showed only increases closer to a single order of magnitude increase. Two factors were deemed important which were debonding of the patch and cracking of the patch, neither of which were taken into account in the finite element analysis. It was noted that debonding of the patch could be improved by more careful attention to the bonding process and it was found that the thickness and uniformity of the layer of epoxy was very important. The application of pressure normal to the bond surface after the attachment also improved the effectiveness of the patching.

Table 71 Results of Repair Tests.

Repair method	Findings from the tests
Hammer peening	Hammer peening of the weld toes could improve the fatigue resistance by a full SN category (e.g. from category E to D). Hammer peening can be an effective repair for shallow surface cracks less than 3 mm deep.
Weld repairs after arc-gouging	Attachments and bracket details: The ratio of the life of the repaired detail to the original was found to be 0.27 (i.e. the remaining life after repair was found to be 27% of the original life).
Weld repairs after arc-gouging with additional hammer peening	Attachments and bracket details: The ratio of the life of the repaired detail to the original was found to be 2.0.
	Longitudinal-to–web frame connections: Cracks that occurred at the one-sided fillet weld between the longitudinal and web frame were described to be caused by a lack of fusion defect, inherent in the connection. The report further described these as best repaired by arc gouging through the fillet weld into the web frame and replacing with a full-penetration groove weld followed by hammer peening the weld toes. The report indicated that the ratio of the life of the repair to the original would then be expected to be 0.86.
Stop holes	Longitudinal-to–web frame connection: Stop holes were found to be an effective technique and all details repaired with stop holes survived more cycles without failure than the original detail. The addition of a fully tensioned high-strength bolt in the stop hole was indicated to further increase the fatigue life.
Doubler plate repair	Attachments and bracket details: Proved to be by far the best repair method. The two repairs of this type reached the same fatigue life as the original and neither repair had any noticeable crack growth.

Source: Modified from SSC (2003). Fatigue strength and adequacy of weld repairs. Ship Structure Committee report no 425. Washington, DC, US.

Overall, this project demonstrated the effectiveness of composite patching of steel plates. Important factors are the quality control of the bonding procedure and optimisation of the geometry and properties of the patch system which are dependent on the properties of the parent plate and how it was fractured (SSC 2015).

8.6 Repair of Steel Structures

8.6.1 Introduction

The most comprehensive reviews of repair of fixed steel structures are found in MSL (1995), MSL (1997a and b), MSL (1999) and MSL (2004). These reports include detailed guidance for evaluation and assessment of the damage and structure and recommendations for selection of repair techniques for various types of damage and anomalies. To a large extent these reports have been used as the basis for the development of standards and recommended practices for the design of repairs, such as ISO 19902 (ISO 2020). The reader is recommended to access these reports and relevant standards for actual design of repairs. Further, DNVGL (2019) provides significant information about the repair of floating steel structures.

Papers have also been published that provide important information about repairs to offshore structures. For example, Sharp (1993) published a review of strengthening and structural repair techniques for ageing fixed steel offshore platforms. This paper reviewed the ability to repair damage and cracks at that time. The methods included welding, clamps and grout filling. The paper also reviewed the integrity of repaired connections as well as recent developments at the time, including friction welding and underwater adhesives. Further, Sharp et al. (1997) also reviewed the techniques used for deep-water repairs in the North Sea in the period 1977–1997, based on the information in the MTD report (MTD 1994). The paper also addressed the specific problems of deep-water repair. It particularly investigated both hyperbaric and wet welding methods, related to research going on at the time on hyperbaric welding at Cranfield University. The paper also commented on the suitability of different techniques for fully diver-less repairs.

Factors such as application and operational requirements, ranges of application, permissible loads and the static and fatigue strength of the repaired structure need to be addressed. The available repair methods will in approximately increasing cost and effort include:

- improvement of existing structure (e.g. grinding, hole drilling, peening, sandblasting, coating);
- local repair (e.g. weld repair, anode replacement, adhesives);
- load reduction (e.g. removal of marine growth, conductors, redundant members);
- strengthening (e.g. clamps, grout filling, reinstate CP); and
- component removal and replacement.

Although the reviews reported earlier in this chapter can provide some indications of the strength of structures repaired by different methods, most repairs will be analysed and designed by the use of advanced non-linear finite element analysis. As background for these analyses, the data from earlier laboratory tests is essential for the calibration of the analysis models of the repairs.

This section first reviews the process of selecting mitigation and repair methods and then continues with a review of different possible repair methods for steel structures. Primarily fixed steel structures are reviewed as repair of semi-submersibles and ship-shaped offshore structures can often be performed in dry condition inside or in dry dock.

8.6.2   Selection of Mitigation and Repair Methods

It is clear from the reviews of previous work discussed earlier in this chapter that the selection of a repair method is strongly linked to the damage or anomaly in need of repair and its cause. For example, if a crack is found, it is important to determine whether this is caused by fatigue or by other mechanisms (e.g. fabrication defect, impact or overload). If fatigue is the cause of the cracking, improvement of the fatigue strength (hammer peening, toe grinding or modification of the detail) will be important, in addition to removing the crack by, for example, a weld repair. The decision on whether to repair and which method of repair to use is reached based on pertinent criteria such as those in Table 72, where:

- the damaged structure is first evaluated and assessed to determine if repair is needed and its urgency in addition to improving the inspection programmes to take into account the damage;
- it is acknowledged that several possible outcomes of these evaluations and assessments are possible, including to repair the structure locally, globally or totally; and
- if repair is found to be needed, the damage type and the damage cause will have influence on the choice along with several other pertinent issues.

All repair methods have their advantages and disadvantages, both in terms of which type of damage they can effectively be used to repair and to what extent they are proven by previous experience or from earlier studies. The MSL studies (MSL 1995, 1997, 1999 and 2004) and the DNVGL

Table 72 Criteria for Selection of Repair Options.

Process	Possible outcomes	Comment and criteria
1) Evaluate structural adequacy and integrity of intact or damaged structure 2) Identify and optimise extent of any repair or strengthening work and the associated urgency 3) Optimise specification of inspection programmes	No strengthening or repair is necessary.	The damaged structure is found to be able to withstand all design load cases.
	No strengthening or repair is needed but a change in the operating and de-manning procedures	The structure is demonstrated to have adequate strength in most design load cases, but: • operational procedures should be changed or implemented to reduce or eliminate exposure to those design events which the structure is not able to withstand (e.g. modify vessel operating procedures), or • de-manning should be implemented to eliminate the presence of personnel and hydrocarbons when exposure to design events which can be forecast (e.g. storm).
	Local strengthening or repair	A method to carry out the work should be selected. There are often several methods that may be relevant to a specific situation and it is important that several methods are considered to ensure that the final choice is the best one based on relevant criteria. These are: • What is the effect of the repair on for example strength and fatigue life and how likely the prescribed result can be achieved • Depth limitations • Tolerance acceptability (e.g. clamps) • Cost, difficulty, available competency, execution timescale and consequences on production Post-installation inspection requirements
	Global strengthening or repair Total strengthening or repair	A method to carry out the work should be selected. There are often several methods that may be relevant to a specific situation and it is important that several methods are considered to ensure that the final choice is the best one based on relevant criteria. These are: • What is the effect of the repair on for example strength and fatigue life and how likely the prescribed result can be achieved • Depth limitations • Tolerance acceptability (e.g. clamps) • Cost, difficulty, available competency, execution timescale and consequences on production • Post-installation inspection requirements

Source: Based on MSL (2004). Assessment of repair techniques for ageing or damaged structures. Project #502. MSL Services Corporation, Egham, Surrey, UK.

(2019) study to a large extent provide valuable information that should be used on these issues. An overview is shown in Table 73.

It should be noted that the most difficult components to repair on a fixed steel offshore structure are the piles and grouted connections. A further important factor is that these components are also

Table 73 Advantages and Disadvantages of Individual Repair Techniques.

Technique	Advantages	Disadvantages	Defect size that can be repaired	Offshore equipment needs	Offshore installation timescale	Onshore fabrication cost	Post-repair inspection requirements	Added weight/load
Dry welding	Universally accepted from a technical standpoint. Applicable to small fatigue cracks in addition to corrosion and dents when combined with patch plates	Requires the construction of either cofferdam or hyperbaric habitat—both being time consuming and expensive. The cofferdam, in particular, will attract high wave loading.	S	H	H	H	M	N
Wet welding	Proven technique and relatively quick method. Applicable to small fatigue cracks in addition to corrosion and dents when combined with patch plates.	Weld properties not as good as dry welds and therefore not accepted in all parts of the world.	S	M	M	N	M	N
Toe Grinding	Assumed to double fatigue life.	Applicable only to small cracks and insufficient fatigue life.	S	L	M	N	M	N
Remedial Grinding	Proven technique. Relatively quick method for arresting fatigue cracks.	Static strength needs to be assessed.	S	L	M	N	M	N
Hammer Peening	Very effective method for increasing fatigue life.	Applicable only to insufficient fatigue life and fatigue life extension.	N	L	M	N	M	N
Stressed Mechanical Clamp	Proven technique for member and joint repair. Can also be used as an end connection to introduce new members into the structure.	Poor tolerance acceptability precludes use around nodal joints or over girth joints between tubular cans. Welds and other protuberances have to be ground flush.	L	M	M	H	H	M
Unstressed Grouted Clamp	Proven technique. High tolerance acceptability. Can be used as an end connection to introduce new members into the structure.	Clamps are relatively long unless they, and the clamped member, are provided with weld beads.	L	M	M	M	M	M
Stressed Grouted Clamp	Proven technique. High tolerance acceptability. Clamps are relatively short. Can be used as an end connection to introduce new members into the structure.	The grout needs to be undisturbed during setting. The strength is achieved only after grout setting and tensioning of the studbolts.	L	M	H	H	H	M
Neoprene-lined Clamp	Some tolerance acceptability. Can be used as an end connection to introduce new members into the structure.	Friction coefficient can be overestimated. Neoprene introduces flexibility, thereby compromising its ability to take up load if alternate load paths exist.	L	M	M	H	M	M

Table 73 (Continued)

Technique	Advantages	Disadvantages	Defect size that can be repaired	Offshore equipment needs	Offshore installation timescale	Onshore fabrication cost	Post-repair inspection requirements	Added weight/load
Grout-filling of Member	Proven technique. Relatively quick method. Applicable mainly for repair of dents, corrosion, inadequate strength and fatigue life.	Weight penalty (structures in seismic regions are especially vulnerable). Complete grout filling may be difficult to achieve.	M	L	L	L	L	H*
Grout-filling of Joints	Proven technique. Relatively quick method. Good for improving both static and fatigue strengths.	Weight penalty (structures in seismic regions are especially vulnerable). Joints with expanded cans, or internal ring stiffening, are more difficult to grout fill.	M	L	L	L	L	H*
Member Removal	Proven technique. Relatively quick method. Only applicable for redundant members if not followed up by member addition	Safety issue if member springs (releasing stored energy) when final ligament severed.	L	M	L	N	N	N
Composites	Lightweight strengthening and repairs are possible. No hot work.	Longevity for underwater use is not yet proven.	M	L	L	M	NA	L

Nomenclature: Generally, H denotes High, M moderate, L low, S small, N none and NA denotes not applicable. H* on weight/load for grout filling indicates that it only applies to weight.
Source: Based on MSL (2004). Assessment of repair techniques for ageing or damaged structures. Project #502. MSL Services Corporation, Egham, Surrey, UK.

very difficult to inspect. It is important, therefore, that sufficient capacity is provided at the design stage such that any expected reduction in strength as a result of ageing effects can be allowed for, recognising the difficulty of repair.

In semi-submersibles and ship-shaped structures repair of fatigue damage is often done by grinding and weld repair, which is much easier if performed in dry conditions (e.g. internally or in a dry dock). A fatigue crack in a semi or ship-shaped structure is often caused by bad detailing in design or fabrication, often related to stiffeners. Remedial measures in these cases often include redesign of the detail to reduce the stress concentration factor or grinding the existing stiffener in order to reduce the effect of the hot-spot stress. Underwater repairs are often possible to avoid on semi-submersibles and ship-shaped offshore structures.

Structural repairs of steel structures are most often required when fatigue damage is found. In addition, corrosion, overload damage and insufficient fatigue life or strength often requires repair. There are a number of different structural repair method available as illustrated in Figure 95 for various types of damage and anomalies.

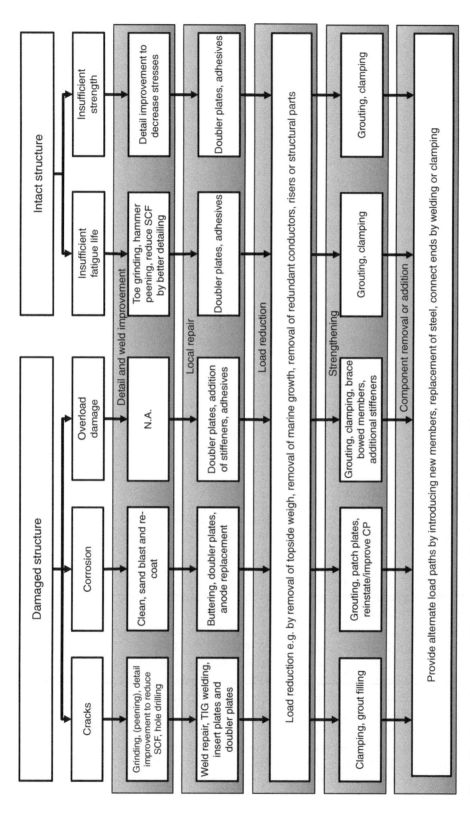

Figure 95 An overview of possible repair and mitigation options (overload damage includes accidental damage from dropped objects and ship impact).

8.6.3   Machining Methods (Grinding)

Machining methods such as burr grinding, disc grinding and water jet gouging can be used as weld improvement methods and to remove cracks (Haagensen and Maddox 2006). These methods require highly skilled operators who should have the relevant training and qualification for the particular task to be undertaken. The final profile after grinding should be thoroughly checked to ensure compliance with recognised standards.

Grinding is a widely used method both for removing fatigue cracks and improving the fatigue strength of welds, see Figure 96. As shown in OTH 89 307 (Department of Energy 1989a), see also Section 8.3.3, the original life of the detail can be more than restored by proper grinding away a defect. The grinding process is normally performed by a rotating burr (typically with a diameter of 12 mm) or by a rotating disc. Disc grinding is effective in removing metal, but the operator needs to be careful not to remove too much material. In addition, the metal may be gouged if the grinding is too severe from operator error. Burr grinders are claimed to be easier to use, but the cutting rate is slower than that of disc grinders. Burr grinding is the preferred method used offshore.

There are mainly three methods of grinding welds:

- weld toe grinding;
- flush grinding of weld cap; and
- weld profiling.

Weld toe grinding will result in a significant improvement in fatigue strength. A circular groove in the transition between weld and base material is created by the grinding tool, as shown in Figure 97.

Weld toe grinding is an effective method for removing shallow fatigue cracks and is likely to be the preferred method for this type of crack, compared to the more costly methods such as welding. DNVGL-RP-C203 (2016) requires the groove to be at least 0.5 mm below the deepest undercut to remove any defects in the area around the weld toe. The depth of the groove should not exceed the smaller of 2 mm or 7% of the plate thickness. After grinding, the groove should be tested preferably with magnetic particle inspection.

Grinding of large fatigue cracks is also possible, and provided that sufficient fatigue strength after grinding can be documented, this is often the preferred solution. An example is tubular joints below the waterline of jackets where the weld repair is a comprehensive and costly operation. An example of geometry after grinding of deep cracks is given in Figure 98.

It is important to point out that toe grinding may not be the optimal solution. A typical example is a fillet weld with a narrow throat thickness (see Figure 99 for illustration of a weld throat). In

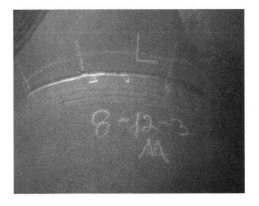

Figure 96   Example of grinding repair. *Source:*
Courtesy of US Bureau of Safety and Environmental
Enforcement (BSEE).

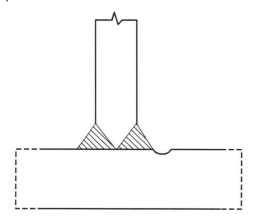

Figure 97 Weld toe grinding. *Source:* Based on DNVGL (2019). Reparasjonsmetoder for bærende konstruksjoner. Report for the Norwegian Petroleum Safety Authority, DNVGL, Høvik, Norway (in Norwegian).

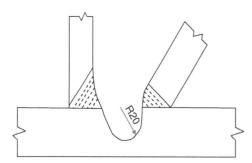

Figure 98 Example of geometry after grinding to remove deep cracks. *Source:* Based on DNVGL (2019). Reparasjonsmetoder for bærende konstruksjoner. Report for the Norwegian Petroleum Safety Authority, DNVGL, Høvik, Norway (in Norwegian).

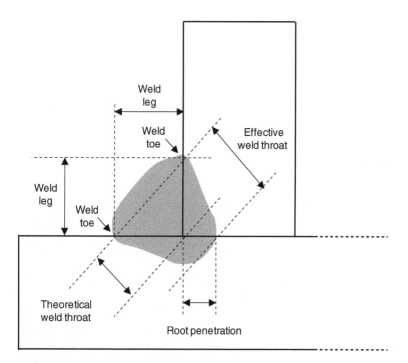

Figure 99 Fillet weld throat terminology.

such cases, there is a likelihood that a large part of the throat cross section is removed by the grinding, which may result in a crack initiation in the weld root (DNVGL 2019). Such cracks are difficult to detect by inspection.

Flush grinding of the weld cap is a method for increasing the fatigue strength of butt welds. The grinding should be carried out as shown in Figure 100 where the transition between plate and weld should be as smooth as possible. The grinding should be made sufficiently deep to remove possible weld defects, such as undercuts.

Weld profiling is done by machining the weld with a given radius as illustrated in Figure 101. Weld profiling will reduce local stress concentrations and thus increase the fatigue strength. The method is not suitable for repair of fatigue cracks. However, it can be used in combination with toe grinding and weld repair. Typical fields of application are tubular joints on jackets and cruciform joints in general.

8.6.4   Re-Melting Methods

Weld dressing methods such as TIG (tungsten inert gas), plasma and laser dressing involve re-melting of the weld in order to remove toe imperfections. Re-melting of the toe region of steel weldments is claimed to give an increase in fatigue strength (DNVGL 2019). TIG welding is a method where the welding material is heated until it melts. Properly designed, this will result in a smoother surface with lower stress concentrations. In addition, the number of weld defects and pores is reduced during the re-melting. However, TIG dressing results in an increased hardness in regions near the weld, which makes the method unacceptable for some applications. The use of weld dressing methods as an improvement method is to the knowledge of the authors not used extensively on offshore structures.

8.6.5   Weld Residual Stress Improvement Methods (Peening)

As mentioned earlier, residual stress modification methods can be used for weld improvement (Haagensen and Maddox 2006). These include:

- mechanical peening methods (hammer peening, needle peening, shot peening, ultrasonic peening);
- mechanical overload methods (initial overloading, local compression); and
- thermal methods (thermal stress relief, spot heating, Gunnert's method).

Figure 100   Illustration of flush grinding of a butt weld.

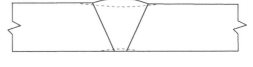

Figure 101   Weld profiling.

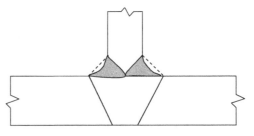

Common to all mechanical peening methods is a tool that deforms the toe plasticly, resulting in compressive residual stresses at the surface of the weld toe. The increase in fatigue strength is primarily due to the introduction of compressive residual stresses preventing the initiation of fatigue cracks. In addition, the geometry after the cold deformation can have lower local stress concentrations compared to an untreated weld, which is also positive for the fatigue strength. The effect of the treatment depends on the level of the residual compressive stresses obtained. As a result, it is important that procedures are developed for the appropriate performance of the methods and that the effect is documented. A general weakness of improvement methods based on superimposed compressive stresses is that these can be reduced or disappear if the weld is subjected to an overload in compression (DNVGL 2019). Machining methods such as toe grinding or profiling are therefore considered to be more robust methods of improvement than peening. Hence, quality control of the performance of peening should be considered to be even more demanding than toe grinding requiring operator's skill and training.

Peening methods for weld improvement are primarily used to extend the fatigue life of welds, and the best results (Haagensen and Maddox 2006) are obtained by a combination of a machining method to modify the weld (e.g. grinding) and a residual stress improvement method (e.g. peening). Peening methods have been used on offshore structures to improve fatigue life.

Mechanical overload methods involve the generation of residual stresses by making the structure yield in compression by mechanical means. Although experiments have shown that this method can increase fatigue life, it has not, to the knowledge of the authors, been used for repair of offshore structures.

Thermal methods include those methods where heat is used to cause plastic deformation, which is cooled instantaneously by spraying a jet of water onto the surface. Since the underlying layers cool more slowly, compression stresses will be created in the surface as these layers cool. To the knowledge of the authors, thermal methods have not been in use in the offshore industry for repairs and particularly not underwater.

8.6.6 Stop Holes and Crack-Deflecting Holes

Drilling of stop holes is a widely used method to prevent the further development of existing cracks and is normally believed to be most effective for the arrest of non-fatigue cracks. The method is primarily intended as a temporary measure, and for ship structures it is recommended only as a temporary repair (SSC 2012). However, it has in some cases been proven to be suitable as a permanent solution (DNVGL 2019). Stop holes are an attractive method for underwater repair and have been used for tubular joints in jacket structures as shown in Figure 102 and for plated structures in ship-shaped structures, although with some variation in their effectiveness as a repair.

The stop hole is normally drilled at the crack tip, and the principle is to transform a crack tip into a blunt notch and as a result reduce the stress concentration, as shown in Figure 103. It is vital that the crack tip is completely removed and as a result the location of the crack tip needs to be verified using appropriate NDT methods prior to repair. The placement of the hole should include sufficient tolerance (twice the thickness of the plate in front of the assumed crack tip has been mentioned as a rule of thumb) to ensure that the actual crack tip is removed. This allows for the possibility that the crack has grown beyond the assumed location. An alternative may be to drill the entire hole in front of the crack tip or use a crack-deflecting hole as shown in Figure 103c.

The stress intensity at the crack tip is removed by the hole if drilled so that it removes the crack tip, but the hole will also introduces a stress concentration (SCF) itself. The $SCF = 1 + 2\sqrt{a/r}$, where a is the length of the crack and r is the radius of the stop hole (Lotsberg 2016). A possible way of sizing the hole would be to require a maximum allowable SCF. If, for example, an SCF of

Figure 102 Example of stop hole repair. *Source:* Courtesy of US Bureau of Safety and Environmental Enforcement (BSEE).

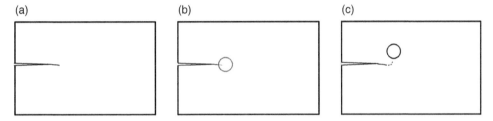

Figure 103 (a) Crack (b) stop hole (c) crack-deflecting hole.

maximum 3 is required, the radius of the hole needs to be equal to the crack length, and if an SCF of 4 is allowed, the radius can be set to half the crack length. This shows that hole drilling is, therefore, more effective for short cracks, as reasonably low SCFs can only be achieved by realistic hole radii. Drilling of small holes to stop long cracks has a reduced effect as the SCF will be significant. Often a ½-inch (12.5 mm) hole is used which will result in an SCF of 5 for a crack of 50 mm, 6.6 for a crack of 100 mm and 9 for a crack of 200 mm. Similarly, for a 1-inch (25mm) hole, the SCFs would be 3.8, 5 and 6.7, respectively. These SCFs are usually considered too high and can obviously be reduced by drilling a larger hole, but this may be challenging for other reasons.

Rolfe and Barsom (1977) and later Fischer et al. (1980) and Fischer et al. (1990) reported experiments and provided a formula for the required radius of a crack arrest hole as a function of the crack length, the yield stress and the cyclic stress range. These two variations of the same formula are shown by Simmons (2013) to be inappropriate for some situations and hence have limited validity range. Simmons (2013) further concluded that calculating an exact radius for a crack-arrest hole is most likely not feasible and that the size should be set in practice to a reasonable value.

A set of practical rules of thumb for crack arrest holes in plates is reported to be:

- for steel plates, ½–1 inch (~ 12.5–25 mm) in thickness the diameter of a stop drilling hole is typically ~ 0.5–2.0 times the plate thickness (SSC 2012);
- the diameter of the crack-arrest hole should be one third the length of the crack (accredited to Dexter by Simmons 2013); and
- the crack-arrest hole must be 4 inches (~100 mm) in diameter and can be permanently effective if the crack is less than 6 inches (~150 mm) long on each side of the transverse stiffener (accredited to Dexter by Simmons 2013).

The second rule of thumb above will provide an SCF of approximately 6, and the third will provide an SCF of approximately 4.5. These SCF values are considered high. The first rule of thumb mentioned above has insufficient information to provide an estimated SCF.

The hole finish after it has been drilled is important to an extent that it is often regarded as more significant than the hole diameter as the remains of burrs or rough edges after the drilling operation can act as crack initiators. All drilled holes should be ground and surfaced to a polished finish, and NDT inspection should be performed upon completion to ensure that the hole circumference is free of defects and that the crack tip has been properly located. These operations can be time consuming, especially underwater.

Several real cases of the use of hole drilling as a repair method have led to an increasing number of holes to be drilled of increasing diameters before there was any evidence of crack arrest. This evidence shows that it is difficult to locate the crack tip accurately in practice and that the necessary hole diameter and hence the SCF is easily underestimated. As shown in Section 8.3.3, the results from fatigue tests after hole drilling for cracks in tubular joints to arrest crack growth were concluded to be partially ineffective based on tests with rather long cracks (approximately 200 mm), a small diameter hole (12.5 mm) and in most cases one hole per crack. A second test was performed in the same test programme with a much shorter crack (approximately 70 mm) and with the same diameter hole, which was more successful. Although these results are not sufficient to prove a relationship between the crack length and the crack arrest hole, they clearly indicate that the findings from plated structures have some relevance also for tubular structures.

As a result, crack-arrest hole drilling to stop or retard crack growth is regarded as an uncertain method and is often only regarded as a temporary repair. However, due to its popularity and simplicity, it is still being used and more work should be performed to verify its effect.

The effect of a stop hole can be improved by the use of bushes or bolts plugged into the holes for creating favourable compressive stresses around the entire hole (Haagensen 1994). It has also been shown that cold forming of the stop hole by expansion by inserting an oversized conical mandrel into the hole to introduce compressive residual stresses can increase the fatigue life of the repair (Lotsberg 2016).

Crack-deflecting holes can also be used and are believed to be more useful for fatigue cracks. These are drilled in the cracked area but at some distance away from the crack with the purpose of altering the stress field and as a result deflect the crack into a lower stressed area. The placement of the hole is very critical and requires a good understanding of the existing stress field and how it is modified by the presence of the hole to gain any benefit (Atteya et al. 2020). The method has been used to some extent and with varied success possibly due to insufficient knowledge of the stress fields.

The use of stop holes to prevent cracks growing along the weld toe can be challenging as the stress concentration of the hole may coincide with the weld. One alternative is to drill a stop hole eccentric to the weld to ensure the stress concentration caused by the hole is away from the weld. Another alternative is to use crack-deflecting holes as shown in Figure 103c.

Drilling of stop holes often requires the use of divers, but ROV tools are understood to have been developed to undertake this task.

8.6.7 Structural Modifications

Fatigue cracks may initiate due to insufficient fatigue strength, typically associated with details with large stress concentrations. In such cases, local modifications may be required in addition to other repair methods to avoid repeated cracks. Modifications to reduce stress concentrations are also common for intact structures with insufficient fatigue life.

Structural modification as a repair method is mostly used on semi-submersible and ship-shaped structures where bracket, flange and stiffener terminations are common. The fatigue cracks initiate due to high local stress concentrations caused by an abrupt termination of the bracket, flange or stiffener towards the plate, particularly where there are close adjacent details that further increase the stress concentration.

The most common modification is to soften the detail by making the transition more gradual. The most efficient result is achieved by grinding the bracket, flange or stiffener toe all the way down to the plate to make the transition as smooth as possible. However, this may not be feasible if the bracket, flange or stiffener is fillet welded to the plate. In such cases the termination of the bracket, flange or stiffener should rather be ground in a way that minimises the impact on the weld itself. Alternatively, a length of the fillet weld (50–100 mm) could be removed and replaced with a full-penetration weld to allow for a smooth transition to be obtained by grinding. If the repair is adequately designed and local stress concentrations minimised, experience shows that these solutions are robust (DNVGL 2019).

Low fatigue strength in such details may also be a result of high nominal stresses, and in such cases a local modification will often not be sufficient to achieve sufficient fatigue strength. Replacement of the plating in the vicinity of the structural detail to a thicker plate may then be a possible method to reduce the nominal stress level (DNVGL 2019). In such cases it may be necessary to weld the insert plate centric with the existing plate to avoid eccentricities that would give rise to bending stresses. Detailed design involving finite element analysis is often needed to verify such repairs.

Closing of openings, such as mouse holes, can also be effective when fatigue cracks are detected or the calculated fatigue is found insufficient. However, it is important to be aware that such openings may be used for functional purposes and closing of such openings may not be optimal.

If a propagating crack has been detected in the early life of the structure, there is then a likelihood that the original design was insufficient. Repair of such cases should include measures to increase the fatigue strength in addition to, for example, a weld repair. Otherwise, a crack may re-initiate at the repaired site. Cold deformation peening methods and detail modifications are examples of methods that can be used to increase fatigue life.

8.6.8 Underwater Welding

Underwater welding offers the opportunity for repair of cracks or the addition of members or attachments (Liu et al. 2010, SSC 1993, Rowe and Liu 1999, Sharp et al. 1997). There are four forms of underwater welding, which are atmospheric, hyperbaric, wet and friction welding. Hyperbaric welding has been used for a number of repairs—the MTD review (MTD 1994) lists 24 repairs using hyperbaric methods. In addition, there were cofferdam repairs (8) and wet welding (3) repairs reported in this review. A number of wet-welding repairs are reported by Reynolds (2010) as described in Section 8.3.15. Cranfield University, in the UK, has been active in research on several types of underwater welding, and for the physics and technology of this welding underwater, the reader is referred to Nixon (1995). The review by Liu et al. (2010) provides considerable information on different weld repair procedures for offshore structures and should be consulted for further information.

The purpose of a weld repair is normally to reinstate the detail to its original state. If an increased fatigue life of the detail is needed, further improvements may be required (e.g. grinding). Repair welding offshore is often done under conditions that are more difficult to control than in a yard and the welding may be of poorer quality. As a result, it is important to develop and use specific accepted welding procedures when performing weld repairs on site (DNVGL 2019).

Repair welding is in addition to grinding the most commonly used method for removing fatigue cracks. Weld methods can be classified into four categories:

- Welding under dry atmospheric pressure. Typical examples are welding on deck areas and in ballast or storage tanks in ship-shaped or semi-submersible structures. Welding can also be done in the dry using a habitat on underwater structures (see below).
- Welding under dry hyperbaric pressure. Welding underwater in an open-bottom habitat where the air pressure is approximately equal to the water pressure. The higher than normal pressure affects the welding process.
- Welding in direct contact with water.
- Friction welding.

Weld repair is best suited for replacing existing welds and it can normally be assumed that the original fatigue strength is restored, assuming a good welding practice. For castings the weld must be ground afterwards to achieve a smooth surface (DNVGL 2019). Weld repair of the base material in rolled and forged components usually leads to a shorter fatigue life compared to the original as demonstrated by the appropriate SN-curves.

Atmospheric Welding

Welding in dry atmospheric pressure is a very widely used repair method. It is especially common for ship-shaped and semi-submersible structures as well as deck structures with relatively easy access. Atmospheric welding can also be used above the splash zone on fixed structures and by the use of cofferdams also in the splash zone and below.

A cofferdam is a watertight structure that surrounds the repair location and is open to the atmosphere. The structure can be open-topped or it can have a closed top with an access shaft to the surface. With this technique welding repair can be carried out using conventional methods providing high-quality welds. There is a significant cost implication in providing the cofferdam, and the method is usually limited to shallow water situations. The MTD review (MTD 1994) lists 3 cofferdam-type repairs, all in shallow water less than 10 m depth. Further, the review also noted problems with maintaining seals against hydrostatic pressure.

Weld repairs to floating structures often involve repairs which are undertaken in the dry at normal pressures, for example in tanks and bulkheads. One option for welds on the outer skin may require a habitat to be installed underwater on the outside which is dewatered. This allows the weld to proceed without contact with the seawater. Examples of such habitats being installed by robots (crawlers) have been undertaken (Equinor 2019). There are significant safety issues for the welder working in a closed environment. This is particularly the case if the tank has been used to store hydrocarbons, fuel or other toxic fluids.

Hyperbaric Welding

Welding under hyperbaric pressure is usually carried out using a purpose-built waterproof chamber (either steel or a flexible material) with an open bottom from which the water has been expelled by admitting gas at the same pressure as the surrounding seawater, see Figure 104.

The pressure in water increases by about one bar for each ten metres of depth. Welding is therefore carried out in a nominally dry environment but at a higher pressure than that on the surface. This higher pressure can have a significant influence on the performance of the welding process, through increasing the arc voltage and reducing arc stability (Richardson 1993). In addition, the increased pressure affects the reaction between gas, slag and metal. Furthermore, the heat loss from the welding process increases due to the increased gas density. As a result, specialised welding procedures should be developed for this method.

Figure 104 Example of a hyperbaric chamber and a diver inside the habitat performing a weld. *Source:* Courtesy of US Bureau of Safety and Environmental Enforcement (BSEE).

There are a number of potential techniques for hyperbaric welding, with Gas Tungsten Arc Welding (GTAW) sometimes called Tungsten Inert Gas (TIG) welding as the front-runner. This uses an arc struck between a tungsten electrode and the work piece as a heat source. GTAW-type welding is often regarded as a complex process, but research has shown that it is a viable and effective technique for hyperbaric welding (Richardson 1993). Cranfield University has undertaken research on developing hyperbaric welding for deep water situations, particularly for pipelines (Hart et al. 2001). Using Cranfield's specialised research equipment (HyperWeld 250), welded joints have been produced at pressures up to 150 bar using the GTAW process. However, to operate this type of welding effectively, special power supply control systems are required in order to optimise process stability.

As noted above, the MTD review (MTD 1994) lists 24 hyperbaric welds in depths ranging from 10 m to below 100 m. Details of a hyperbaric welded repair of a through-thickness crack on BP's Magnus platform at a depth of 182 m, close to the seabed, are given in Sharp et al. (1997) and *Offshore Engineer* (1990). This was the repair of a closure weld joining a diagonal tubular brace to the stub of a leg on a jacket structure. A decision was taken to use hyperbaric welding as the loads were considered too large for a clamped-type repair. The preparation of a work habitat was required, but this was in a difficult position as it was at the junction of an inclined member and between two pile guides. The defective weld was cut out using abrasive water jetting, cutting though the full thickness of the 45 mm wall thickness of the member. This thickness made significant demands on the pre-heating and post-weld heating requirements needed to maintain weld quality and minimise hydrogen-induced cracking. An induction heating system was deployed to provide full circumferential heating of the repaired region. The hyperbaric chamber consisted of two parts, using both steel and a high-strength fabric. A service chamber was linked to the main work chamber, which had facilities for the subsea welding and pre-heating modules, as well as life support. The habitat system was tested using a full-scale mock-up at a base in Aberdeen. The repair was successfully completed in 1990, but the operator reported difficulties in finding qualified welder divers at the time.

Wet Welding

Wet welding is carried out in direct contact with water as shown in Figure 105. The advantage is that there is no need to build a habitat around the area to be repaired. However, a significant disadvantage is that the quality of the weld is poorer compared to the corresponding weld under dry atmospheric pressure. Wet welding has been reported to be used in the offshore industry (Liu et al. 2010) with welds of sufficient quality (AWS D3.6 class A quality). Section 9.3.10 has details of actual repairs undertaken by wet welding mainly in the GoM. However, the use on both the UK and Norwegian continental shelf is very limited.

Wet welding offers a simpler and less costly solution as a repair method, but fast cooling rates occur in the weld metal. This can produce hard and brittle structures in the weld and heat-affected zones (HAZ) with only modest ductility and impact toughness, particularly in the case of higher carbon content steels more typical of North Sea structures.

The wet welding procedure leads to dissociation of water by the arc. The rapid interaction between oxygen and the more readily oxidisable elements in the molten weld metal leads to removal of oxygen creating a hydrogen-rich environment. This hydrogen diffuses readily through any slag layer to dissolve in the molten weld metal. As the weld cools, the hydrogen begins to diffuse away. However, the hydrogen that remains may be sufficient to cause cracking in the weld metal or in the heat-affected zone adjacent to the weld.

A semiautomatic wet welding process 'water curtain welding' (with the action of a conical water jet containing a gas shield) and a flux-cored wire welding method (without a gas shield) has been used with some success. The former has been reported as being able to produce high-quality welds (Hoffmeister et al. 1983). However, on balance the quality of welds has generally not reached that standard achieved with the other welding processes. As a result, the technique is often restricted to repairs where structural strength is not a major concern such as connecting anodes to non-structural elements and adding weld beads to members as part of a clamp-type repair.

Friction Welding

The friction-welding technique was developed by TWI and is applicable for underwater repair by being able to add, for example, studs and attachments (TWI 2020). It is a solid-phase process which involves rotating a stud at high speed whilst pressure is applied forcing the stud onto the metal substrate. The friction between the stud tip and the substrate causes the metal surfaces to heat and a thin layer of metal to flow plasticly under pressure without melting to the periphery of the weld,

Figure 105 Example of a diver performing a wet weld. *Source:* Courtesy of US Bureau of Safety and Environmental Enforcement (BSEE).

removing impurities from the interface. Pressure is maintained for a few seconds after the rotation is stopped, producing a solid-phase forged weld with a fine grain structure.

Friction welding underwater can cause rapid cooling of the weld and may lead to poor weld properties such as excessive hardness with potential cracking. It has been found that the use of a polymer sleeve to form a shroud around the weld region protects it from the surrounding water and therefore reduces the problem (TWI 2020) and can give acceptable welds.

It is reported by Gibson et al. (2010) that quality welds can be made by friction welding both in shallow water and at depth. It is claimed to be a relatively simple process to operate, and divers can be rapidly trained and qualified in using the process. It can also be adapted for remote operations with ROVs.

8.6.9 Doubler Plates

The use of doubler plates can be an effective reinforcement method where it is reasonable to increase the thickness of the plate or member. A thorough analysis of the design is needed to make sure that the double plate is effective in carrying loads. IACS (1996) recommends machining slots in the doubler plate and the plate should be welded to the parent material both on the edges and through the slots to achieve effective stress transfer to the doubler plate. Doubler plate repair is also used on fixed steel offshore structures as shown in Figure 106.

8.6.10 Removal of Structural Elements

Removal of structural elements may in some cases be the preferred solution for a repair. As a result of advanced analysis techniques, of members that can be removed rather than repaired have become readily available. The purpose of such a measure can be put into three main categories:

- removal of secondary elements to avoid local stress concentrations;
- removal of structural elements or parts of a structural element to change global or local load effects; and
- removing a structural element to be replaced by a new one.

Removal of a secondary element may be relevant for structural elements primarily included due to requirements in early temporary installation phases. An example of this is lifting lugs. Such

Figure 106 Example of doubler plate repair to a jacket structure. *Source:* Courtesy of US Bureau of Safety and Environmental Enforcement (BSEE).

details can have low fatigue strength if they are located in an area subjected to high cyclic loads. Removal may serve as an effective measure.

Removal of structural elements or parts of a structural element may in some cases be efficient for reducing external loads on the structure. After several years of operation, riser or other appurtenances may no longer be in use. Removal of these structural elements (including support structures) will reduce the hydrodynamic loads and potentially have a positive effect on the fatigue strength locally. In addition, the likelihood of fatigue crack initiation in the transition between the support structure and the main support structure is removed.

Removing the conductor guide frames on fixed offshore structures may be a valuable repair for older structures, such as those designed before about 1981 (MSL 2004). The understanding of how conductors behave under internally applied loads has improved, and there is now an inherent capacity for these early types of conductors to span greater distances (MSL 2004). Other items which can be removed from fixed offshore structures include redundant caissons, conductors and risers, launch rails, boat landings, bumpers, pad-eyes and pile guides.

Removing a structural element to be replaced by a new one may also be used as an effective modification. Examples of such repairs include a jacket on the Norwegian continental shelf where a crack was detected at the tubular joint between the original horizontal brace and the pile sleeve. The repair was performed by removing the original brace, and a new brace was installed by the use of clamps.

Cutting techniques to remove the member include mechanical cutting (cutter, wire saw, abrasive water jet and diamond wire), thermal cutting (oxy-acetylene, oxy-hydrogen, oxy-arc, thermic and ultra-thermic lance, plasma arc, pyronol), explosives and electro-chemical cutting (MSL 2004).

8.6.11 Bonded-Type Repairs

Bonded-type repairs are increasingly being used in many industry sectors to bond ("glue") materials of various types to a dry substrate (base material). For bonding to wet surfaces, special materials are required, and some of these have been developed for repairs on ships.

Bonded types of repair consist of a fibre-reinforced laminate patch (bonded patch) that is attached to the base material with an adhesive (see Figure 107). Bonded repairs can also involve attaching a metal or composite plate to the base material. The most used bonded repair in the offshore industry is resin repair involving a layer of fibre-reinforced composite material. The fibre-reinforcement is typically glass fibre, carbon fibre or aramid fibre. Such repairs can provide a structural reinforcement and should be designed to allow the transfer of loads from the damaged

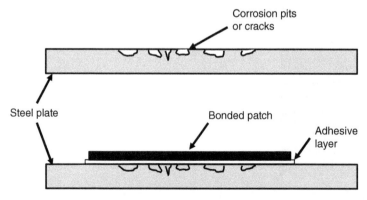

Figure 107 Illustration of bonded patch repair (before and after repair).

material into the fibre-reinforced composite repair. The laminate needs to be designed to carry these loads and the adhesive to transfer the forces between the steel structure and the laminate.

Failure modes to be considered in the design are fractures of the laminate, fractures of the adhesive compound and delamination of the laminate. DNVGL-RP-C301 (DNV GL 2015) provides requirements and guidelines for the use of laminating and bonding with composites in repairs and provides guidelines on where such repair can be used and how to obtain a repair of sufficient quality.

The repair method is only recommended for non-critical structural elements (DNVGL 2019). However, the method can also be used on a critical structural element where the damage is considered minor. MSL (2004) describes the use of bonded repairs to caissons but is not clear on whether this was performed underwater or above water.

As previously described in Section 8.3.4, a UK Department of Energy–funded programme was undertaken to develop adhesives for application to wet surfaces. This proved successful in developing high-strength tube-to-tube connections with a resin filler. A company (Wessex Resins) further developed the technology. However, given the early emphasis on structural repairs to offshore steel structures using, for example, a sleeve-type joint, the resin technology has not so far been taken up by the industry. Grout is the preferred filler, despite its lower strength when used in a sleeve joint. Grouting is an established offshore technique used for connecting the pile and jacket, and the use of similar methods for grouted repairs is an obvious step. There is the possibility of using resin techniques for repairs to other parts of an offshore structure such as damaged caissons. These caissons house large pumps for providing seawater for cooling and fire services. Caissons have suffered damage from galvanic action between the pump and caisson leading to severe localised corrosion and holes in the wall. Another cause of damage is depletion of the original anodes when corrosion can occur if the anodes are not renewed.

8.6.12 Structural Clamps and Sleeves

Structural clamps and sleeves are used as a repair method primarily in fixed steel offshore structures. Clamps and sleeves consist of two or more parts which are assembled around the structural element (member or joint).

A clamp is used to repair or strengthen a tubular member or joint within an existing structure by providing an alternative load path. A clamp typically weighs between 0.5 and 10 tonnes (MSL 1995). A sleeve surrounds and is connected to a tubular member providing an alternative parallel load path.

Clamps and sleeves have several applications as a repair method. The most common are:

- strengthening of fatigue and impact-damaged members or joints (cracks, dents and bows etc.);
- strengthening of members or joints with insufficient strength; and
- support for a new structural element added to an existing structure.

The main aim of a structural clamp and sleeve repair is to transfer the loads and stresses into the clamp or sleeve from the area of the damaged or under-strength area of the structure. One of the critical factors in achieving this is being able to obtain a close connection between the clamp or sleeve and the existing steelwork such that stress transfer can be effective. Ways to achieve this have included using friction directly or a grout or an elastic membrane (e.g. neoprene) to fill the interface between the original steelwork and the clamp or sleeve. The stress transfer can be made more effective by stressing the clamp or sleeve directly onto the tubular section by using long stud-bolts (MSL 2004). A schematic diagram of a clamp repair of a K joint is shown in Figure 108.

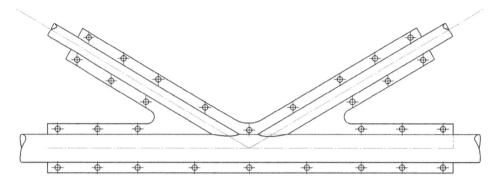

Figure 108 Schematic diagram of a repair of a tubular K joint using bolted connections.

Figure 89 shows a bolted clamp used to strengthen a connection on the Viking AD platform as discussed previously in Section 8.3.11. This clamp was recovered from the platform following decommissioning which allowed tests to be carried out on the components such as the bolts and grout.

Different clamp and sleeve technologies exist distinguished by their load transfer methods and requirements for allowing for geometric tolerances in relation to the existing structural element. The most common types include:

1) Mechanical friction clamps (prestressed mechanical clamps), which are a clamp with a body of two or more segments stressed together by bolts to provide the load path in the clamp. The outer segments need to be formed with strict tolerances to fit the existing structure. The coefficient of friction between the two steel surfaces is as a result an important factor for the strength of the repair. Application of mechanical friction clamps is reported to include strengthening of one or more brace members (static strength and fatigue) and connecting members at an inclined angle into an existing structure. The use of mechanical friction clamps is often not seen as suitable for repair of joints as the geometry of a joint is rather complex and the strict tolerances required are difficult to obtain. Guidance for static strength and fatigue analysis are provided in ISO 19902 (ISO 2020) and OTH 88 283 (Department of Energy 1988), presented in Section 8.3.2.

2) Grout-filled clamps (split sleeve clamps) are clamps in which the split sleeves are closed by pre-tightened bolts leaving an annular space between the clamp and the existing structure for the injection of grout. Grout-filled clamps are therefore unstressed. The use of grout reduces the need for strict tolerances. Loads are transferred from the existing structure to the clamp through the grout by shear and compressive loads. Weld beads (shear keys) can be placed around the circumference of the clamp and member or joint to increase the shear capacity (DNVGL 2019). Placing weld beads on an existing underwater member or joint would require underwater welding (such as wet welding). Filling the clamp with grout is regarded as a standard underwater operation, but there may be individual features that require large-scale testing prior to the repair to verify the execution procedures. Such issues may include filling of particularly large volumes, placement of grout piping and the contractor's experience with materials and methods. Ensuring sufficient sealing at, for example, the lower end of the clamp may also be a challenge. Observations of insufficient filling have been noted when inspecting clamps from decommissioned structures (DNVGL 2019), highlighting the importance of having a verified grout-filling method that ensures sufficient quality of the repair. Applications are reported (D.En. 1988) to include strengthening of one or more brace members at tubular joints and connecting a new brace member into the structure. Formulae for static strength and stress concentration factors are provided in ISO 19902

(ISO 2020) and OTH 88 283 (Department of Energy 1988), presented in Section 8.3.2. The strength of unstressed grouted clamps is lower than mechanical friction clamps. As a result, unstressed grouted clamps often need to be made longer to achieve sufficient capacity.

3) Stressed grout-filled clamps (prestressed grouted clamps) are similar to grout-filled unstressed clamps in many ways. Grout is placed in the annular space between the clamp and the joint and allowed to reach a predefined strength prior to the application of an external stressing force normal to the steel-grout interface applied by tightening the bolts. Stressed grouted clamps share the advantages of both friction clamps and unstressed grouted clamps with regards to load-bearing capacity. Similar to unstressed grout-filled clamps, the requirements for tolerances are less strict compared to mechanical friction clamps. Application are reported to include strengthening of one or more brace members at tubular joints against static or fatigue loading and connecting a new brace member at an inclined angle into an existing structure. This type of clamp has been used extensively on the Norwegian continental shelf and the experience has generally been good (DNVGL 2019).

4) Stressed neoprene-lined clamps (prestressed lined clamps) are similar to mechanical friction clamps except that a layer of elastomer material, for example neoprene, is applied between the clamp and the structural element. The strength is determined by the external bolt loads and the coefficient of friction between the elastomer layer and the steel surface. The use of an elastomer-liner provides less stringent tolerance requirements compared to friction clamps (see Section 8.3.12).

An example of a clamp repair is shown in Figure 109.

Similar solutions to the four types of clamp repairs are available for sleeve-type connections. A mechanical sleeve connection relies on the friction between the two tubulars for load transfer. The outer tubular has two or more segments stressed together to generate a force normal to the friction surface. A grouted sleeve connection is formed by injecting cementitious material (grout) into the

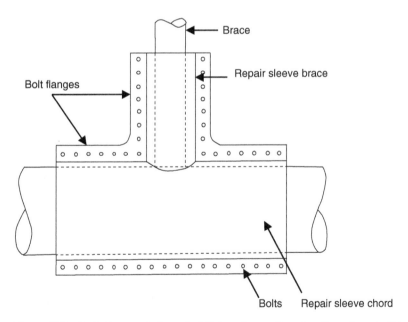

Figure 109 Typical grouted clamp for T joints. *Source:* Based on Department of Energy (1988). Offshore Technology Report OTH 88 283 Grouted and mechanical strengthening and repair of tubular steel offshore structures. Prepared by Wimpey Offshore Engineers and Constructors LTD (R.G. Harwood and E.P.

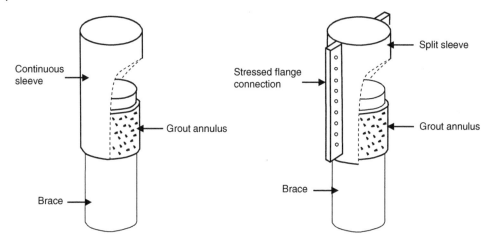

Figure 110 Grouted sleeve connections. *Source:* Based on Department of Energy (1988). Offshore Technology Report OTH 88 283 Grouted and mechanical strengthening and repair of tubular steel offshore structures. Prepared by Wimpey Offshore Engineers and Constructors LTD (R.G. Harwood and E.P. Shuttleworth) for the Department of Energy. HMSO, London, UK.

annular space between the tubulars. Two examples of grouted sleeves are shown in Figure 110. Both consist of a stressed grouted sleeve connection in which grout is placed into the annular space between the tubulars and allowed to reach a predefined strength prior to the application of an external stressing force normal to the steel-grout interface.

At present, various forms of conductor guide securing clamps are also in use such as tie down clamps, tension clamps, compression clamps and lobster clamps (see e.g. DNVGL 2019).

Structural repairs by clamps and sleeves have performed satisfactorily over many years but require regular inspection particularly if bolts have been used to connect the two halves of a repair. Bolt tightening is regular maintenance routine for many of these clamps and sleeves.

These repairs have a disadvantage in increasing the wave loading, a factor which needs to be taken into account during design of the repair and the assessment of the global structure. In addition, the cost of using clamps is high because of the required underwater working and support vessels.

As already mentioned, the tolerances of a clamp design are important, particularly for mechanical friction clamps. At the time of writing this book, the standard method is to use photogrammetry of the underwater structural part to be repaired. This provides an accurate 3D model that is used as the basis for the design of the clamp to ensure optimal tolerances.

Formulae for static strength and stress concentration factors are provided in ISO 19902 (ISO 2020) and OTH 88 283 (Department of Energy 1988), presented in Section 8.3.2.

8.6.13 Grout Filling of Members

Grout filling of damaged or understrength members or joints (filling the tubulars with cementitious material) can increase their static strength, see Figure 111. Grout filling of tubular members is a common repair and strengthening method for offshore structures, particularly for older jacket structures.

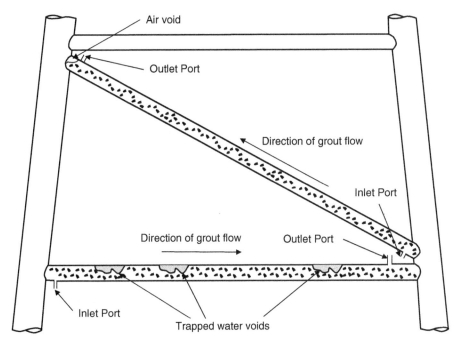

Figure 111 Void formation during grout filling. *Source:* Based on MSL (2004). Assessment of repair techniques for ageing or damaged structures. Project #502. MSL Services Corporation, Egham, Surrey, UK.

It has been shown that filling of grout can lead to full recovery of the strength of damaged members as the grout will prevent further local buckling deformation inwards and the capacity against global buckling will increase. The best load transfer to the grout is through the contact stress at each end of the brace. As a result, grout filling as a repair method is most effective in braces subjected to compressive loads and requires complete filling of the member and cavities are reduced to a minimum, as shown in Figure 111. This may be difficult for horizontal braces and braces with internal ring stiffeners. The filling procedure, including material selection, inlet and outlet points and equipment should be verified using full-scale testing prior to the repair (DNVGL 2019). Load transfer through friction or bonding between the steel and the grout cannot normally be assumed (DNVGL 2019).

Filling of grout will increase the stiffness and mass of the member and thus change the load distribution in the structure. Hence, it is important that the effect of load distribution due to grout filling is documented through reanalysis of the structure.

Several research programmes (Department of Energy 1988, Department of Energy 1989b, Department of Energy 1992, Rickles et al. 1992, Rickles et al. 1997, Hebor 1994, Ricles et al. 1995 and Patterson and Ricles 2001) have been undertaken to provide design information for such repairs, including both fatigue and static strength properties. These have included testing of repaired and grout-filled components in the laboratory and analytical work.

ISO 19902 (ISO 2020) provides design formulae for fully grouted intact members and fully grouted dented tubular members with dent depths less than 0.3 times the diameter and 10 times the thickness of the member.

8.6.14 Grout Filling of Tubular Joints

According to ISO 19902 (ISO 2020), two varieties of grouted joints are commonly used in practice, namely a grouted chord (as shown in Figure 112) and grouting in the annulus of a double-skin joint. As a repair method, grout filling is typically performed by grout filling the chord. The strength of a tubular joint increases by grout filling the chord as it strengthens the wall and as a result limits local bending and ovalisation of the cross section. Increased capacity is achieved both against fatigue and static stress. As previously described in Section 8.3.6, grout filling of existing undamaged nodes can also be used to extend the fatigue life provided the dominant load condition is axial (Department of Energy 1992).

Grout filling of tubular joints is a relatively simple and effective method of repair. However, the effect can be reduced as the capacity of the joint will in a repaired state be limited by the strength of an incoming brace. Many of the limitations related to the use of grout in braces will also apply to repairs of tubular joints. The influence on stiffness and load re-distribution should be carefully considered in order to achieve the best possible repair.

In terms of fatigue, the introduction of grout considerably reduces the stress generated under axial and out-of-plane bending conditions, thereby reducing the SCFs around the brace-to-chord weld intersection. It has also been shown that the failure modes of grout-stiffened nodes were not altered from those of conventional nodes.

8.6.15 Installation of New Structural Elements

In some cases, the most effective repair method is to introduce new structural elements. For example, additional braces can be inserted in order to increase the strength of a jacket to withstand increased wave forces. These may be attached by welding if located above water, while below water it may be appropriate to use clamps to attach the new elements.

For floating structures, additional stiffeners may be required for a damaged stiffened plate or plates with insufficient capacity. This will require access to the plate area with damage or reduced capacity in order to perform repair work. The repair strategy will depend on the extent of damage and the structural arrangement. Stiffened plates are to a large extent used as

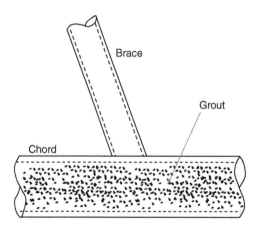

Figure 112 Grout-filled joint. *Source:* Based on MSL (2004). Assessment of repair techniques for ageing or damaged structures. Project #502. MSL Services Corporation, Egham, Surrey, UK.

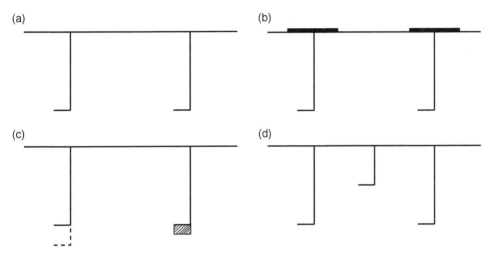

Figure 113 Reinforcement methods for stiffened plates, (a) indicates the original stiffened plate, (b) indicates the use of doubler plates, (c) indicates methods for strengthening the stiffeners, (d) indicates additional stiffeners to achieve an increased effective flange-width.

load-bearing components in marine structures, such as the hull girder of a ship and the pontoons of a semi-submersible. In addition, stiffened plates are used in the decks of offshore platforms. In a majority of cases repair of such stiffened plates is possible to perform above water or inside a floating structure, which is outside the scope of this book. An overview of repair methods to stiffened plates can be found in DNVGL (2019), Paik and Thayamballi (2007) and in Paik and Melchers (2008). Underwater repair is used if needed and then typically in the form of underwater welding.

The stresses in the outer skin of floating structures due to direct wave load are largely determined by the ratio of stiffener spacing (s) to plate thickness (t), the so-called s/t ratio. If the stiffener spacing is too large, the use of intermediate stiffeners or transverse plates are possible solutions to reduce the stress level. However, the stiffeners are normally attached by fillet welds, and the size of the welds is often not sufficient for this to be considered as an alternative. The use of intermediate stiffeners or transverse stiffeners is illustrated in Figure 113d.

A brief overview of repair and strengthening methods in situations when these plates are found to have insufficient capacity is provided in Table 74, based on DNVGL (2019) and Paik and Melchers (2008).

Possible reinforcement methods for the web of girders are described in chapter 9 of DNV-RP-C201 (2010).

8.6.16 Summary of Steel Repairs

Based on the previous studies reviewed in this chapter, an overview of repair methods including a short description is given for different types of damage, including:

- cracks and insufficient fatigue life in Table 75;
- overload damage and insufficient strength in Table 76;
- corrosion in Table 77; and
- other types of damage in Table 78.

Table 74 Repair and Reinforcement of Stiffened Plates.

Cause of the insufficient capacity	Temporary repair	Permanent repair
Insufficient design or fabrication or increase in loading (see Figure 113 for examples)	Doubler plates may be used but may lead to fatigue cracking.	Intermediate stiffeners to achieve an increased effective flange-width. Reinforcement of the stiffener flange. Reinforcement of the web of girders
Corrosion of plates or stiffeners	Repair by sandblasting and repainting. Doubler plates over the affected area. Stronger member may be used to support weakened stiffeners and composite repair may be used	Crop and renew (replacement of panels or part of panels). Cutting, weld sequence, edge preparation and alignment should be done in accordance with fabrication standards for new builds.
Fatigue cracks in plates or stiffeners	Crack arresting holes with or without expansion bolts may be used as a temporary repair but insert plates should be used as soon as possible. Weld repairs are also often used but should be used in combination with grinding or hole drilling to reduce the stresses.	Rewelding of cracks, replacing the cracked plate by suitable inserts, modifying details, increase scantlings. Latent defects should be veed out by grinding or arc gouging, prepared and rewelded.

Table 75 Repair Methods for Cracks and Insufficient Fatigue Life in Steel Structures

Repair method	Description and discussion
Grinding: • grinding of welding toe; • plane grinding of butt weld; • weld profiling.	The preferred method of crack repair if the crack is not too deep, especially in contained areas (tanks); weld profiling will reduce local stress concentrations and thus increase fatigue strength. Toe grinding is an effective method for removing shallow fatigue cracks, and it has been shown to significantly improve the fatigue strength. Plane grinding of the weld is a method to increase the fatigue strength of a butt weld. The methods are not suitable for repair of deep fatigue cracks but can be used in combination with weld repair.
Weld repair in: • atmospheric dry conditions; • hyperbaric conditions; • wet conditions.	The preferred method when fatigue cracks are long, deep or through thickness. Placing the weld in a low-stress area should be chosen to avoid low fatigue life of a repair. Execution includes grinding the area down to uncracked material and to a geometry suitable for a repair weld. Qualified welding procedures are vital, especially inside welding on the outer skin plates (with the sea on the other side) and for hyperbaric welding. Wet welding is to some extent also used but not accepted in all offshore regions.
Shot-, needle-, hammer- and ultrasonic peening	Methods to deform the weld toe plastically and, as a result, to introduce compressive residual stresses in the surface of the material. A weakness of this improvement method is that strict procedures need to be followed and the residual compressive stresses can be reduced or disappear after the structure has been subjected to a possible compressive overload. Peening has been reported to be used on the Norwegian continental shelf in order to increase fatigue life of details.

Table 75 (Continued)

Repair method	Description and discussion
Drilling stop holes	A method to prevent the further development of existing cracks and can be used also for stopping the growth of through-thickness cracks. Primarily used as a temporary repair method but also as a permanent repair. It has been reported to be used on nodes of fixed jacket structures, inner plating and outer skin of floating structures.
	The long-term effect is not yet established.
	A possible problem is that the crack continues to grow beyond the stop hole, and to avoid this, the location of the stop hole is vital. The effect of a stop hole is best for short cracks and large-diameter stop holes.
	The effect of stop holes can be enhanced by introducing compressive residual stresses around the hole (tapered bolt, inserting a sleeve in the hole and by other specialized tools).
Modification of the detail	Used to improve fatigue life of details with insufficient life and on floating structure for cracks that initiate at the termination of brackets, flanges and stiffeners. The purpose of the modification is to reduce the local stress concentration of the detail and especially at the weld. Often used in combination with a weld repair.
Insert plates	Used if the nominal cyclic stress is high. Parts of an existing plate and any stiffeners are cut out and replaced with a thicker plate to reduce the nominal cyclic stress level.
Doubler plates	Another possible method of reducing the nominal stress level. This is a well-established method and guidelines for execution are given in standards.
Closing of openings	Closing openings (e.g. penetrations for stiffeners) can be an effective measure.
Addition of new structural elements	Can be used to reduce stresses in highly stressed components. On fixed steel offshore structures these added members are often connected by the use of clamps; see below.
	Addition of stiffeners in floating structures is a useful method if the distance between stiffeners in the outer skin is too large. The method is used to reduce the stresses locally in the longitudinal weld between stiffeners and plates.
Member and nodal strengthening	Clamps and sleeves are particularly useful for repair of insufficient fatigue life and significant cracks (e.g. severance). Clamps and sleeves can be stressed or unstressed, mechanical, grouted or elastomer lined.
Removal of secondary structural elements	Cracks or low fatigue life related to secondary structural elements that are no longer in use (e.g. pad eyes for lifting) may be rectified by removing these.
Removal of primary structural elements	Primary purpose of this method is to reduce overall loading (e.g. wave loading on unused horizontal frames).
Removal and replacement of structural elements	Found to be useful for severed or near-severed jacket braces. The remains of the existing brace are removed and a new brace is inserted, possibly at a slightly different location.
Adhesives • composite (CRP); • steel.	A useful method for adding material (steel or composite) over cracked section. Repair by adhesives is preferred to avoid hot work (welding). In general, underwater adhesive repairs are not widely used and the experience and possible limitations are not well documented.

Table 76 Repair Methods Suggested for Overload Damage and Insufficient Strength in Steel Structures.

Repair method	Description and discussion
Weld repair in: • atmospheric dry conditions; • hyperbaric conditions; • wet conditions.	Mostly used in combination with other repair methods (e.g. to weld doubler plates or for addition of new members). Can also be used to reconnect ends of severed braces.
Grouting	Grouting will increase the capacity of the cross section at a dent and will also increase the axial and bending capacity in general. In addition, grouting has been used to increase damping and alter the eigen-frequency by adding mass. This repair method is suitable for damage both above and below water.
Addition of braces	A method to reduce loading in, for example, a brace or to reduce the buckling length of a brace.
Addition of stiffeners to plates	Method used in cases where there is insufficient cross-sectional area.
Doubling plates	Method used in cases with insufficient cross-sectional area or dented members.
Strengthening of existing stiffeners	Method used in cases with insufficient bending capacity in the stiffener. The method includes the addition of a stiffened area (doubling plates on stiffener web or additional stiffener on the existing stiffener).
Addition of new structural elements	Can be used to reduce loading in highly stressed components, as for repair methods for cracks and insufficient fatigue life, see Table 75.
Member and nodal strengthening	Clamps can be used as a repair method for structural elements or nodes damaged by buckling, dents and for structural elements that have experienced large plastic deformations due to overload. Clamps can also be used to strengthen structural elements with insufficient strength and can be used to attach new members to a structure.
Adhesives • composite; • steel.	A useful method for adding material (steel or composite) over damaged or under-strength sections, see also Table 75.

8.7 Repair of Corrosion and Corrosion Protection Systems

8.7.1 Introduction

Most offshore installations experience some degree of corrosion, even when protective systems are in place. Typical cases are due to damaged coatings or poor control of the cathodic protection voltage. A thorough assessment prior to the implementation of mitigating measures is important.

One possible outcome may be that the corroded component or area still has sufficient strength for all relevant limit states. In such cases it will be sufficient to prevent further corrosion by cleaning, sandblasting and re-applying a protective coating. If the corrosion damage is more severe, then remedial measures are required. In general methods for mitigation of corrosion and corrosion protection systems may include repair of damaged coatings, replacement of material and repair or replacement of the corrosion protection system as further reviewed in this section.

Table 77 Repair of Corrosion Damage in Steel Structures.

Repair method	Description and discussion
Cleaning, abrasive blasting and reinstating protective coating	A recommended method if the strength of the structure is still sufficient taking into account the level of corrosion.
Replacement of steel	Used for heavily corroded components to reinstate strength. Inserts are used either to return the component to its original state or to reinforce by increasing the thickness of the plate.
Weld repair in: • atmospheric dry conditions; • hyperbaric conditions; • wet conditions.	Buttering (welding of strings) to replace lost material. This is considered a good alternative in cases where the area of reduced thickness is limited, for example for corrosion in welds.
Slab plates	Used in areas of reduced thickness to increase the load-carrying capacity primarily as a temporary measure.
Grout filling	Grout filling can increase the capacity of a corroded cross section, see also Table 76.
Addition of new structural members	Can be used to provide alternative load path for significantly corroded components, see also Table 75. Addition or replacement of stiffeners in floating structures is useful if stiffeners are significantly corroded.
Enhance cathodic protection	Anode replacement, adding sled or bracelet anodes.
Resins • composite • steel	A useful method for adding material (steel or composite) over corroded sections or patch, see also Table 75.

Table 78 Repair Methods for Other Types of Damage for Steel Structures.

Anomaly Defect or change	Repair method (temporary and permanent)
Wear	Grout member, joint or sleeve. Weld patch plates. Member removal or load reduction. Provide alternative load path by adding new members or connecting ends by welding or clamping.
Marine growth	Marine growth cleaning by water jet
Settlement and subsidence	Lifting or straightening by jacking. Removal of lower decks or lower deck equipment to reduce wave loading.
Scour	Sand filling, rock dumping.
Material embrittlement (hydrogen embrittlement)	Reduce cathodic potential overprotection, if possible, to avoid further cracking. Cracks may be repaired as for fatigue cracks (Table 7). Replacement may be required in serious cases.
Fabrication faults	Treat as fatigue cracks, corrosion or insufficient fatigue or static strength.
Under-design	Treat as fatigue cracks, corrosion or insufficient fatigue or static strength.

8.7.2 Repair of Damaged Coatings

Coatings are applied for corrosion control as they act as barriers that isolate the steel from the corrosive environment. Coatings can be applied in the construction yard or offshore although the latter is more expensive and often less efficient. Typical coatings used offshore include coal tar epoxy (as noted in Section 3.1.2 for the West Sole WE platform) and epoxy resins. Coatings on fixed steel structures are more normally applied on members in the splash zone while on floating structures large parts of the external hull structure are coated.

Coatings are inspected as part of an inspection programme and may require repair or replacement depending on the level of damage. Several scales exist (ISO 2016) for measuring the degree of either blistering, rusting, flaking or cracking which may occur with coatings. Pinpoint rusting or spot rusting is one of the more common types of coating failure, often as a result of ageing. Assessing the level of damage to a substrate during inspection can be difficult without removing the coating, which can be expensive and will require substantial remedial work.

Coating repairs need to include proper cleaning of the surface for marine growth (brushes or water jet), sandblasting of the surface to remove the damaged coating and to ensure good adhesion before re-applying a protective coating. Although coatings exist that can be applied manually underwater with good adhesion, a habitat extends the ability to perform the repair more effectively. ROV-created habitats exist but are primarily used for pipeline repairs.

Replacement of coatings in the splash zone has access difficulties and requires a coating to be applied to a wet surface. In order to obtain good quality adhesion special coatings are required (Belzona 2020).

Coatings and anodes are also a feature of corrosion protection for ballast tanks in floating structures. Breakdown of these coatings can lead to localised corrosion and loss of watertight integrity. Proper re-installment of coatings in such areas is an important mitigating measure.

8.7.3 Replacement of Corroded Material

A more comprehensive measure is replacement of material. This typically implies that components or plate areas need to be partly or completely replaced with new material or with local reinforcements. Methods commonly used include replacement of steel by insert plates, buttering and doubler plates. These methods are used when an area is so severely corroded that it is considered that complete replacement is the most appropriate solution.

Insert plates are primarily used to return the component to its original state but thicker insert plates than the original can be used to increase the plate thickness thereby obtaining a larger corrosion allowance for further operation.

Weld beads can be added to replace lost material (buttering). This is considered a good alternative in cases where the area of reduced thickness is limited such as pitting corrosion and where corrosion is limited to areas adjacent to welds.

Doubler plates can be welded to a plate or member to increase thickness. The use of doubler plates is primarily considered as a temporary measure but can occasionally be used as permanent strengthening.

As previously mentioned, corroded tubular members may be strengthened by grout filling, sleeves or clamps. Removal of the damaged member or section and replacement by a new member is also possible.

8.7.4 Repair or Replacement of the Corrosion Protection System

Corrosion protection systems are used on most offshore structures by means of sacrificial anodes or impressed currents systems. Sacrificial anodes are designed with an expected life commensurate with the design life, typically 25 years.

Sacrificial anodes are designed to provide the expected current demand of the system over the expected life. As part of the inspection plan, the anode condition and the cathodic potential are checked. Depending on the remaining anode material, replacement may be required. Figure 40, in Section 4.3.10, shows a series of drawings of sacrificial anodes with increasing loss of material. A partially depleted anode from the West Sole WE platform is shown in Figure 12. At some stage, such as the severely pitted example in Figure 40, the loss of material results in a less negative protection potential and therefore less efficient cathodic protection. At this stage, anode replacement is required to maintain efficient protection.

Individual anodes can be replaced, but this can involve a considerable amount of underwater working, often requiring divers. ABS has produced guidance on retrofitting of sacrificial anodes (ABS 2018). These include:

- Hanging impressed current anode systems; these have been successfully applied to offshore floating storage vessels which had originally used sacrificial anodes for the hull. In this system the anode is freely suspended from locations above water either from the feed cable or a strain member. The anode is often weighted and contained within a frame. This can be a viable approach for shallow water structures with a relatively short life expectancy (< 5 years) and a moderate to high current demand. Anode failure is expected on a regular basis but with very low replacement cost.
- Gravity anode sleds, which are designed to sit on the sea floor at some distance from the structure and are connected electrically to the protected structure.
- Buoyant anode sleds, which are high-capacity impressed-current cathodic protection systems. They utilise impressed-current titanium-anode rods housed in buoyant floats. These are similar to gravity sleds but with the anode elements held up in the seawater by means of buoys. The advantage of the buoyant sled is that the critical elements can move freely if hit by falling debris and the overall sled structure is much lighter.
- Anode pods, which are aluminium anode systems arranged in stable, self-contained "pods" and are ideal for replacing depleted anodes on mature assets. The pods are deposited on the seabed and connected electrically to the protected asset using a clamp tieback system to any local tubular member or flange. However, this system can result in less efficient spread of the protection compared to adding conventional individual anodes.
- Anode links, which are cost-effective cathodic protection retrofit systems consisting of 3 to 15 anodes cast directly onto a heavy-duty wire rope. The anode link is attached electrically and mechanically above the waterline allowing the string to hang in the seawater with at least two anodes situated in the mud. This solution is typically suitable for any structure in less than 30 meters (90 feet) of water. Anode links are available to increase cathodic protection capacity by securely fitting around pipes. The two half segments can be welded together (most likely above water) or mechanically using bolted connections. Each segment should be electrically connected to the pipe (Bahadori 2014).

As earlier mentioned, corrosion protection systems are designed with an expected life commensurate with the design life. Life extension, therefore, may require assessment of these systems and consideration of any required mitigation to meet the extended life.

8.8 Repair of Mooring Systems

Repair is required for mooring systems with significant damage as indicated in Section 3.8.4, Table 18. The most common types of damage requiring repair are fatigue cracking and corrosion. However, also overload and mechanical damage, particularly at the fairlead has been reported as

Figure 114 Kenter shackle. *Source:* HSE 2006.

causes leading to repair or replacement. In recent years, the increasing use of high-strength chain (for example R5) has led to several cases of brittle fracture.

The typical repair methods for damaged mooring lines are:

- replacement of a section length (chain or wire rope);
- replacement of individual chain links (e.g. Kenter links);
- stud tightening or mechanically refitted (in studded chain); and
- grinding of chain links to remove cracks, gouges and other surface defects (excluding weld cracks) provided that the resulting reduction in link diameter does not exceed 5% and the cross-sectional area, due to abrasion, wear and grinding is at least 90% of the original nominal area (IACS 2010).

Weld repair is not allowed (IACS 2010) on the higher-strength chains, such as the now common R4 and R5 chains.

The term "discard criteria" is a useful guide for the stage at which to replace a mooring line. Guidance is provided in for example API RP-2I (API 2008), based on the degree of damage such as cracks in chains and broken wire rope. Where chain links have become significantly damaged and have met the discard criteria, IACS (2010) states that such links should be removed and replaced with joining shackles, such as a Kenter-link as shown in Figure 114. These shackles should be capable of passing through fairleads and windlasses in the horizontal plane. It is also recognised that these joining shackles have much lower fatigue lives than ordinary chain links and, hence, as few as possible should be used. On average it is recommended that such shackles should be separated by at least 122 m.

8.9 Repair of Concrete Structures

8.9.1 Introduction

Repair is typically required for damage as described in Chapter 3 and when strengthening of the structure is needed. These typical damage types to concrete offshore requiring repair include:

- corrosion of reinforcement leading to cracking, spalling and delamination (splash zone);
- vessel impact and dropped object damage;
- insufficient design or fabrication (detailing, material, cover to reinforcement); and
- corrosion around embedded steelwork, including prestressing anchorages.

Some of the defects observed on offshore concrete structures are reported to result from the breakdown of repairs from the time of construction. This may indicate that the long-term properties of these had not been optimal. This may be related to the workmanship, method and material for these repairs at the time of construction.

The repair of concrete structures underwater represents a challenging task, particularly due to the harsh environmental conditions associated with working underwater and in the splash zone. In order to determine the need for repair, a thorough evaluation or assessment of the present condition of the structure (with the damage present) is an essential first step. This should be based on comprehensive documentation of the cause and extent of damage in addition to the relevant strength requirements for the repaired structure. If repair is found to be needed, a broad evaluation of possible repair techniques should be undertaken to determine the most effective solution. An assessment should also be made whether a temporary strengthening is needed until the repair is complete.

In addition, proper quality assurance of the repair work is important to provide a sufficient repair. Repair procedures (including equipment) should be verified using full-scale tests that represent relevant conditions such as environment, size and temperature. These tests should also be used to qualify personnel to perform the job.

The American Concrete Institute (ACI) has prepared a guide to underwater repair of concrete (ACI 1998) which covers an overview of typical damage types, inspection methods, evaluation of this damage, preparation for the repair, repair materials and methods in addition to inspections of the completed repairs. This has together with the previous studies presented in Section 8.4 formed the basis for this chapter.

Another ACI report (ACI 2016) has also been developed to provide design engineers with a recommended practice for the evaluation of the damage and deterioration for concrete structures (primarily land-based structures). In addition to the design of appropriate repair and rehabilitation strategies, this code provides minimum requirements for the assessment, repair and rehabilitation of existing structural concrete buildings including structural components. Similarly, a European standard (EN 2005) for concrete repair of land-based concrete structures specifies requirements for the identification, performance (including durability) and safety of products and systems to be used for the structural and non-structural repair of concrete structures. It covers repair mortars and concretes and systems to replace defective concrete and to protect reinforcement, necessary to extend the service life of a concrete structure showing deterioration. Its applicability to offshore structures is limited.

Typical repair methods for concrete structures related to these types of damage include:

- concrete material replacement (cementitious or epoxy):
  - spraying (in splash zone and dry areas);
  - hand mortaring;
  - concrete casting into formwork by use of grout or pre-placed (pre-packed) aggregate followed by cementitous material injection.
- injection methods (e.g. cracks or voids);
- re-instating damaged reinforcement and prestressing; and
- steel sleeve repair (normally not useful for large-scale offshore components).

The various repair methods require preparation of the base material or substrate in order to obtain a good quality repair. This includes for example removal of chloride-infected concrete and also the removal of concrete attached to the reinforcement to achieve good adhesion. Previous studies on repairs have concluded that published information comparing long-term properties of different repair methods under similar conditions (HSE 2003 a, b, c and d) is limited.

Table 79 Causes and Consequences of Damage and Underwater Repair Solutions.

Cause of damage	Consequence of damage	Repair solution
Vessel impact	Water seeping through cracked wall	Epoxy resin and caulking used to seal damaged area externally; a 200 mm section of wall was removed internally and recast. A coffer dam was placed against the outer wall and the outer 200 mm of wall then removed and recast. Resin injection points were cast into the new concrete to ensure a good bond between the old and new material.
Faulty construction joint	Leak manifested several years after construction. Repaired 10 years after construction.	Initial repair attempts using resin injection were unsuccessful. Optic fibre examination showed voids within 1200 mm wall. Voids were injected with cementitious grout, with epoxy grout being used for the final contact with the old concrete.
Leaky grouted prestressing duct	Water leakage (depth of 90 m)	Stopped by injecting resin from the dry end.
Dropped object	Damage to a 500 mm thick cell roof slab at a depth of 80 m.	Deep hole (300 mm in depth) was formed in the concrete, with water flowing through slab. Repair made by prepacking aggregate within the hole, covering the hole with a steel plate and injecting grout to restore the original concrete profile.
Cracks in external shear walls	Water ingress	Resin was injected into cracks. In some cases, the crack was jacked apart before the resin was placed so that on removal of the jack the resin was compressed into the crack void. Ballast was added to the structure to prevent the cracks reopening under wave loading
Scour hole at pipeline entry point	Hole extending 4 m beneath the structure	Void was grouted and subsequently protected by rock dumping.

An overview of underwater repairs to offshore concrete structures is shown in Table 79 and also as reported by OceanStructures (2009) in Table 15.

The following sections present an overview of the most common repair methods used on offshore concrete structures. Each repair method is evaluated in relation to its maturity, quality, reliability and long-term properties.

8.9.2   Choice of Repair Method

In selecting a repair method there should be a well-defined requirement for the properties to be achieved, such as capacity or durability. The repair solution, including the method and materials, should be selected based on engineering requirements, the cause of the defect, performance requirements for the completed repair and previous experience with the method proposed and its complexity. The following options are available when damage is found:

- monitor for a given period before considering repair;
- prevent or reduce further degradation;
- re-analyse the design capacity, with possible downgrading of the function;
- reinforce or repair; and
- rebuild or replace all or part of the concrete structure.

Qualification of the proposed method and personnel prior to planned repair work is necessary. The reason why extra focus is given to design for the repair methods for concrete structures, versus steel structures, is because it is more difficult to test quality and workmanship after the repair (DNVGL 2019). Hence, sufficient time should be made available for such preparation and quality control. Large-scale testing to verify the method may be required together with some initial materials testing as part of the development depending on the method selected, its maturity and repair material. The inspection and control regime for repair work should comply with the requirements of the standard used in the original design.

8.9.3 Concrete Material Replacement

Damaged concrete material or concrete that covers corroded reinforcement may have to be removed and replaced. This can involve chiselling, pneumatic hammer, high-pressure water jets or pneumatic or hydraulic-powered saws to remove the relevant material and different ways of replacing the material either by cementitious materials (hand mortaring, casting and dry and wet spraying) or by the use of epoxy resins. An important consideration is that the concrete structure itself is usually subjected to permanent compressive forces due to self-weight and prestressing during the execution of the repair and as a result, large repairs may not contribute to carry these permanent loads.

Due to the difficulty of the use of these methods proper planning and prior testing are normally required. The repair should include (DNVGL 2019):

- removal of marine growth, corrosion products, loose or chloride-infected concrete;
- the boundary of the spalled area should be cut back to a depth of about 10–20 mm depending on the amount of damage;
- cleaning and surface preparation;
- avoidance of spillage on nearby areas prepared for repair, for example when using spraying;
- operating conditions such as temperature and humidity and for spraying the orientation on the base material (vertical surface or under a horizontal surface) and the thickness of repair (possible need for layering); and
- possible finishing by epoxy application over cementitous repairs.

These repair methods may be used to fill cavities, cracks and other damage, such as impact damage to the domes of the roofs of underwater concrete storage tanks, as described in OTH 90 318 (Department of Energy 1990). This study also assessed the efficiency of different materials for repairing this type of damage.

The OTH 90 318 report and the report by DNVGL (2019) listed the typical properties of repair materials for cementitous and epoxy types of repair needed to be considered in the selection of suitable material. The key factors with respect to the repair's design and durability and the integrity of the structure after repair are:

- operating time and pumpability;
- compatibility with basic concrete material (e.g. temperature coefficient of expansion, permeability);
- low permeability to seawater;
- suitability for the environment (temperature limits, relative humidity, exposure);
- volume stability (shrinkage);
- long-term properties (environmental and cyclic loading);
- adhesion to base material; and
- strength and strength development in tension and compression.

There are many available mortar products suitable for repair of concrete structures. According to DNVGL (2019), cement-based materials which can be used underwater include pre-mixed, packaged or mortar mixed at the workplace and methods mostly relevant for above water and possibly in the splash zone such as hand-applied mortar, dry sprayed mortar mixture and wet spraying of a concrete mix. The compressive strength of the repair material should be determined using core samples. It is also recognised to be difficult to achieve the same strength as the base material (particularly for high-strength concrete) if a full load-bearing repair is needed.

Surface treatment techniques to remove damaged material prior to repair commonly used on concrete offshore structures are water or mechanical chiselling. Several studies and publications have previously documented that water chiselling generally produces less micro-scratches in the surface than mechanical chiselling and is therefore preferable for achieving good adhesion to the base material (HSE 2003c). Water chiselling is, therefore, a preferred alternative for achieving good adhesion to the base material as well as durability. In addition, water chiselling provides an uneven surface which is favourable for adhesion to the base material.

The typical methods for applying cementitious materials are shown and described in Table 80. Underwater concrete casting type of repair can be performed by using:

- a cofferdam installed around the damage;
- sealing the outer surface of the damage with, for example, epoxy and then the casting can be performed from inside a dry tower; and
- injection into formwork as shown in Figure 115.

Table 80 Typical Methods for Applying Cementitious Materials.

Method	Description
Spraying	Spraying is a widely used method for repair of offshore concrete structures above water or on internal shafts. The method is often used where there are defects such as peeling and delamination in the concrete due to reinforcement corrosion which often occurs in areas where there have already been defects from construction. Spraying is an effective method of repairing larger areas with extensive damage, typically on the shafts, where access is difficult. In many cases, dry spraying is the chosen repair method above water and preferably above the splash zone, partly because of its positive historical experience and its operational advantages over other methods.
Hand mortaring	Hand mortar is a repair method that is widely used in a number of repairs, particularly above water or on internal shafts. For repairs to existing concrete structures this method is mainly used for smaller and more individual defects. The quality of the repair by this method is very dependent on the craftsmanship, especially for upper horizontal surfaces. The advantage of this method, compared to dry spraying, is that the pre-packaged dry mix is mixed with a certain amount of water before application, thus allowing better control over the properties and quality of the applied material. Compacting the material through the full thickness of the repair can be challenging and the method should therefore be subject to close control during execution. In preparation for repair using hand mortaring, a hand chisel is often used to remove for example loose and chloride-infected concrete to prepare for the mortar repair.
Concrete casting using formwork	Concrete casting with formwork is often used for extensive repairs associated with deep defects and multiple defects with a significant amount of exposed reinforcement. It can be used underwater, in the splash zone and above water.
	Concrete casting can be combined with additional reinforcement installation and prestressing. Prestressing is used if there is a need to ensure that the concrete repair is to be subjected to compressive forces and to satisfy water-tightness requirements.

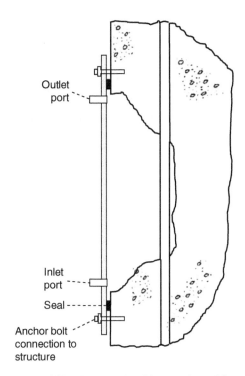

Figure 115 An example of formwork used for concrete repair.

A liquid and pumpable concrete or mortar is required. In the past, a "two-stage" process was used, involving aggregate being installed in a mould and a mortar being pumped in to fill the void to form a monolithic repair. More recently premixed micro-cement concrete, often with rough aggregate, has been used for repairs with formwork (DNVGL 2019). Due to its large size, this type of repair will typically place high demands on compatibility with the surrounding structure, pre-treatment and roughness in castings and factors such as modulus of elasticity and temperature properties.

8.9.4 Injection Methods

Injection of a cementitious or epoxy material is a versatile and historically widely used repair material on offshore concrete structures. The material is injected into voids such as cracks, fabrication joints (to seal a leaking joint), delaminations, cavities in walls and in prestressing cable ducts with incomplete filling or leaks. According to OTH 90 318 (Department of Energy 1990), cementitious grouts were shown to have difficulty in penetrating cracks of surface width of 10 mm and penetration was impossible below 5 mm. Hence, for cracks less than 10 mm in width, epoxy resins would be the repair material of choice.

A liquid two-component epoxy may be applied through hoses connected to injection tubes or nipples, which are placed in the concrete to fill voids which will be too narrow to enable a cementitious material filling to be effective. Epoxy material is not typically used to fill voids in large volumes due to challenges with heat generation during the curing of the epoxy, creep properties as well as constraints in the material's shelf life.

There are a number of limitations in using epoxy injection as a repair method. When epoxy is injected into delaminations in walls or into through-thickness cracks, it can be challenging to seal the surfaces, typically with an epoxy resin before injection starts. It can also be challenging to completely fill the void. For example, on a vertical surface the injection begins at the lowest point and continues until the epoxy begins to ooze out of the port above, which is the visual sign that the void has been filled to that level. Failure to seal the surface may result in loss of epoxy material and compromise the efficiency of the operation. Other issues concerning the use of epoxy materials are:

- limited cure life for some of the materials on the market limiting the use to only small volume cavities;
- Influence of temperature can have a significant impact on the material's cure life; this is important underwater with typical North Sea temperatures;
- the influence of relative humidity can have a significant impact on the properties of the final product; and
- injection at too high pump pressures can cause defects in the concrete.

Epoxy coatings are widely used on offshore concrete structures for surface protection, particularly in the splash zone and above water. The primary use is to improve the durability of the concrete in areas with low cover, fabrication joints and penetrations as well as protection of completed repairs. The typical applications and experience with epoxy as surface protection are (DNVGL 2019):

- Full-coverage coating applied as a repair method. There exists good experience with such coatings to minimise chloride penetration. As a result, a complete membrane has been subsequently applied to, for example, the Troll A platform after 10 to 15 years in operation. Similar full-coverage coatings were applied to the Statfjord A platform during construction with good results, as discussed earlier.
- Local application of a coating is often used in smaller areas to compensate for low concrete cover and to protect the repair material, steel attachments, penetrations and as a membrane for any repairs performed.

Surface protection with an epoxy coating is sensitive to the design and material selection as well as operational temperature and relative humidity. Adequate testing of the method and material should therefore be performed prior to the application. In addition, cleaning and preparation of the concrete surface is essential for achieving good adhesion. The application of epoxy coatings often requires specialist contractors with a good knowledge of both materials and workmanship.

8.9.5 Repair of Reinforcement and Prestressing Tendons

Corroded reinforcement may in many cases require repair. Removal of the corrosion products can be performed with high-pressure water jets and abrasive blasting, but the back surfaces of the reinforcement can be challenging to clean. If the cross section of the reinforcing steel has been reduced significantly, the addition of new reinforcing bars may be necessary. This will require the original reinforcement to be exposed beyond the corroded section by a distance equal to the required design length for the lap splice. New reinforcement can be added by splicing (e.g. by tying) or welding new bars to the undamaged material, typically with an overlap of at least 100 mm for welded connections and more for tied bars. Welding underwater will often be difficult for reinforcement.

Major impact damage may also include exposure of the reinforcement which may have experienced damage requiring similar repair. In addition, prestressing tendons may also be exposed and damaged. As mentioned in Section 8.4.2, there are three different methods where fractured prestressing tendons are replaced (Department of Energy 1988c). The first involves the use of "flat jacks" which produce a similar effect in maintaining compressive stresses in the concrete. The second method utilises Macalloy bars which can accommodate the tensile stresses in the ultimate load condition. The third method uses a steel "splint" which is capable of taking tensile loads.

8.9.6 Summary of Concrete Repairs

Typical repair systems for offshore concrete platforms are shown in Table 81.

Table 81 Anomalies and Possible Evaluation Methods for Concrete Structures.

Cause of damage	Types and description	Suggested repair method
Corrosion	Spalling and delamination (above water and in the splash zone) as result of reinforcement corrosion as well as potential loss of reinforcement area	Removal of concrete cover to enable cementitious mortars to be used to fill the damaged area. Sandblasting of reinforcement to remove any corrosion products and a bonding coating can be applied to the damaged area.
		Above water, grout spraying (wet and dry) potentially followed by introducing an epoxy coating can be used. Alternatively, re-casting concrete with or without additional reinforcement is an option for larger areas of damage.
		In the splash zone, water-tolerant epoxy mortars can be suitable for small areas of damage. For larger areas of damage, formwork may be required to hold the repair material (grout or cement) in position whilst it is setting.
	Seepage of corrosion products at cracks under water as result of reinforcement corrosion	Removal of concrete cover, sandblasting of reinforcement and replacement of material. CP system should also be considered for improvement.
	Chloride ingress above water causing reinforcement corrosion and potential loss of reinforcement area	Removal of concrete cover, sandblasting of reinforcement and grout spraying (wet and dry), potentially followed by introducing an epoxy cover
	Corrosion of steel attachments (e.g. pre-tension cable at anchorages, steel support attachments) and potential loss of prestressing capacity	For minor damage, sand blasting to remove corrosion and epoxy filling. For major damage involving potential loss of local prestressing capacity, flat jack, Macalloy bars and steel splint repair may be required. Reinstall of the anchorage may be also required.
Overload and accidental damage (vessel impact and dropped object)	Local impact damage to towers resulting in reduction in concrete section area and possible water ingress	Sealing (epoxy resin) of damaged area if minor. For larger damage, removal and replacement of the concrete by use of cementitous materials is required. Repair of damaged reinforcement and prestressing may also be required.
	Local impact damage to tops of storage cells from dropped objects resulting in reduction in concrete section area and possible water ingress	Replacement of concrete by injecting grout or cementitious material into formwork (with or without preplaced aggregate) to restore the original concrete profile. Typical formwork could be a steel plate covering the damaged area (as shown Figure 115.
		If reinforcement is damaged, treat as mentioned above.

(Continued)

Table 81 (Continued)

Cause of damage	Types and description	Suggested repair method
Wear and tear	Abrasion (local wear) and erosion resulting in minor reduction in concrete section area	Filling of the area by cementitous or epoxy material if needed.
Marine growth	Increased loading	Marine growth cleaning by water jet. Care needs to be taken not to damage the concrete structure.
Settlement and subsidence	Settlement and subsidence	Removal of lower decks or lower deck equipment to reduce loading.
Scour and build-up of drill cuttings	Scour around the base of the platform possibly leading to instability	Sand or grout filling of voids followed by rock dumping to protect the repair.
Material deterioration	Loss of the concrete material due to aggressive agents (sulfate, chloride)	Removal and replacement of infected material and possibly add epoxy coating.
	Cracking due to shrinkage (after construction)	Sealing of cracks by epoxy or grout injection.
	Cracking, scaling and crumbling due to freeze and thaw (above water)	Sealing of cracks by epoxy or grout injection Re-casting concrete (with or without additional reinforcement)
	Strength loss of the concrete material due to sulfate producing bacteria (primarily in oil-storage tanks)	Often this anomaly is accepted partly due to the enormous difficulty in accessing the interior of the tank. However, no real data exists on the strength loss due to sulfate reducing bacteria.
Fabrication fault	Defective construction joints, minor cracking, surface blemishes, remaining metal attachments, patch repairs from construction, low cover to the reinforcement possibly leading to spalling in the splash zone and possible water ingress	In case of significant water leakage from defective construction joints, repair by injecting epoxy filling or cementitious grout may be required. The other types of faults may require repair in line with similar damage types mentioned above in this table.
Under-design	Cracking due to low reinforcement or prestressing, crushing of concrete in rare cases, failure in the junction between shafts and cells	Repair may be needed if loss of strength is significant using techniques listed above.

8.10 Overview of Other Mitigation Methods

As shown in Figure 61 in Section 7.1 there were several possible outcomes of an evaluation and assessment outlines. These included:

- acceptable as-is;
- specific inspection;
- load reduction;
- strengthening or repair; and
- change of operational procedures.

In addition to inspection, load reduction, strengthening and repair, there are other mitigation methods that can fall under the heading of "change of operational procedures". These are often used when strengthening and repair are difficult or too expensive or a short-term solution is needed. These mitigation methods are generally classified in standards such as API RP-2SIM (API 2014), NORSOK N-006 (Standard Norge 2015) and NORSOK N-005 (Standard Norge 2017). These methods generally consist of exposure reduction (exposure to personnel and the environment) and likelihood reduction. The following mitigations can be used for exposure reduction:

- production shut down in severe events;
- de-manning (temporarily in forecasted events or permanently);
- operational limitation changes linked to redefined safe operating conditions (e.g. visiting vessel sizes, sea-states, weight, centre of gravity, draft, trim, ballast, mooring pretension);
- reduction of the environmental consequences of a structural failure (including such as sub-surface safety valves, reducing hydrocarbon storage, removing or re-routing gas flow lines, plugging of abandoned wells); and
- storm and hurricane preparedness (evacuation prior to severe events, safe shut in, secure loose objects and equipment that could become projectiles).

In some standards, post-event assessment by identifying and inspecting critical members and joints are mentioned as a possible mitigation. For likelihood reduction such standards typically describe inspection, load reduction, strengthening and repair activities as described earlier.

Regulation in some parts of the world allows for the use of risk reduction processes based on an ALARP approach (As Low As Reasonable Practicable). In this, a comparison is made between the benefits of additional risk reduction measures beyond the minimum required by regulation and the cost of undertaking these measures. This approach is relevant to the assessment of existing damaged structures. For example, depending on the circumstances de-manning under severe storm conditions would represent an appropriate risk reduction.

Decommissioning or re-use of installations is a final option if repair, strengthening and other mitigation methods fail to provide a safe structure. Often the platform will stop production and be left in a "cold-state" or "light-house mode" until the actual decommissioning takes place. In this "cold-state", inspection and repair may still be necessary as the structure needs to retain sufficient integrity to allow for a safe removal. Re-use includes, for example, the addition of wind turbines onto the structure which is being considered. The US rigs to reef programme made use of toppled structures on the seabed to create a habitat for marine life.

8.11 Bibliographic Notes

Section 8.6 is to some extent based on DNVGL report for Petroleum Safety Authority (DNVGL 2019) and MSL (2004). Section 8.9 is to some extent based on ACI (1998) and DNVGL (2019).

References

ABS (2018), "Guidance Notes on Cathodic Protection of Offshore Structures", American Bureau of Shipping, New York, US.

ACI (1998), "Guide to Underwater Repair of Concrete ACI 546.2R-98", Reported by ACI Committee 546. American Concrete Institute.

ACI (2016), "ACI 562-16 Code Requirements for Assessment, Repair, and Rehabilitation of Existing Concrete Structures and Commentary", American Concrete Institute.

API (2008), *API RP-2I In-Service Inspection of Mooring Hardware for Floating Structures*, Third Edition, American Petroleum Institute, 2008.

API (2014), API RP-2SIM *Recommended Practice for Structural Integrity Management of Fixed Offshore Structures*, American Petroleum Institute, 2014.

Atteya, M., Mikkelsen, O., Dimitrios, G.P. and Ersdal, G. (2020), "Crack Arresting with Crack Deflecting Holes in Steel Plates", In Proceedings of the ASME 39th International Conference on Ocean, Offshore and Artic Engineering (OMAE 2020), Fort Lauderdale, Florida, US. Paper number OMAE2020-19358.

Bahadori, A. (2014), *Cathodic Corrosion Protection Systems—Guide for Oil and Gas Industries*, Gulf Professional Publishing, Elsevier Waltham, Massachusetts, US.

Belzona (2020), "Solutions for Splash Zones", https://www.belzona.com/en/focus/splash_zones.aspx, Accessed on 23rd June 2020.

Browne, R.D. (1993), "Underwater Repair", In Allen, R.T.L., Edwards, S.C. and Shaw, D.N. (1993), *Repair of Concrete Structures*, CRC Press.

Bruin, W.M. (1995), "Assessment of the Residual Strength and Repair of Dent-Damaged Offshore Platform Bracing", Master Thesis presented at Lehigh University, Bethlehem, Pennsylvania, US.

BSSE (2020). https://www.bsee.gov/site-page/master-list-of-tap-projects-0 . Visited 29th June 2020.

BSI (1984), *BS 6319-8:1984 Testing of Resin and Polymer/Cement Compositions for Use in Construction. Method for the Assessment of Resistance to Liquids"*, British Standardisation Institute. London, UK.

BSI (1990), *BS 6319-3:1990 Testing of Resin and Polymer/Cement Compositions for Use in Construction. Methods for Measurement of Modulus of Elasticity in Flexure and Flexural Strength*, British Standardisation Institute, London, UK.

BSI (2005), *BS 5400-5 Steel, Concrete and Composite Bridges—Code of Practice for Design of Composite Bridges*, British Standardisation Institute, London, UK.

Campbell-Allen, D. and Roper, H. (1991), *Concrete Structures—Materials, Maintenance and Repair*, Longman Singapore Publishers, Singapore.

Clarke, J.D., Sharp, J.V. and Bowditch M. (1986), "An Underwater Adhesive-Based Repair Method for Offshore Structures", Conference on Inspection, Maintenance and Repair, Aberdeen, 1986.

Clarke, S. (2020), Private communication.

Department of Energy (1977), *First Edition Guidance Notes for the Design and Construction of Offshore Structures*, Department of Energy, HMSO, London, UK.

Department of Energy (1984), *Third Edition Guidance Notes for the Design and Construction of Offshore Structures*, Department of Energy, HMSO, London, UK.

Department of Energy (1985), *Offshore Technology Report OTH 84 202 Grouted Repairs to Steel Offshore Structures,* Performed by Wimpey Laboratories for the Department of Energy. HMSO, London, UK.

Department of Energy (1987a), *Offshore Technology Report OTH 87 259 Remaining Life of Defective Tubular Joints—An Assessment based on Data for Surface Crack Growth in Tubular Joints Tests*, Prepared by Tweed, J.H. and Freeman, J.H., 1987 for the Department of Energy, HMSO, London, UK.

Department of Energy (1987b), *Offshore Technology Report OTH 87 278 Remaining Life of Defective Tubular Joints: Det of Crack Growth in UKOSRP II and Implications*, Prepared by Tweed, J.H. for Department of Energy, HMSO, London, UK.

Department of Energy (1988), *Offshore Technology Report OTH 88* 283 *Grouted and Mechanical Strengthening and Repair of Tubular Steel Offshore Structures*, Prepared by Wimpey Offshore Engineers and Constructors LTD (R.G. Harwood and E.P. Shuttleworth) for the Department of Energy. HMSO, London, UK.

Department of Energy (1988b), *Offshore Technology Report OTH 88 289 Grouts and Grouting for Construction and Repair of Offshore Structures—A Summary Report*, HMSO, London, UK.

Department of Energy (1988c), *Offshore Technology Report OTH 87 250 Repair of Major Damage to the Prestressed Concrete Tower of Offshore Structures*, Prepared by Wimpey Laboratories for the Department of Energy, HMSO, London, UK.

Department of Energy (1989), *OTH 89 307 Fatigue Performance of Repaired Tubular joints*, Prepared by the Welding Institute (P.J. Tubby) for the Department of Energy, HMSO, London, UK.

Department of Energy (1989b), *Offshore Technology Report OTH 89 314 Residual and Fatigue Strength of Grout Filled Damaged Tubular Members*, HMSO, London, UK.

Department of Energy (1989c), *OTH 87 248 Concrete in the Ocean Programme—Coordinating Report on the Whole Programme*, HMSO, London, UK.

Department of Energy (1990), *Offshore Technology Report OTH 90 318 Assessment of Materials for Repair of Damaged Concrete Underwater*, Prepared by University of Dublin (S.H. Perry) and Imperial College of Science and Technology (J.M. Holmyard) for the Health and Safety Executive (HSE), HMSO, London, UK.

Department of Energy (1992), *Offshore Technology Report OTH 92 368 Fatigue Life Enhancement of Tubular Joints by Grout Injection*, HMSO, London, UK.

Department of Energy (1992b), *OTH 89 298 Scaling of Underwater Concrete Repair Materials*, Prepared by University of Dublin (S.H. Perry) and Imperial College of Science and Technology (J.M. Holmyard) for the Health and Safety Executive (HSE), HMSO, London, UK.

DNVGL (2015), *DNVGL-RP-C301 Design, Fabrication, Operation and Qualification of Bonded Repair of Steel Structures*, DNVGL, Høvik, Norway.

DNVGL (2016), *Recommended Practice DNVGL-RP-C203 Fatigue Design of Offshore Steel Structures*, DNVGL, Høvik, Norway.

DNVGL (2019), *Reparasjonsmetoder for bærende konstruksjoner*, Report for the Norwegian Petroleum Safety Authority, DNVGL, Høvik, Norway (in Norwegian).

El-Reedy, M.A. (2019), *Assessment, Evaluation, and Repair of Concrete, Steel, and Offshore Strucutres*, CRC Press, Taylor & Francis Group, Boca Raton, Florida, US.

Ellinas, C.P. (1984), "Ultimate Strength of Damaged Tubular Bracing Members", *Journal of Structural Engineering*, ASCE, Vol. 110, No. 2.

EN (2005), *EN 1504-3:2005 Products and Systems for the Protection and Repair of Concrete Structures—Definitions, Requirements, Quality Control and Evaluation of Conformity—Part 3: Structural and Non-Structural Repair*, European Standardisation organisation.

Energo (2006), "Assessment of Fixed Offshore Platform Performance in Hurricanes Andrew, Lili and Ivan", MMS Project no 549, Energo 2006.

Energo (2007), "Assessment of Fixed Offshore Platform Performance in Hurricanes Katrina and Rita", MMS project no 578, Energo 2007.

Energo (2010), "Assessment of Fixed Offshore Platform Performance in Hurricanes Gustav and Ike", MMS project no 642. Energo 2010.

Equinor (2019), "Integritetsstyring på en aldrende FPSO med begrenset POB - Utvikling av sprekker, samsvar med analyser og utfordrende utbedringer", Presentation at PSA Structural Meeting, August 2019 (In Norwegian).

Fisher, John W., Barthelemy, B. M., Mertz, D. R., and Edinger, J. A. (1980), "Fatigue Behavior of Full-Scale Welded Bridge Attachments", National Cooperative Highway Research Program (NCHRP) Report 227, National Transportation Research Board, Washington, DC, US.

Fisher, John W., Jin, Jain, Wagner, David C., and Yen, Ben T. (1990), "Distortion-Induced Fatigue Cracking in Steel Bridges". National Cooperative Highway Research Program (NCHRP) Report 336, National Transportation Research Board, Washington, DC, US.

Gibson, D., Paculba, N. and Grey, I. (2010), "Friction Stud Welding Underwater in the Offshore Oil and Gas Industry", In *International Workshop on the State of the Art Science and Reliability of Underwater Welding and Inspection Technology*, Edited by Liu, S., Olson, D.L., Else, M., Merritt, J. and Cridland, M. November 17–19, 2010. ABS, Houston, Texas, US.

Haagensen, P.J. and Maddox, S.J. (2006), "IIW Recommendation on Post Weld Improvement of Steel and Aluminium Structures", IIW Report XIII-1815-00, International Institute of Welding.

Haagensen, P.J. (1994), *Methods for Fatigue Strength Improvements and Repair of Welded Structures,* OMAE1994, ASME.

Hamakareem, M.I. (2020), "Repair of Underwater Concrete Structures—Methods and Procedure", *The Constructor*, https://theconstructor.org/concrete/repair-underwater-concrete.

Hart, P.R., Richardson, I.M. and Nixon, J.H. (2001), "The Effects of Pressure on Electrical Performance and Weld Bead Geometry in High Pressure GMA Welding", *Welding in the World* 45 (11-12) pp 25–33, London, UK.

Hebor, M.F. (1994), Residual Strength and Epoxy-Based Grout Repair of Corroded Offshore Tubular Members, Lehigh University, Bethlehem, Pennsylvania, US.

Hoffmeister, H., Küster, K. and Schafstall, H-G. (1983), "Weld Joint Properties of Medium Strength Steels after Underwater Wet MIG-Welding by the Water Curtain Process", Proc. Second Int. Conf. on Offshore Welded Structures, 16–18 November 1982, London, UK, Published by The Welding Institute, 1983, pp. 17-1–17-8.

HSE (1997), *Offshore Technology Report OTO 96 030 A Review of Adhesive Bonding for Offshore Structures*, Prepared by Professor M.J. Cowling, Glasgow Marine Technology Centre, Glasgow University for the Health and Safety Executive (HSE), HMSO, London, UK.

HSE (1997), *Offshore Technology Report OTO 96 057 Repairs of Offshore Installations—Background to Section 60 of the Guidance Notes*, Prepared by Techword Services for the Health and Safety Executive (HSE), HMSO, London, UK.

HSE (2002a), *Offshore Technology Report 2000/057 Assessment of Strengthening Clamp from the Viking Offshore Platform: Phase I*, Prepared by the Health and Safety Laboratory for the Health and Safety Executive (HSE), HMSO, London, UK.

HSE (2002b), *Offshore Technology Report 2000/058 Assessment of Strengthening Clamp from the Viking Offshore Platform: Phase II*, Prepared by the Health and Safety Laboratory for the Health and Safety Executive (HSE), HMSO, London, UK.

HSE (2002c), *Research Report 031 Development of Design Guidance for Neoprene-Lined Clamps for Offshore Application Phase II,* Prepared by MSL Engineering Limited for the Health and Safety Executive (HSE), London, UK.

HSE (2003a), *RR 175 Field Studies of the Effectiveness of Concrete Repairs Phase 1 Report: Desk Study and Literature Review*, Prepared by Mott MacDonald Ltd for the Health and Safety Executive (HSE), HMSO, London, UK.

HSE (2003b), *RR 184 Field Studies of the Effectiveness of Concrete Repairs Phase 3 Report: Inspection of Sites, Sampling and Testing at Selected Repair Sites*, Prepared by Mott MacDonald Ltd for the Health and Safety Executive (HSE), HMSO, London, UK.

HSE (2003c), *RR 185 Field Studies of the Effectiveness of Concrete Repairs Phase 3a Report: An Investigation of the Performance of Repairs and Cathodic Protection (CP) Systems at the Dartford West Tunnel*, Prepared by Mott MacDonald Ltd for the Health and Safety Executive (HSE), HMSO, London, UK.

HSE (2003d), *RR 186 Field Studies of the Effectiveness of Concrete Repairs Phase 4 Report: Analysis of the Effectiveness of Concrete Repairs and Project Findings*, Prepared by Mott MacDonald Ltd for the Health and Safety Executive (HSE), HMSO, London, UK.

HSE (2003e), *RR053 Ship/Platform Collision Incident Database (2001)*, Prepared by Serco Assurance for the Health and Safety Executive (HSE), HSE Books, Sudbury, UK.

IACS (1996), *Shipbuilding and Repair Quality Standard 47, 1996 (Rev. 1, 1999) Part B*, International Association of Classification Societies (IACS).

IACS (2010), *Guidelines for the Survey of Offshore Chain Cable in Use, No. 38, 2010*. International Association of Classification Societies (IACS).

ISO (2016), *ISO 4628-3:2016 Paints and Varnishes—Evaluation of Degradation of Coatings*, International Organization for Standardization.

ISO (2020), *Petroleum and Natural Gas Industries—Fixed Steel Offshore Structures*, International Organization for Standardization.

Liu, S., Olson, D.L., Else, M., Merritt, J. and Cridland, M. (2010), "International Workshop on the State of the Art Science and Reliability of Underwater Welding and Inspection Technology", November 17–19, 2010. ABS, Houston, Texas, US.

Loh, J.T, Kahlich, J.L. and Broekers, D.L. (1992), "Dented Tubular Steel Members", EXXON Production Researh Company Internal Report, Houston, Texas, US.

Loh, J.T. (1991), "Grout-Filled Undamaged and Dented Tubular Steel Members", EXXON Production Research Company Internal Report, Houston, Texas, US,

Lotsberg, I. (2016), *Fatigue Design of Marine Structures*, Cambridge University Press, 2016.

MSL (1995), "Strengthening, Modification and Repair of offshore Installations—Final Report for a Joint Industry Project," MSL Document No. C11100R243, MSL Engineering Limited, MSL House, Sunninghill, Ascot, UK. Also published as MMS TAP project report no. 189.

MSL (1997a), "JIP—Strengthening, Modification and Repair Techniques for Offshore Platforms—Phase II—Demonstration Trials of Diverless Strengthening and Repair Techniques—Final Report", MSL Document no. C15800R025 Rev1, MSL Engineering Limited, MSL House, Sunninghill, Ascot, UK. Also published as MMS TAP project report no. 273aa.

MSL (1997b), "Development of Grouted Tubular Joint Technology for Offshore Strengthening and Repair—Draft Final Report", MSL Document no. C14100R020 Rev1, MSL Engineering Limited, MSL House, Sunninghill, Ascot, UK. Also published as MMS TAP project report no. 273ab.

MSL (1999), "Subsea Structural Repairs Using Composite Materials—Final Report for a Joint Industry Project", MSL Engineering Limited, Doc. Ref. C20200R008, June 1999.

MSL (2004), "Assessment of Repair Techniques for Ageing or Damaged Structures", Project #502. MSL Services Corporation, Egham, Surrey, UK.

MTD (1994), *Review of Repairs of Structures and Pipelines*, MTD report no. 94-102, Marine Technology Directorate, London, UK.

Nichols, N. and Harif, H.M. (2014), "Use of Platform Response Measurements from On-Line Monitoring (OLM) System to Verify the Effectiveness of Structural Repairs and Managing On-going Structural Integrity", OTC Asia Paper 24947-MS.

Nichols, N. and Khan, R. (2017), "Remediation and Repair of Offshore Structures", In *Encyclopedia of Maritime and Offshore Engineering*, Edited by John Carlton, Paul Jukes and Yoo-Sang Choo. John Wiley & Sons Inc.

Nixon J.H. (1995), *Underwater Repair Technology*, Gulf Professional Publishing, 1995.

Ocean Structures (2009), *Ageing of Offshore Concrete Structures-Report for Petroleum Safety Authority Norway*, Laurencekirk, Scotland.

Ostapenko, A., Berger, T.W., Chambers, S.L. and Hebor, M.F. (1996), "Corrosion Damage—Effect on Strength of Tubular Columns with Patch Corrosion", ATLSS Report No. 96.01. ATLSS Engineering Reseach Center, Fritz Engineering Laboratory Report No. 508.5, Lehigh University, Bethlehem, Pennsylvania, US. Also published as MMS report 277.

Paik, J.K. and Thayamballi, A.K. (2007), *Ship-Shaped Offshore Installations – Design, Building and Operation*, Cambridge University Press, Cambridge, UK.

Paik, J.K. and Melchers, R.E. (2008), *Condition Assessment of Aged Structures*, Woodhead Publishing.

Patterson, D. and Ricles, J.M (2001), "Structural Integrity Assessment of Corrosion-Damaged Offshore Tubular Braces Subjected to Inelastic Cyclic Loading", ATLSS Report No. 01-09, Lehigh University, Bethlehem, Pennsylvania, US. Also published as MMS report 101.

Reynolds, T.J. (2010), "Service History of Wet Welded Repairs and Modifications", In International Workshop on the State-of-the-Art Science and Reliability of Underwater Welding and Inspection Technology, November 17–19, 2010, Houston, Texas, US. Edited by Stephen Liu and David L. Olson.

Richardson, I.M. (1993), "The Influence of Ambient Pressure on Arc Welding Processes", Int. Institute of Welding Doc. 212-828-93, Proc. Of IIW Study Group 212 Seminar, pp. 43–68, Delftse Uitgevers Maatschappij.

Ricles, J., Gillum, T. and Lamport, W (1992), "Residual Strength and Grout Repair of Dented Offshore Tubular Bracing", ATLSS Report No. 92-14, Lehigh University, Bethlehem, Pennsylvania, US.

Ricles J. M., Bruin W. M., Sooi T. K., Hebor M. F. and Schonwetter P. C. (1995), "Residual Strength Assessment and Repair of Damaged Offshore Tubulars", OTC: Offshore Technology Conference Houston, Texas, US.

Ricles, J.M., Bruin, W.M. and Sooi, T.K. (1997), "Repair of Dented Tubular Columns—Whole Column Approach", ATLSS Reports, Lehigh University, Bethlehem, Pennsylvania, US. Also published as MMS TAP report 101.

Rolfe, S. T. and Barsom, John M. (1977), *Fracture and Fatigue Control in Structures: Applications of Fracture Mechanics*, First Edition, Prentice Hall.

Rowe, M. and Liu, S. (1999), "Global—JIP—Phase II Final Report", *Published as MMS TAP report* 263.

Sharp, J.V. (1993), "Strengthening and Structural Repair of Ageing North Sea Platforms: A Review", OMAE 1993.

Sharp J.V., Nixon J.H., Billingham J. and Richardson I. (1997), "A Review of the Technology of Deepwater Repairs for Offshore Structures", Boss Conference, 1997.

Sharp J.V., Stacey, A. and Birkinshaw, M. (2002), "Application of Risk Based Acceptance Criteria to Offshore Structural Integrity Management", ERA conference, London, 2002.

Simmons, G.G., "Fatigue Enhancement of Undersized Drilled Crack-Arrest Holes", PhD Thesis at University of Kansas, US.

SSC (1969), "Recommended Emergency Welding Procedure for Temporary Repairs of Ship Steels,". Ship Structure Committee report no 195. Washington, DC, US.

SSC (1993), "Underwater Repair Procedures for Ship Hulls (Fatigue and Ductility of Underwater Wet Welds)", Ship Structure Committee Report no 370, Washington, DC, US.

SSC (2000), "SSC-416 Risk-Based Life Cycle Management of Ship Structures", Ship Structure Committee Report no 416, Washington, DC, US.

SSC (2003), "Fatigue Strength and Adequacy of Weld Repairs", Ship Structure Committee Report no 425, Washington, DC, US.

SSC (2005), "Design Guidelines for Doubler Plate Repairs of Ship Structures", Ship Structure Committee Report no 443, Washington, DC, US.

SSC (2012), "Review of Current Practices of Fracture Repair Procedures for Ship Structures", Ship Structure Committee Report no 462, Washington, DC, US.

SSC (2015), "Strength and Fatigue Testing of Composite Patches for Ship Plating Fracture Repair", Ship Structure Committee Report no 469, Washington, DC, US.

Standard Norge (2015), NORSOK N-006 *Assessment of Structural Integrity for Existing Offshore Load-Bearing Structures, edition1; March 2009*, Standard Norge, Lysaker, Norway.

Standard Norge (2017), NORSOK N-005 *In-Service Integrity Management of Structures and Maritime Systems. Edition 2, 2017*, Standard Norge, Lysaker, Norway.

TWI (2020), https://www.twi-global.com/technical-knowledge/job-knowledge/friction-stir-welding-147, Visited 30th June 2020.

UEG (1978), "Underwater Inspection of Offshore Installations—Guidance for Designers", CXJB Underwater Engineers Report UR10, CIRIA Underwater Engineering Group, London, UK.

Wessex Ressins (2020), https://www.wessex-resins.com/, Visited 28 June 2020.

9

Conclusions and Future Possibilities

9.1 Overview of the Book

Offshore energy production of oil, gas and wind from a large number of fixed and floating structures is important for the world's economy. These structures are subject to damage and degradation such as corrosion, fatigue and impact and the continued safety of these is essential. In addition, the first offshore oil and gas structures date back to the 1940s and many of these existing structures are now getting old and have exceeded their original design life.

Underwater inspection and repair are key processes in ensuring these structures continue to operate safely and effectively. These activities have developed and improved significantly since the early offshore structures were installed. Initially divers played an important part, but with the development of ROVs and more recently AUVs, the activity has been done remotely. Standards and recommended practices have also developed for inspection and repair, particularly with the introduction of the ISO 19900 (ISO 2007), API RP-2A (API 2014) and NORSOK N-001 (Standard Norge 2012), as reviewed in Chapter 2 of this book.

Over the years, particularly in the period 1980 to 2000, many research and development programmes were initiated, mainly in the UK and US, providing valuable information on inspection, repair and related activities. The results of these programmes have been reviewed in this book to provide the reader with a background to enable current practice and standards to be better understood. In addition, the purpose of including these has also been to avoid duplication of research.

Inspection methods range from GVI and FMD for the detection of larger defects underwater, to CVI and NDE methods such as EC, MPI and ACFM that are available for the detection of small damage and anomalies, such as fatigue cracks. There is sufficient evidence that current inspection methods are able to detect damage and anomalies at a sufficiently early stage if used effectively. Such early detection is required to enable repair before more significant damage develops which could lead to structural collapse. The evaluation and assessment of structure with identified damage and anomalies has also improved significantly, enabling repair decisions to be made with more confidence.

A range of repair options have been developed with the ability to mitigate most typical damage occurring in offshore structures and have been extensively used in practice over many years. This experience has enabled operators to select the most appropriate among the several repair techniques available. In the early years of the offshore industry, such repairs needed to be undertaken by divers, but the use of remote vehicles to undertake repairs has developed significantly and this is expected to continue.

Underwater Inspection and Repair for Offshore Structures, First Edition.
John V. Sharp and Gerhard Ersdal.
© 2021 John Wiley & Sons Ltd. Published 2021 by John Wiley & Sons Ltd.

Structural monitoring is, in principle, a valuable technique for gaining continuous data on the condition and behaviour of structures. Natural frequency monitoring was tried in the early days of North Sea developments but proved to be limited in being able to detect only large defects in most offshore fixed structures with the technology available at that time. More recently, structural monitoring methods and the related data processing abilities have improved significantly and are now able to detect and locate defects at a much smaller scale. These techniques have significant benefits in continuous damage detection and can provide damage alerts for the increasing number of ageing offshore structures. These techniques are increasingly being used and are expected to be further developed in combination with digital technologies and robots.

9.2 Emerging Technologies

The loading and response on offshore structures are at present, to an increasing extent, monitored in real-time and all these data are stored and available for the structural integrity engineer. Real-time detection of damage can be obtained and enhanced by post-processing of monitoring data. These data can be linked directly to the integrity management database in order to give improved insight into the safety of the structure and to detect adverse structural changes in real-time. As a result, the structural engineer may be able to respond even more timely to mitigate or repair the structure.

Appropriate (real-time) data processing algorithms can link these measurements to models that recognise damage and anomalies and predict the condition of the structure. In many cases these analyses have been done retrospectively, requiring manual transfer of data to the shore and this time lag limits the value of such structural health monitoring systems. However, the ideal is that these analyses are done in real-time. More recently, monitoring devices that include wireless collection and transfer of data to the shore have been introduced making it possible to provide such real-time damage detection. Monitoring records are stored remotely (e.g. cloud) where they are easily accessible for any structural engineer involved.

This monitoring data can be used to calibrate structural analysis models (modal frequencies from accelerometer monitoring and stress levels from measured stresses) before significant damage occurs. Computer algorithms can be established so that regression analysis can be performed, estimating the relationship between different parameters (load and responses). Based on such a setup, the analysis can be used to predict anomalies and damage in a structure more or less in real time. One possible method is the machine learning algorithm (recurrent neural network) that can be used to establish the necessary methods to detect such anomalies and damage. Neural networks are also a computer technique that is commonly used for feature extraction, pattern recognition and classification. Its capability of recognising patterns in a database is especially useful. The neural network's ability to learn the correlations between the input and output can be a very useful method for identifying changes in the structural behaviour.

When using these emerging technologies, it is vital to have a clear idea of:

- which failure modes of the structure are critical;
- which damage and anomalies will precede such failure modes; and
- how and what to monitor to gain early information about such damage and anomalies.

A benefit is that the results from traditional inspection and surveillance can be used in these computerised algorithms as an addition to the continuous monitoring data.

The emergence of advanced monitoring techniques and wireless sensors (Internet of Things) can influence how structures are managed in the future. Embedding intelligent sensors and

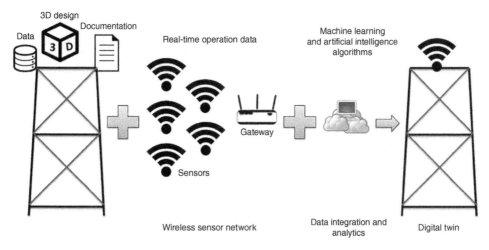

Figure 116   Illustration of a digital twin.

computer algorithms into managing offshore structures is already underway. 3D computer aided design (CAD) models with additional data related to structural specifications, the required inspection intervals, cost estimation and other structural integrity management data are already in use. In addition to the 3D model, the implementation of real-time data from sensors and data processing are components needed for creating a so-called digital twin, as illustrated in Figure 116.

The digital twin is designed to replicate a physical structure and improve its operational efficiency, enabling predictive maintenance by utilising sensors. For example, a fixed offshore steel structure may have a 3D model that includes all data and the documents from fabrication, installation and operation to allow for an easy retrieval of information for the structural integrity engineer. Further, if the structure has a sufficient number of sensors, these data can be used to train a machine learning algorithm to understand the behaviour of the structure and predict damage and anomalies. However, with the large uncertainties that exist, for example in the soil data, the measured motions and strains are unlikely to be identical to those predicted by a structural analysis program (e.g. finite element program). As a result, the structural analysis models may need to be updated based on the measured motions and strains in order to minimise the discrepancy between the digital twin and the real structure (Thygesen et al. 2018).

The updating of the structural analysis model can be performed by system identification methods (see e.g. Chantzi and Papadimitriou 2016), which are a process of identification of the natural frequencies and associated mode shapes along with the damping properties of the real structure and using these as a basis for the digital twin model updating (Thygesen et al. 2018). In addition, the load models can be similarly calibrated which will even more increase the accuracy between the digital twin and the real structure. Several engineering houses are developing the necessary software to perform such analysis, e.g. Rambøll and DNV GL. Digital twins are at the time of writing in use for a select number of platforms, for example Apache's Beryl Alpha platform (*Oil & Gas News*, June 8, 2020).

Autonomous underwater robots to monitor offshore structures offer a further promising addition to traditional inspection techniques and structural monitoring. Robot-based autonomous monitoring systems allow for more frequent inspections at a potentially lower cost. Recent advancements in robotic systems, already in use on bridge structures and electricity pylons, have led to the development of so-called swarm robots that can adapt their functionality dynamically and are expected to bring a radical change in the existing structural monitoring techniques in the near future.

Underwater inspection and integrity management of offshore structures in the future may be performed by structural monitoring and autonomous underwater robots that detect damage and anomalies. These may be supported by computer algorithms that predict the location and severity of the findings. A follow-up repair using autonomous robots is, however, a more difficult step but developments are also expected in this field.

9.3   Final Thoughts

Based on some 70 years of experience with offshore structures in many different areas of the world, the standards for design, fabrication, installation and operation of these are well developed. This conclusion is borne out by the absence of any recent major accidents involving the structure at the time of writing this book. However, many offshore structures are now ageing and being life extended, which would be expected to lead to an increasing amount of damage and deterioration. This is, in particular, already seen on some floating structures but should also be expected to apply to other types of structures in due course. This anticipated increase in damage and deterioration should be leading to increased inspections to ensure continuous safe performance of these structures and will realistically also lead to an increased amount of repairs.

Inspection methods and their deployment are at present well developed and capable of detecting most expected damage. However, in the need to minimise costs the inspection execution is continuously optimised using longer intervals between inspections and lower cost inspection methods (e.g. FMD). The challenge of these approaches is to deal with situations where, for example, multiple failures occur or failure occurs between inspections. The consequence of this is that damage may develop into a more serious event that may place the platform at an unnecessary risk. Some structural engineers are becoming increasingly aware of these challenges but unfortunately, in the authors view, this is not in general reflected in the industry as a whole. The development of autonomous robots, monitoring techniques and digital technology may offer a more cost-effective and possibly safer route for managing structural integrity.

Taking into account what we have discussed in this book, we feel it would be useful to recommend which key factors should be taken into account when making a structure more suitable for inspection and repair. At the design stage these are:

- ensuring that the design allows for all safety-critical parts of the structure to be inspectable and repairable;
- building in appropriate and durable monitoring systems enabling early detection of damage; and
- ensuring that sufficient redundancy and ductility is built in to provide continuing integrity when damage begins to accumulate due to ageing.

During operation these are:

- increased use of monitoring systems enabling early detection of damage;
- sharing information with other operators of similar structures; and
- keeping data about the structure accessible and well organised, for example by the use of digital twins and related damage identification algorithms.

The continued safety of offshore structures is an increasingly important issue with underwater inspection and repair being acknowledged key activities, particularly with the increasing need for life extension of offshore structures. Ultimately, the safety of personnel working on these structures is paramount and the awareness of damage types and anomalies, their detection and the possible repair of these plays a vital part in ensuring continued safety. It is hoped that this book will contribute to this understanding so that the structures are managed appropriately and safely.

References

API (2014), API RP-2A *Recommended Practice for Planning, Design and Constructing Fixed Offshore Platforms*, API Recommended practice 2A, 22nd Edition, American Petroleum Institute, 2014.

Chatzi, E. and Papadimitriou, C. (2016), *Identification Methods for Structural Health Monitoring*, Springer.

ISO (2007), *ISO 19902:2007 Petroleum and Natural Gas Industries—Fixed Steel Offshore Structures*, International Organization for Standardization.*Oil & Gas News*, June 8, 2020.

Thygesen, U.T., Jepsen, M.S., Vestemark, J., Dollerup, N. and Pedersen, A. (2018), "The True Digital Twin Concept for Fatigue Re-Assessment of Marine Structures", In Proceedings of the ASME 2018 37th International Conference on Ocean, Offshore and Arctic Engineering, OMAE2018, June 17–22, 2018, Madrid, Spain. OMAE2018-77915.

Standard Norge (2012) *NORSOK N-001: Integrity of Offshore Structures, Rev. 8, September* 2012, Standard Norge, Lysaker, Norway.

Index

Underwater Inspection and Repair for Offshore Structures, First Edition.
John V. Sharp and Gerhard Ersdal.
© 2021 John Wiley & Sons Ltd. Published 2021 by John Wiley & Sons Ltd.